AF361897

Antimicrobial Resistance in
Veterinary Science

Antimicrobial Resistance in Veterinary Science

Guest Editor

Nahla O. Eltai

Basel • Beijing • Wuhan • Barcelona • Belgrade • Novi Sad • Cluj • Manchester

Guest Editor
Nahla O. Eltai
Microbiology Department
Biomedical Research Centre
Qatar University
Doha
Qatar

Editorial Office
MDPI AG
Grosspeteranlage 5
4052 Basel, Switzerland

This is a reprint of the Special Issue, published open access by the journal *Antibiotics* (ISSN 2079-6382), freely accessible at: www.mdpi.com/journal/antibiotics/special_issues/3925VUMT49.

For citation purposes, cite each article independently as indicated on the article page online and using the guide below:

Lastname, A.A.; Lastname, B.B. Article Title. *Journal Name* **Year**, *Volume Number*, Page Range.

ISBN 978-3-7258-3354-2 (Hbk)
ISBN 978-3-7258-3353-5 (PDF)
https://doi.org/10.3390/books978-3-7258-3353-5

Contents

Article

Variation in Pen-Level Prevalence of BRD Bacterial Pathogens and Antimicrobial Resistance Following Feedlot Arrival in Beef Calves

Jennifer N. Abi Younes [1], John R. Campbell [1], Simon J. G. Otto [2], Sheryl P. Gow [3], Amelia R. Woolums [4], Murray Jelinski [1], Stacey Lacoste [1] and Cheryl L. Waldner [1,*]

[1] Department of Large Animal Clinical Sciences, Western College of Veterinary Medicine, University of Saskatchewan, Saskatoon, SK S7N 5B4, Canada; jennifer.abiyounes@usask.ca (J.N.A.Y.)

[2] School of Public Health, University of Alberta, Edmonton, AB T6G 1C9, Canada; simon.otto@ualberta.ca

[3] Canadian Integrated Program for Antimicrobial Resistance Surveillance, Public Health Agency of Canada, Saskatoon, SK S7L 0Z2, Canada

[4] Department of Pathobiology and Population Medicine, College of Veterinary Medicine, Mississippi State University, Mississippi State, MS 39762, USA

* Correspondence: cheryl.waldner@usask.ca

Abstract: Antimicrobials are crucial for treating bovine respiratory disease (BRD) in beef feedlots. Evidence is needed to support antimicrobial use (AMU) decisions, particularly in the early part of the feeding period when BRD risk is highest. The study objective was to describe changes in prevalence and antimicrobial susceptibility of BRD bacterial pathogens at feedlot processing (1 day on feed (1DOF)), 12 days later (13DOF), and for a subset at 36DOF following metaphylactic antimicrobial treatment. Mixed-origin steer calves ($n = 1599$) from Western Canada were managed as 16 pens of 100 calves, receiving either tulathromycin ($n = 1199$) or oxytetracycline ($n = 400$) at arrival. Deep nasopharyngeal swabs collected at all time points underwent culture and antimicrobial susceptibility testing (AST). Variability in the pen-level prevalence of bacteria and antimicrobial susceptibility profiles were observed over time, between years, and metaphylaxis options. Susceptibility to most antimicrobials was high, but resistance increased from 1DOF to 13DOF, especially for tetracyclines and macrolides. Simulation results suggested that sampling 20 to 30 calves per pen of 200 reflected the relative pen-level prevalence of the culture and AST outcomes of interest. Pen-level assessment of antimicrobial resistance early in the feeding period can inform the evaluation of AMU protocols and surveillance efforts and support antimicrobial stewardship in animal agriculture.

Keywords: bovine respiratory disease; feedlot; bovine; antimicrobial resistance; antimicrobial use; longitudinal; prevalence; pen; sampling

Citation: Abi Younes, J.N.; Campbell, J.R.; Otto, S.J.G.; Gow, S.P.; Woolums, A.R.; Jelinski, M.; Lacoste, S.; Waldner, C.L. Variation in Pen-Level Prevalence of BRD Bacterial Pathogens and Antimicrobial Resistance Following Feedlot Arrival in Beef Calves. *Antibiotics* **2024**, *13*, 322. https://doi.org/10.3390/antibiotics13040322

Academic Editor: Nahla O. Eltai

Received: 10 February 2024
Revised: 28 March 2024
Accepted: 29 March 2024
Published: 2 April 2024

1. Introduction

Antimicrobials are essential for maintaining health and welfare and preventing economic losses in animal production systems. In North American cattle feedlots, parenteral antimicrobials are most frequently used to manage bovine respiratory disease (BRD), the leading cause of morbidity and mortality [1]. While management strategies such as preconditioning, vaccination at arrival, and reducing stress can better prepare cattle for the transition from farm to feedlot, an estimated 39% of calves entering Western Canadian feedlots remain at high risk for developing BRD, and these means alone have been insufficient to adequately manage the disease [2–5]. As a result, antimicrobials continue to be necessary and have proven effective for BRD control [6].

However, efficient strategies for using laboratory tools to inform antimicrobial use (AMU) decisions for BRD treatment and control are lacking. The need for evidence to target AMU is growing with the global awareness of antimicrobial resistance (AMR), which threatens the efficacy of antimicrobials as well as the health of humans, animals, and the environment [7,8]. Additionally, a better understanding of AMR in common bacterial pathogens in the development of BRD management and treatment protocols is essential for antimicrobial stewardship [9,10]. A reduction in the availability of antimicrobials to treat BRD in feedlot cattle would be detrimental to animal welfare and the productivity of the beef industry [11].

The control and treatment of BRD in feedlots is complicated due to its polymicrobial and multi-etiologic nature. Consistently identified risk factors include the placement of young and lighter-weight calves, vaccination status at arrival, recent and abrupt weaning, prolonged transport time, commingling animals from different origins, and inclement weather [12–14]. These stressors are thought to suppress the respiratory immune system, increasing the animal's susceptibility to contagious pathogens as well as opportunistic viral and bacterial infection, ultimately resulting in respiratory disease. Principal bacterial agents implicated in BRD include three members of the *Pasteurellaceae* family: *Mannheimia haemolytica*, *Pasteurella multocida*, and *Histophilus somni*. Not only have these bacteria been consistently recovered from clinical cases but cattle from which *M. haemolytica* was recovered at arrival were more likely to become ill within 10 days [15].

A 2023 study by Smith et al. [16] collected data from 25 U.S. commercial feed yards, representing 4.4 million cattle on feed, with the objective of determining the temporal distributions of first BRD treatment. The results agree with others showing that BRD incidence is often greatest during the first few weeks on feed [12,17]. However, studies exploring bacterial changes in pathogen and AMR prevalence over time during the early feeding period (<14 days) are only recently gaining interest [18–20]. While macrolides are the most effective antibiotics for metaphylaxis to reduce the incidence of BRD, longitudinal changes in bacterial prevalence and AMR after feedlot arrival are less commonly reported than samples collected at arrival [21–23]. Consequently, feedlot veterinarians face the challenge of lacking readily available AMR data to inform antimicrobial choices in calves requiring first treatment for BRD. Instead, antimicrobial drug choices are made based on prior experience, animal history, historical treatment records, and historical data on therapeutic effectiveness.

To address the aforementioned challenges related to AMU in feedlots, veterinarians and managers need practical strategies to support laboratory-based antimicrobial decision-making. These strategies should align with the World Health Organization (WHO) recommendations for using laboratory-based tools to select antimicrobials for treatment to promote prudent AMU and monitor AMR in food-producing animals [7]. Commercial feedlots can house thousands of cattle, making it impractical to sample each individual animal due to the associated time, resources, and costs involved. However, as feedlot cattle are managed as groups within pens [6], a subset of animals per pen could be sampled to estimate the frequency of the bacterial pathogens with AMR of interest.

Sampling cattle at feedlot arrival provides insights into incoming levels of AMR, facilitating the effective monitoring, surveillance, and identification of intervention points both pre- and post-arrival. However, factors such as stress, commingling, and environmental contamination lead to modifications in the upper respiratory microbiome and affect the prevalence of bacteria and AMR after arrival [24,25]. Additionally, cattle at a high risk of developing BRD are typically administered metaphylactic antimicrobial therapy within the first few days of feedlot arrival to decrease the pathogen burden and reduce the risk of pen-level disease [6]. Metaphylaxis may induce selective pressure and result in greater proportions of AMR bacteria [26–30], with differences in the nasopharyngeal microbiota of treated cattle up to 60 days on feed [23,31]. Thus, in cattle administered long-acting injectable antimicrobials for metaphylaxis, sampling after a post-metaphylactic interval (PMI) could provide more meaningful data on the pathogen and AMR prevalence that occurs during the early feeding period with the greatest BRD risk, and could better inform future antimicrobial treatment decisions.

The overall goal of this study was to leverage an opportunity to sample all calves from auction-sourced feedlot pens at 1 and 13 days on feed (DOF) to determine the prevalence of selected BRD pathogens and clinically relevant AMR targets. This study is unique because of the large number of calves (n = 1599) sampled over a two-year period [19,20,22]. Calves in each purchased lot were auction-sourced from multiple farms and housed in pens of 100 head, which more closely approximated commercial feedlot practices and comingling risk in Western Canada compared to pen sizes from previous reports [19,20,32,33]. Complete culture and antimicrobial susceptibility data from all calves in larger pens at two consistent time points across different years and metaphylaxis protocols expand on other studies evaluating changes in the first few weeks on feed [19,22,34].

The first objective was to describe the prevalence and variability of BRD pathogens and AMR among pens, metaphylaxis groups, and years during the early feeding period, including 1DOF, 13DOF, and 36DOF, for fall-placed auction market calves at Western Canadian feedlots. The second objective was to compare the prevalence of BRD pathogens and associated AMR recovered from calves at 1DOF with those observed at 13DOF to evaluate the additional insights gained from sampling following the PMI.

2. Results

2.1. Study Population

Radio Frequency Identification (RFID) tag numbers were scanned for each calf. In 2020, calves were sourced from 292 unique herds of origin (derived from the first 12 digits of the calves' 15-digit RFID tag) (as described in Section 4.2) (Table 1). The number of unique herds of origin for 2020 calves purchased for the feedlot varied from 30–81 herds per pen, suggesting a high risk of pathogen exposure due to commingling. The calves sampled in 2021 were less diverse, with 208 unique herds of origin and 12–38 unique herds per pen. At 1DOF, the mean calf weight in 2020 was 253 kg (556 lbs) (range: 211–291 kg (464–640 lbs)) (Table 1). Lighter calves were targeted in 2021, resulting in a mean calf weight of 225 kg (496 lbs) (range: 160–315 kg (351–694 lbs)).

Table 1. Descriptive summary of calves at time of processing including unique herd of origin (using first 12 digits of RFID tag), calf weight (kg), and standard deviation (standard dev.).

Year	Pen	Unique Herds of Origin	Avg. Weight (kg)	Standard Dev. (kg)
	1	34	257	13
	2	81	243	10
	3	48	254	11
	4	43	253	13
2020	5	31	260	9
	6	38	262	11
	7	41	256	13
	8	30	239	13
	Combined	292	253	14
	9	31	223	18
	10	36	229	14
	11	33	222	14
	12	35	229	10
2021	13	23	220	18
	14	12	223	20
	15	38	228	13
	16	38	230	12
	Combined	208	225	15

Of the 1600 steers purchased for this study, one calf was recumbent at the time of initial feedlot processing (as outlined in Section 4.3) in 2020 and removed, resulting in 1599 cattle sampled at 1DOF from October to December. Three calves died prior to 13DOF (one calf in November of 2020 and two in November of 2021), leaving a total of 1596 calves sampled at 13DOF. At 36DOF, 310 calves were sampled. Antimicrobial susceptibility data were not available from the laboratory for four samples in 2021 (two at 1DOF, one at 13DOF, one at

36DOF), resulting in the following total samples cultured and tested for susceptibility: 1597 at 1DOF, 1595 at 13DOF, and 309 at 36DOF.

For all pens combined, 8.1% of calves (n = 130) were treated for BRD within 45DOF. Cohorts receiving tulathromycin metaphylaxis observed similar BRD incidences, with 3.5% (n = 28) of calves treated in 2020 and 3.3% (13 calves) treated in 2021 for tulathromycin-treated cohorts. In contrast, in 2021, oxytetracycline-treated cohorts observed a higher incidence of 22.3% of calves (n = 89) being treated for BRD. Total mortalities in 2020 included three calves (0.4%) succumbing to BRD and one case of bloat. In 2021, mortality attributed to BRD increased slightly to 1.0% (eight calves), with 0.6% (five calves) from pen 16. Other mortalities in 2021 included four cases of bloat (0.5%) and one calf (0.1%) euthanized due to neurologic symptoms that also had lung lesions.

2.2. Differences in Bacterial Recovery between Years and Metaphylaxis Options

The proportion of calves that were culture-positive for bacteria of interest (*M. haemolytica*, *P. multocida*, and *H. somni*) varied among pens during the study (1DOF, 13DOF, and 36DOF) and between sampling years (Tables 2 and 3).

Table 2. Percentage of calves from each pen cohort from which *M. haemolytica*, *P. multocida*, and *H. somni* were recovered at 1DOF (n = 799 calves) at on-arrival processing and before metaphylaxis, 13DOF (n = 798), and 36DOF (n = 80) for the 2020 study population.

Metaphylaxis Drug	Pen	Sampling Time	No. Calves	Recovery Rates of BRD Pathogens from Calves		
				M. haemolytica	*P. multocida*	*H. somni*
tulathromycin	1	1DOF	100	26%	32%	4%
		13DOF	100	77%	2%	0%
		36DOF	10	80%	20%	30%
tulathromycin	2	1DOF	100	35%	47%	3%
		13DOF	100	84%	7%	1%
		36DOF	10	70%	10%	50%
tulathromycin	3	1DOF	100	47%	48%	6%
		13DOF	100	70%	13%	0%
		36DOF	10	80%	40%	30%
tulathromycin	4	1DOF	100	47%	55%	4%
		13DOF	100	80%	21%	4%
		36DOF	10	80%	20%	60%
tulathromycin	5	1DOF	100	33%	67%	12%
		13DOF	100	14%	11%	2%
		36DOF	10	80%	10%	40%
tulathromycin	6	1DOF	99	33%	63%	7%
		13DOF	99	9%	8%	5%
		36DOF	10	70%	20%	30%
tulathromycin	7	1DOF	100	28%	70%	12%
		13DOF	99	30%	11%	12%
		36DOF	10	70%	30%	100%
tulathromycin	8	1DOF	100	15%	70%	29%
		13DOF	100	57%	8%	11%
		36DOF	10	70%	20%	50%

Table 3. Percentage of calves from each pen cohort from which *M. haemolytica*, *P. multocida*, and *H. somni* were recovered at 1DOF (*n* = 800 calves) at on-arrival processing and before metaphylaxis, 13DOF (*n* = 798 calves), and 36DOF (*n* = 230 calves) for the 2021 study population.

Metaphylaxis Drug	Pen	Sampling Time	No. Calves	Recovery Rates of BRD Pathogens from Calves		
				M. haemolytica	*P. multocida*	*H. somni*
oxytetracycline	9	1DOF	100	60%	23%	6%
		13DOF	100	61%	35%	10%
		36DOF	30	33%	33%	70%
oxytetracycline	10	1DOF	100	49%	29%	0%
		13DOF	100	52%	60%	5%
		36DOF	30	13%	57%	73%
oxytetracycline	11	1DOF	100	40%	35%	4%
		13DOF	100	47%	18%	19%
		37DOF	30	47%	20%	70%
tulathromycin	12	1DOF	100	58%	39%	7%
		13DOF	100	25%	7%	4%
		36DOF	30	63%	13%	70%
tulathromycin	13	1DOF	100	47%	29%	13%
		13DOF	100	17%	14%	21%
		36DOF	30	67%	27%	53%
tulathromycin	14	1DOF	99	48%	51%	8%
		13DOF	99	33%	14%	6%
		36DOF	30	50%	23%	73%
tulathromycin	15	1DOF	100	43%	56%	9%
		13DOF	99	44%	12%	9%
		36DOF	30	43%	20%	40%
oxytetracycline	16	1DOF	100	43%	73%	7%
		12DOF	99	61%	41%	13%
		30DOF	20	20%	35%	60%

Irrespective of year, for all calves sampled at 1DOF, prior to the administration of metaphylactic antimicrobials, 41% (CI: 35–46%) were culture-positive for *M. haemolytica*, 49% (CI: 41–57%) for *P. multocida*, and 8% (CI: 6–11%) for *H. somni* (Figure 1a–c). However, the proportion of calves that were culture-positive for *M. haemolytica* at 1DOF was higher in 2021 (49% of calves [CI: 42–55%]) compared to 2020 (33% of calves [CI: 27–39%]) (Table 4). In contrast, *P. multocida* was recovered from more calves in 2020 (56% [CI: 46–66%]) than in 2021 (42% [CI: 32–52%]) at 1DOF. The recovery of *H. somni* at 1DOF did not differ between 2020 (9% [CI: 5–14%]) and 2021 (7% [CI: 4–11%]).

Table 4. Differences in recovery of *M. haemolytica*, *P. multocida*, and *H. somni* at arrival to feedlot and prior to metaphylaxis administration (1DOF) between year 2021 and year 2020, reported as population-averaged odds ratios (ORs) and 95% confidence intervals (CIs), accounting for clustering at the pen level. *n* = 1599 calves.

Bacteria	OR 2021 vs. 2020	95% CI	*p*-Value
M. haemolytica	1.9	1.3, 2.8	<0.001
P. multocida	0.6	0.3, 0.97	0.04
H. somni	0.7	0.4, 1.5	0.42

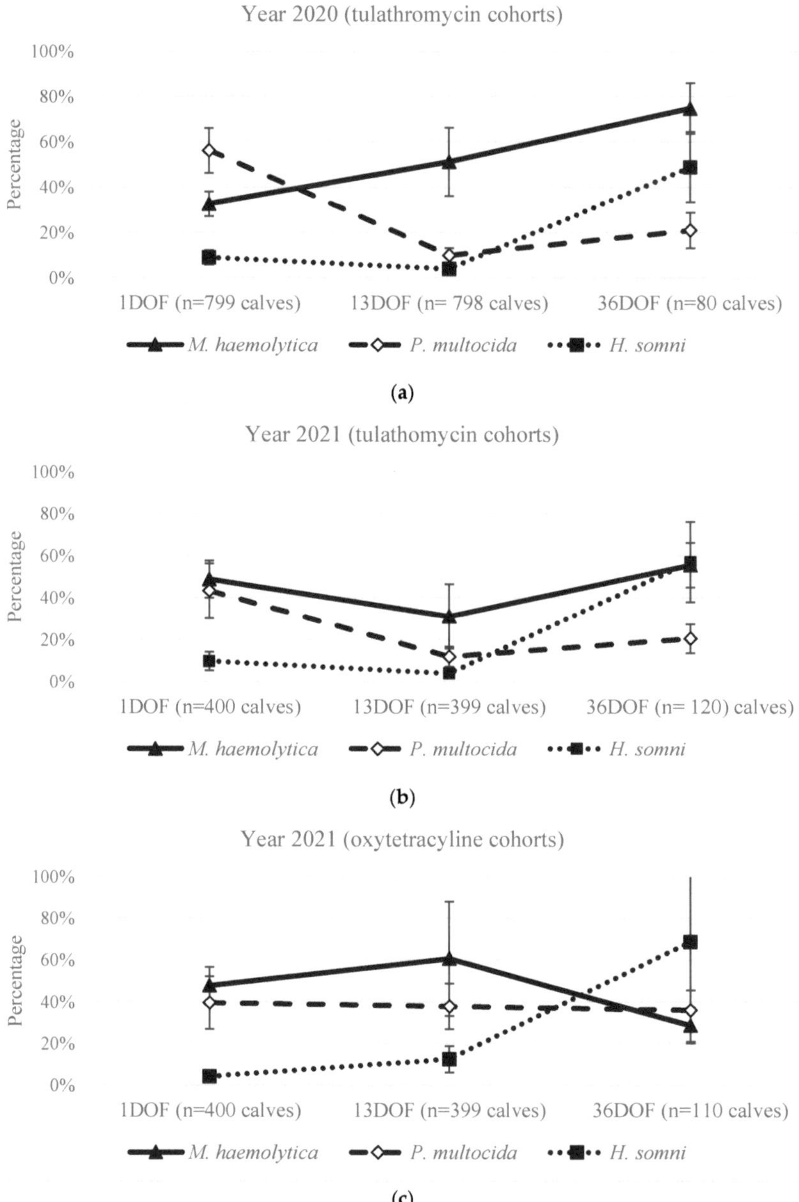

Figure 1. Percentage (%) of calves from which *M. haemolytica*, *P. multocida*, and/or *H. somni* were recovered at 1DOF, 13DOF, and 36DOF for (**a**) 2020 pen cohorts that received tulathromycin metaphylaxis, (**b**) 2021 pen cohorts that received tulathromycin metaphylaxis, and (**c**) 2021 pen cohorts that received oxytetracycline metaphylaxis. Samples at 1DOF were collected during on-arrival processing and before metaphylaxis. Percentages displayed are population-averaged with 95% confidence intervals, adjusted for clustering at the pen level using mixed effects logistic regression. For pen 16, sampling at 30DOF, prior to mass treatment, was used for analysis.

When adjusting for year, metaphylaxis antimicrobial, and clustering at the pen level, there were no significant differences in the recovery of *M. haemolytica* at 13DOF across sampling years and metaphylactic treatment groups (Table 5). *P. multocida* was around five times more likely to be cultured from calves at 13DOF that had received metaphylactic oxytetracycline in 2021 compared to calves from either year that received tulathromycin (Table 5). For calves that received tulathromycin on arrival, there was no difference ($p = 0.55$) in *P. multocida* recovery at 13DOF between years (Table 5). *H. somni* was 3.3 times more likely to be cultured at 13DOF from calves in 2021 treated with oxytetracycline at arrival than calves in 2020 treated with tulathromycin (Table 5). There was no difference in the recovery of *H. somni* at 13DOF between 2020 and 2021 tulathromycin-treated calves (*p*-value: 0.06) or between year 2021 tulathromycin- and oxytetracycline-treated calves (*p*-value: 0.67) (Table 5).

Table 5. Differences in recovery of *M. haemolytica*, *P. multocida*, and *H. somni* at 13DOF between years and metaphylaxis administered (year 2020/tulathromycin, year 2021/tulathromycin, and year 2021/oxytetracycline), reported as population-averaged odds ratios (ORs) and 95% confidence intervals (CIs), accounting for clustering at the pen level. $n = 1596$ calves.

Bacteria	Year/Metaphylaxis Comparison	OR	95% CI	*p*-Value
M. haemolytica	2021/tulathromycin vs. 2020/tulathromycin	0.4	0.1, 1.3	0.12
	2021/oxytetracycline vs. 2020/tulathromycin	1.2	0.3, 4.0	0.82
	2021/oxytetracycline vs. 2021/tulathromycin	3.1	0.7, 13	0.12
P. multocida	2021/tulathromycin vs. 2020/tulathromycin	1.2	0.6, 2.4	0.55
	2021/oxytetracycline vs. 2020/tulathromycin	5.5	2.9, 9.3	<0.001
	2021/oxytetracycline vs. 2021/tulathromycin	4.4	2.1, 9.3	<0.001
H. somni	2021/tulathromycin vs. 2020/tulathromycin	2.6	0.6, 2.4	0.06
	2021/oxytetracycline vs. 2020/tulathromycin	3.3	1.2, 3.7	0.018
	2021/oxytetracycline vs. 2021/tulathromycin	1.3	0.4, 3.7	0.67

2.3. Differences in Bacterial Recovery over Time within Year and Metaphylaxis Options

The probability of recovering *M. haemolytica* more than doubled from 33% (CI: 27–39%) at 1DOF before metaphylaxis to 75% (CI: 64–84%) at 36DOF post metaphylaxis administration in 2020 (Figure 1a). While there was substantial pen-to-pen variation at 13DOF (Table 2), calves in 2020 were more likely to have *M. haemolytica* isolated at 13DOF than at 1DOF and at 36DOF compared to 13DOF (Table 6).

In contrast, the recovery of *M. haemolytica* from pens receiving tulathromycin in 2021 decreased from 1DOF to 13DOF (49% [CI: 40–58%] to 31% [CI: 16–52]), followed by an increase from 13DOF to 36DOF (Figure 1b; Table 6). For calves receiving oxytetracycline at arrival in 2021, the recovery of *M. haemolytica* decreased from 1DOF to 36DOF (Figure 1c; Table 6).

For calves receiving tulathromycin at arrival, *P. multocida* recovery decreased between 1 and 13DOF from 56% (CI: 46–66%) to 10% (CI: 7–14%) in 2020 and from 44% (CI: 30–58%) to 12% (CI: 7–19%) in 2021 (Figure 1a, b; Table 6). *P. multocida* then increased in the tulathromycin groups from 13DOF to 36DOF in both 2020 and 2021 (Figure 1a, b; Table 6).

In contrast, the recovery of *P. multocida* from calves receiving oxytetracycline at arrival in 2021 did not significantly change between 1DOF (40% [CI: 27–54%]) and 36DOF (36% [CI: 27–47%]) (Figure 1c; Table 6).

Table 6. Pairwise comparisons from the repeated measures and multilevel logistic regression models for the likelihood of bacterial recovery (*M. haemolytica, P. multocida, H. somni*) from calves within pens at 1DOF at on-arrival processing and before metaphylaxis, 13DOF, and 36DOF, stratified by sampling year and metaphylactic antimicrobial administered. Differences reported population-averaged odds ratios (ORs) and 95% confidence intervals (CIs) conditioned on pen and calf levels. n = 1599 calves at 1DOF, 1596 at 13DOF, and 310 at 36DOF.

Bacteria	Year/Metaphylaxis	DOF Comparison	OR	95% CI	*p*-Value
M. haemolytica	2020/tulathromycin	13DOF vs. 1DOF	2.4	1.9, 2.9	<0.001
		36DOF vs. 1DOF	6.8	4.0, 12	<0.001
		36DOF vs. 13DOF	2.8	1.7, 4.8	<0.001
	2021/tulathromycin	13DOF vs. 1DOF	0.5	0.3, 0.6	<0.001
		36DOF vs. 1DOF	1.3	0.9, 1.9	0.19
		36DOF vs. 13DOF	2.9	1.9, 4.3	<0.001
	2021/oxytetracycline	13DOF vs. 1DOF	1.3	1.0, 1.7	0.041
		36DOF vs. 1DOF	0.5	0.3, 0.7	<0.001
		36DOF vs. 13DOF	0.3	0.2, 0.5	<0.001
P. multocida	2020/tulathromycin	13DOF vs. 1DOF	0.1	0.1, 0.1	<0.001
		36DOF vs. 1DOF	0.2	0.1, 0.3	<0.001
		36DOF vs. 13DOF	2.3	1.3, 4	0.004
	2021/tulathromycin	13DOF vs. 1DOF	0.2	0.1, 0.2	<0.001
		36DOF vs. 1DOF	0.3	0.2, 0.5	<0.001
		36DOF vs. 13DOF	1.9	1.2, 3.2	0.01
	2021/oxytetracycline	13DOF vs. 1DOF	0.9	0.7, 1.2	0.65
		36DOF vs. 1DOF	0.9	0.6, 1.3	0.55
		36DOF vs. 13DOF	0.9	0.6, 1.4	0.75
H. somni	2020/tulathromycin	13DOF vs. 1DOF	0.5	0.3, 0.7	<0.001
		36DOF vs. 1DOF	9.8	5.7, 17	<0.001
		36DOF vs. 13DOF	22	12, 40	<0.001
	2021/tulathromycin	13DOF vs. 1DOF	1.1	0.7, 1.7	0.70
		36DOF vs. 1DOF	13	7.7, 22	<0.001
		36DOF vs. 13DOF	12	7.1, 20	<0.001
	2021/oxytetracycline	13DOF vs. 1DOF	2.7	1.6, 4.6	<0.001
		36DOF vs. 1DOF	44	23, 84	<0.001
		36DOF vs. 13DOF	16	9.3, 27	<0.001

H. somni recovery increased over time, averaging 8% at 1DOF and 13DOF and sharply increasing to 58% at 36DOF (Figure 1a–c). However, recovery varied from 1DOF to 13DOF across metaphylaxis options (Table 6). In year 2020 tulathromycin-treated pens, calves were less likely to have *H. somni* at 13DOF than at 1DOF. In 2021, there was no significant difference for tulathromycin-treated calves; however, the recovery of *H. somni* was greater in oxytetracycline-treated calves at 13DOF (Table 6).

2.4. Bacterial Co-Isolation Patterns in Years 2020 and 2021 at 1DOF and 13DOF

Most calves with BRD bacteria detected had a single species isolated at both 1DOF and 13DOF (Table 7). Irrespective of year and metaphylaxis antimicrobial administered, the most common co-isolation pattern observed at 1DOF was *M. haemolytica* and *P. multocida* (15%), followed by *P. multocida* and *H. somni* (3%) and *M. haemolytica* and *H. somni* (0.9%) (Table 7). At 13DOF, these proportions changed only slightly, with 8% of calves having *M. haemolytica* and *P. multocida*, 2% having *M. haemolytica* and *H. somni*, and 1% having *P. multocida* and *H. somni*. Very few calves had all three bacteria isolated concurrently.

Table 7. Number of calves with bacterial co-isolation patterns recovered at 1DOF at on-arrival processing and before metaphylaxis and 13DOF stratified by sampling year and metaphylactic antimicrobial administered.

Year	Meta. [1]	Time Point	No. Calves	Neg. Culture	MH	PM	HS	MH + PM	MH + HS	PM + HS	MH + PM + HS
								Number (%) of Calves with Bacterial Co-Isolation Pattern			
2020	Tula	1DOF	799	183 (23%)	133 (17%)	286 (36%)	26 (3%)	120 (15%)	6 (0.8%)	40 (5%)	5 (0.6%)
2020	Tula	13DOF	798	322 (40%)	366 (46%)	32 (4%)	19 (2%)	43 (5%)	10 (1%)	4 (0.5%)	2 (0.3%)
2021	Tula	1DOF	400	83 (21%)	117 (29%)	94 (24%)	19 (5%)	69 (17%)	7 (12%)	8 (2%)	3 (0.8%)
2021	Tula	13DOF	399	219 (55%)	102 (26%)	26 (6.5%)	27 (7%)	12 (3%)	4 (1%)	8 (2%)	1 (0.3%)
2021	Oxy	1DOF	400	102 (26%)	131 (33%)	96 (24%)	5 (1%)	54 (14%)	2 (0.5%)	5 (1%)	5 (1%)
2021	Oxy	13DOF	399	84 (21%)	133 (33%)	71 (18%)	13 (3%)	64 (16%)	15 (4%)	11 (3%)	8 (2%)
All Years	All Groups	1DOF	1599	368 (23%)	381 (24%)	476 (30%)	50 (3%)	243 (15%)	15 (0.9%)	53 (3%)	13 (0.8%)
All Years	All Groups	13DOF	1596	625 (39%)	601 (38%)	129 (8%)	59 (3%)	119 (8%)	29 (2%)	23 (1%)	11 (0.7%)

[1] Antimicrobial used for metaphylaxis (Meta.): Tula, tulathromycin; Oxy, oxytetracycline; MH, *M. haemolytica*; PM, *P. multocida*; HS, *H. somni*.

In general, calves that received metaphylaxis treatment with tulathromycin had a substantial decrease in overall bacterial recovery from 1DOF to 13DOF (OR: 2.91; 95% CI: 2.45–3.47; *p*-value: <0.001), with an average of 22% of calves with a negative culture result at 1DOF and 45% at 13DOF (Table 7). Within this overall pattern, the recovery of *M. haemolytica* increased between 1DOF and 13DOF in 2020, although not in 2021 (Figure 1). This contrasted with the results for oxytetracycline-treated calves, for which the average number of animals with negative culture results remained relatively stable over time (OR: 0.77; 95% CI: 0.56, 1.08; *p*-value: 0.13) (Table 7).

2.5. Differences in Antimicrobial Susceptibility of Bacteria between Years at 1DOF

The overall crude (unadjusted) prevalence of calves exhibiting resistance to the tested antimicrobials among the BRD pathogens of interest at 1DOF was low (Tables 8 and 9). The population-averaged prevalence, accounting for clustering by pen, was 7% (CI: 5–10%) for 2020 and 5% (CI: 4–8%) for 2021 (*p*-value: 0.23). Frequency tables for minimum inhibitory concentrations are provided in Supplementary Materials S1.

Table 8. Number (%) of calves with bacteria interpreted as resistant to select antimicrobials [1] at 1DOF at on-arrival processing and before metaphylaxis, 13DOF, and 36DOF for the 2020 study population based on CLSI minimum inhibitory concentration breakpoints (all calves received metaphylactic tulathromycin at arrival).

Time Point	No. Calves	Bacteria [2]	AMP	DANO	FLOR	SPECT	TET	GAM	TILD	TILM	TUL
			Number (%) of Calves with Isolates Resistant to Select Antimicrobials [1]								
1DOF	799	MH	1 (0.1%)	0	0	0	0	1 (0.1%)	0	0	1 (0.1%)
1DOF	799	PM	15 (2%)	0	0	31 (4%)	28 (4%)	0	0	NI	0
1DOF	799	HS	0	NI	0	1 (0.1%)	0	0	4 (0.5%)	NI	6 (0.8%)
13DOF	798	MH	2 (0.3%)	0	1 (0.1%)	0	1 (0.1%)	351 (44%)	3 (0.4%)	25 (3%)	341 (43%)
13DOF	798	PM	7 (0.9%)	0	0	10 (1%)	8 (1%)	1 (0.1%)	1 (0.1%)	NI	1 (0.1%)
13DOF	798	HS	0	NI	0	0	0	1 (0.1%)	2 (0.3%)	NI	2 (0.3%)
36DOF	80	MH	2 (2.5%)	0	0	0	1 (1.3%)	44 (55%)	1 (1.3%)	7 (8.8%)	40 (50%)
36DOF	80	PM	0	0	0	3 (3.8%)	3 (3.8%)	0	0	NI	0
36DOF	80	HS	0	NI	0	0	0	0	0	NI	0

[1] Antimicrobials for which isolates were tested: AMP, ampicillin; DANO, danofloxacin; FLOR, florfenicol; SPECT, spectinomycin; TET, tetracycline; GAM, gamithromycin; TILD, tildipirosin; TILM, tilmicosin; TUL, tulathromycin. No resistance observed for penicillin, ceftiofur, or enrofloxacin. [2] MH, *M. haemolytica*; PM, *P. multocida*; HS, *H. somni*. NI = not interpretable, CLSI breakpoints not available.

Table 9. Number (%) of calves with bacteria interpreted as resistant to select antimicrobials [1] at 1DOF at on-arrival processing and before metaphylaxis, 13DOF, and 36DOF for the 2021 study population based on CLSI minimum inhibitory concentration breakpoints, stratified by metaphylactic antimicrobial administered at arrival.

| Meta. [2] | Time Point | No. Calves | Bacteria [3] | Number (%) of Calves with Isolates Resistant to Select Antimicrobials [1] | | | | | | | | |
				AMP	PEN	DANO	SPECT	TET	GAM	TILD	TILM	TUL
Tula	1DOF	399	MH	2 (0.5%)	0	2 (0.5%)	0	1 (0.3%)	0	1 (0.3%)	2 (0.5%)	0
		399	PM	9 (2%)	0	0	5 (1%)	3 (0.8%)	0	0	NI	0
		399	HS	0	0	NI	1 (0.3%)	0	0	0	NI	0
	13DOF	399	MH	0	0	0	0	18 (5%)	27 (7%)	38 (10%)	43 (11%)	27 (7%)
		399	PM	6 (2%)	0	0	0	0	0	0	NI	0
		399	HS	1 (0.3%)	0	NI	1 (0.3%)	1 (0.3%)	0	0	NI	0
	36DOF	119	MH	1 (0.8%)	0	0	0	23 (19.3%)	10 (8.4%)	12 (10.1%)	12 (10.1%)	10 (8.4%)
		119	PM	0	0	0	0	0	0	0	NI	0
		119	HS	3 (2.5%)	1 (0.8%)	0	0	6 (5.0%)	0	0	NI	0
Oxy	1DOF	399	MH	0	1 (0.3%)	0	0	1 (0.3%)	0	0	0	0
		399	PM	8 (2%)	0	0	11 (3%)	9 (2%)	0	0	NI	0
		399	HS	0	0	NI	0	1 (0.3%)	0	0	NI	0
	13DOF	398	MH	1 (0.3%)	2 (0.5%)	0 (0%)	0	3 (0.8%)	3 (0.8%)	3 (0.8%)	3 (0.8%)	3 (0.8%)
		398	PM	3 (0.8%)	0	0 (0%)	53 (13%)	52 (13%)	0	0	NI	0
		398	HS	0	0	NI	0	14 (4%)	0	0	NI	0
	36DOF	110	MH	0	0	0	0	0	0	0	0	0
		110	PM	2 (1.8%)	1 (0.9%)	0	15 (13.6%)	17 (15.5%)	0	0	NI	0
		110	HS	1 (0.9%)	0	NI	0	11 (10%)	0	0	NI	0

[1] Antimicrobials for which isolates were tested: AMP, ampicillin; PEN, penicillin; DANO, danofloxacin; SPECT, spectinomycin; TET, tetracycline; GAM, gamithromycin; TILD, tildipirosin; TILM, tilmicosin; TUL, tulathromycin. No resistance observed for ceftiofur, florfenicol, or enrofloxacin. [2] Antimicrobial used for metaphylaxis (Meta.). [3] MH, *M. haemolytica*; PM, *P. multocida*; HS, *H. somni*. NI = not interpretable, CLSI breakpoints not available.

In 2020 at 1DOF, the most commonly observed resistance was to tetracycline (4% of calves) or spectinomycin (4% of calves) with *P. multocida* (Table 8). For all three species, <2% of calves had isolates resistant to ampicillin, tulathromycin, gamithromycin, or tildipirosin. No calves had bacteria with resistance to penicillin, ceftiofur, danofloxacin, enrofloxacin, florfenicol, or tilmicosin.

In 2021 at 1DOF, 2% of calves had ampicillin- or spectinomycin-resistant *P. multocida* (Table 9). Tetracycline resistance was observed in 0.8% of calves in the tulathromycin-treated cohorts and 2% in the oxytetracycline-treated cohorts. Less than 1% of calves had *M. haemolytica* or *H. somni* with any AMR. No calves had bacteria with resistance to enrofloxacin, florfenicol, or tulathromycin.

2.6. Differences in Antimicrobial Susceptibility of Bacteria between Years at 13DOF

The population-averaged prevalence of calves with an organism resistant to at least one antimicrobial at 13DOF, adjusted for pen, was 41% (CI: 23–62%) for year 2020 tulathromycin-treated calves, 15% (CI: 5–37%) for 2021 tulathromycin-treated calves, and 20% (CI: 7–46%) for 2021 oxytetracycline-treated calves. There were no significant differences in the prevalence of calves with at least one AMR pathogen at 13DOF between year/metaphylaxis groups (2021/tulathromycin vs. 2020/tulathromycin, p-value: 0.062; 2021/oxytetracycline vs. 2020/tulathromycin, p-value: 0.18; 2021/tulathromycin vs. 2021/oxytetracycline, p-value: 0.65).

The highest prevalence of AMR was observed for tulathromycin- or gamithromycin-resistant *M. haemolytica* isolated from calves in 2020 at 13DOF (Table 10). Calves were more likely to have tulathromycin- or gamithromycin-resistant *M. haemolytica* isolates from 2020 tulathromycin-treated pens than either 2021 tulathromycin-treated or 2021 oxytetracycline-treated pens (Table 10). There were no significant differences between years for the frequency of recovery of tilmicosin-, tildipirosin-, or tetracycline-resistant *M. haemolytica* (p-values of 0.36, 0.20, and 0.52, respectively).

Table 10. Differences in antimicrobial resistance (AMR) patterns from the repeated measures multilevel logistic regression models at 13DOF between years and metaphylaxis administered (year 2020/tulathromycin, year 2021/tulathromycin, and year 2021/oxytetracycline), reported as population-averaged odds ratios (ORs) and 95% confidence intervals (CIs), accounting for clustering at the pen-level. n = 1595 calves for which susceptibility data were available.

AMR Outcome of Interest	Pairwise Comparison of Year/Metaphylaxis	OR	95% CI	p-Value
M. haemolytica tulathromycin	2021/tulathromycin vs. 2020/tulathromycin	0.1	0.03, 0.7	0.02
	2021/oxytetracycline vs. 2020/tulathromycin	0.06	0.01, 0.4	0.004
	2021/oxytetracycline vs. 2021/tulathromycin	0.4	0.05, 3.2	0.39
M. haemolytica gamithromycin	2021/tulathromycin vs. 2020/tulathromycin	0.1	0.03, 0.7	0.018
	2021/oxytetracycline vs. 2020/tulathromycin	0.06	0.01, 0.4	0.004
	2021/oxytetracycline vs. 2021/tulathromycin	0.4	0.05, 3.1	0.39
P. multocida tetracycline *	2021/tulathromycin vs. 2020/tulathromycin **	0.2	0, 0.90	0.08
	2021/oxytetracycline vs. 2020/tulathromycin	15	6.9, 37	<0.0001
	2021/oxytetracycline vs. 2021/tulathromycin **	83	19, ∞	<0.0001

Post-hoc Wald test for significance of differences between year and metaphylaxis options was not different for the recovery of any pathogen with AMR or *M. haemolytica* with resistance to tildipirosin, tilmicosin, or tetracycline. * Zero calves with tetracycline resistance at 13DOF from year 2021 tulathromycin-treated pens. Exact logistic regression used. ** Median unbiased estimate reported.

Tetracycline-resistant *P. multocida* was not identified at 13DOF in any calves in the 2021 tulathromycin-treated pens, compared to 13% with tetracycline-resistant isolates in the 2021 oxytetracycline-treated calves at 13DOF (Table 9). Tetracycline-resistant *P. multocida* was more prevalent at 13DOF in 2021 oxytetracycline-treated calves than either the 2020 or 2021 tulathromycin-treated calves (Table 10).

2.7. Within-Year Comparison of Bacterial and Antimicrobial Susceptibility from 1DOF to 36DOF

For the 2020 study population, the population-averaged prevalence of tulathromycin-resistant *M. haemolytica* increased over time from 0.1% (CI: 0.02–0.9%) at 1DOF to 34% (CI: 17–57%) at 13DOF (*p*-value: <0.001; Tables 8 and 11). The prevalence of calves with gamithromycin-resistant *M. haemolytica* also increased from 0.1% (CI: 0.02–0.9) at 1DOF to 35% (CI: 17–58) at 13DOF (*p*-value: <0.001; Tables 8 and 11). Both tulathromycin- and gamithromycin-resistant *M. haemolytica* also increased from 1DOF to 36DOF (Table 11). There were no significant differences over time for tildipirosin-resistant *M. haemolytica*, while resistance to tilmicosin increased across all time points (Table 11).

In 2021, there were no calves with tulathromycin- or gamithromycin-resistant *M. haemolytica* at 1DOF (CI: 0–0.005%) in either metaphylaxis group (Table 9). By 13DOF, the population-averaged prevalence of tulathromycin-resistant *M. haemolytica* in the year 2021 increased to 7% (CI: 2–22%) for tulathromycin-treated calves but only to 0.8% (CI: 0.5–15%) for oxytetracycline-treated calves (Table 11). In 2021 tulathromycin-treated cohorts, gamithromycin-, tildipirosin-, and tilmicosin-resistant *M. haemolytica* increased from 1DOF to 13DOF and from 1DOF to 36DOF, but the difference between 13DOF and 36DOF was not significant. In contrast, there were no significant differences over time for the oxytetracycline-treated cohorts (Table 11).

Tetracycline-resistant *M. haemolytica* also increased in 2021 tulathromycin-treated calves across all time points (Table 11), but there were no significant differences for the oxytetracycline-treated cohorts. The recovery of *P. multocida* with tetracycline resistance decreased from 1DOF to 13DOF in 2020 (Table 12) but subsequently rebounded from 13DOF to 36DOF (*p*-value: 0.05). There were no differences over time in 2021 tulathromycin-treated calves. For oxytetracycline-treated calves in 2021, the recovery of tetracycline-resistant *P. multocida* increased significantly from 1DOF to 13DOF as well as from 1DOF to 36DOF, with no change between 36DOF and 13DOF (Table 12).

Table 11. Pairwise comparisons from the repeated measures, multilevel logistic regression models for the likelihood of a calf within a pen having *M. haemolytica* with antimicrobial resistance (AMR) to tulathromycin, gamithromycin, tilmicosin, tildipirosin, or tetracycline across time for each year and metaphylaxis option, reported as population-averaged odds ratios (ORs) and 95% confidence intervals (CIs), accounting for clustering at the pen level. $n = 1595$ calves for which susceptibility data were available.

AMR Outcome of Interest	Year/Metaphylaxis	DOF Comparison	OR	95% CI	*p*-Value
Tulathromycin	2020/tulathromycin	13DOF vs. 1DOF	151	36, 638	<0.001
		36DOF vs. 1DOF	206	46, 915	<0.001
		36DOF vs. 13DOF	1.4	0.9, 2.1	0.14
	2021/tulathromycin *	13DOF vs. 1DOF **	41	9.1, ∞	<0.001
		36DOF vs. 1DOF **	50	10, ∞	<0.001
		36DOF vs. 13DOF	1.3	0.5, 3.4	0.67
	2021/oxytetracycline	13DOF vs. 1DOF **	3.9	0.6, ∞	0.25
		36DOF vs. 1DOF	.	.	.
		36DOF vs. 13DOF **	0.9	0, 6.2	0.96
Gamithromycin	2020/tulathromycin	13DOF vs. 1DOF	160	38, 672	<0.001
		36DOF vs. 1DOF	260	58, 1151	<0.001
		36DOF vs. 13DOF	1.6	1.1, 2.5	0.02
	2021/tulathromycin *	13DOF vs. 1DOF **	41	9.1, ∞	<0.001
		36DOF vs. 1DOF **	50	10, ∞	<0.001
		36DOF vs. 13DOF	1.3	0.5, 3.4	0.66
	2021/oxytetracyclin	13DOF vs. 1DOF **	3.9	0.6, ∞	0.25
		36DOF vs. 1DOF	.	.	.
		36DOF vs. 13DOF **	0.90	0, 6.2	0.96
Tildipirosin	2020/tulathromycin *	13DOF vs. 1DOF **	3.9	0.6, ∞	0.25
		36DOF vs. 1DOF **	9.9	0.5, ∞	0.18
		36DOF vs. 13DOF	3.3	0.06, 42	0.64
	2021/tulathromycin	13DOF vs. 1DOF	10	3.0, 34	<0.001
		36DOF vs. 1DOF	11	3.0, 37	<0.001
		36DOF vs. 13DOF	1.0	0.7, 1.6	0.85
	2021/oxytetracycline *	13DOF vs. 1DOF **	3.9	0.6, ∞	0.12
		36DOF vs. 1DOF	.	.	.
		36DOF vs. 13DOF **	0.9	0, 6.2	0.48
Tilmicosin	2020/tulathromycin *	13DOF vs. 1DOF **	37	8.1, ∞	<0.001
		36DOF vs. 1DOF **	103	20, ∞	<0.001
		36DOF vs. 13DOF	3.0	1.0, 7.4	0.04
	2021/tulathromycin	13DOF vs. 1DOF	6.6	2.9, 15	<0.001
		36DOF vs. 1DOF	6.4	2.6, 16	<0.001
		36DOF vs. 13DOF	1.0	0.6, 1.5	0.84
	2021/oxytetracycline *	13DOF vs. 1DOF **	3.9	0.6, ∞	0.25
		36DOF vs. 1DOF	.	.	.
		36DOF vs. 13DOF **	0.9	0, 6.2	0.96

Table 11. *Cont.*

AMR Outcome of Interest	Year/Metaphylaxis	DOF Comparison	OR	95% CI	*p*-Value
Tetracycline	2020/tulathromycin *	13DOF vs. 1DOF **	1.0	0.05, ∞	1.00
		36DOF vs. 1DOF **	10	0.5, ∞	0.18
		36DOF vs. 13DOF	10	0.1, 1000	0.35
	2021/tulathromycin	13DOF vs. 1DOF	7.2	1.9, 27	0.001
		36DOF vs. 1DOF	27	6.3, 115	<0.001
		36DOF vs. 13DOF	3.7	2, 7	<0.001
	2021/oxytetracycline *	13DOF vs. 1DOF	3.0	0.2, 159	0.62
		36DOF vs. 1DOF **	3.6	0, 69	1.57
		36DOF vs. 13DOF **	0.9	0, 6.2	0.96

* Exact logistic regression equation performed. · odds ratio (OR) was not estimable. ** Indicates a median unbiased estimate.

Table 12. Pairwise comparisons from the repeated measures, multilevel logistic regression models for the likelihood of a calf within a pen having *P. multocida* with tetracycline resistance or spectinomycin resistance (antimicrobial resistance (AMR)) across time points for each year and metaphylaxis option, reported as population-averaged odds ratios (ORs) and 95% confidence intervals (CIs), accounting for clustering at the pen level. *n* = 1595 calves for which susceptibility data were available.

AMR Outcome of Interest	Year/Metaphylaxis	DOF Comparison	OR	95% CI	*p*-Value
Tetracycline	2020/tulathromycin	13DOF vs. 1DOF	0.4	0.2, 0.7	0.001
		36DOF vs. 1DOF	1.1	0.4, 2.7	0.83
		36DOF vs. 13DOF	2.9	1, 8.2	0.052
	2021/tulathromycin *	13DOF vs. 1DOF **	0.3	0, 1.7	0.25
		36DOF vs. 1DOF **	0.9	0, 5.8	0.91
		36DOF vs. 13DOF	·	·	·
	2021/oxytetracycline	13DOF vs. 1DOF	5.1	2.8, 9.3	≤0.001
		36DOF vs. 1DOF	6.0	2.9, 12.6	≤0.001
		36DOF vs. 13DOF	1.2	0.7, 1.9	0.50
Spectinomycin	2020/tulathromycin	13DOF vs. 1DOF	0.41	0.2, 0.7	0.001
		36DOF vs. 1DOF	1.02	0.4, 2.6	0.97
		36DOF vs. 13DOF	2.5	0.9, 6.9	0.084
	2021/tulathromycin *	13DOF vs. 1DOF **	0.1	0, 0.8	0.06
		36DOF vs. 1DOF **	0.5	0, 2.8	0.54
		36DOF vs. 13DOF	·	·	·
	2021/oxytetracycline	13DOF vs. 1DOF	4.3	2.5, 7.5	≤0.001
		36DOF vs. 1DOF	4.3	2.2, 8.3	≤0.001
		36DOF vs. 13DOF	1.0	0.6, 1.6	0.97

* Exact logistic regression equation performed. · odds ratio (OR) was not estimable. ** Indicates a median unbiased estimate.

2.8. Pen-Level Clustering of Bacterial Recovery and Antimicrobial Resistance

For all calves at 1DOF, differences in the recovery of bacteria of interest between pens measured by the intraclass correlation coefficient (ICC) were low at 0.029 (CI: 0.011–0.078) for *M. haemolytica*, 0.09 (CI: 0.04–0.18) for *P. multocida*, and 0.10 (CI: 0.04–0.23) for *H. somni*. In comparison, the proportion of variance explained by clustering at the pen level increased substantially for the recovery of *M. haemolytica* at 13DOF to 0.24 (CI: 0.13–0.39). The ICCs for *P. multocida* at 13DOF (0.067, CI: 0.025–0.16) and *H. somni* (0.16, CI: 0.06–0.35) were relatively stable compared to 1DOF.

Likewise, the prevalence of tulathromycin-resistant *M. haemolytica* was low (0.1%) at 1DOF, making the ICC negligible. At 13DOF, while conditioning on year and metaphylactic drug, the proportion of total variance explained by pen increased substantially to 0.72 (CI: 0.46–0.89). A similar increase in pen-level variation was observed for calves with tetracycline-resistant *M. haemolytica* where, again, the ICC was negligible at 1DOF but increased to 0.74 (CI: 0.29–0.95) at 13DOF. For calves with tetracycline-resistant *P. multocida*, the ICC at 1DOF was 0.078 (CI: 0.012–0.35), increasing to 0.39 at 13DOF (CI: 0.16–0.68).

2.9. Sample Size Estimates Generated from Simulation Models

The accuracy and precision of sample size estimates resulting from simulation studies are detailed in Supplementary Materials S2, Figure S1. The findings support the recommendation of sampling 20 to 30 calves per pen of 200 calves. This sample size range enabled the differentiation of a low, moderate, or high prevalence of calves with BRD pathogens exhibiting antimicrobial resistance.

3. Discussion

The results of this study support the antimicrobial stewardship efforts of the beef feedlot industry by providing evidence of pathogen and AMR variability over time and between pens. Currently, feedlot cattle are managed in groups at the pen level and antimicrobial decisions are based on experience, available history of the incoming cattle, studies on protocol effectiveness, and limited AMR surveillance reports. Sampling each individual animal prior to AMU might be an ultimate end goal of targeted antimicrobial decisions [7]; yet, it is currently neither feasible nor practical for large commercial feedlots. Instead, our simulation model supports that evidence-based laboratory data on individual pens could be generated by sampling a subset of 20 to 30 animals per pen of 200 calves at arrival for cattle not receiving metaphylaxis or shortly after the PMI in calves where metaphylaxis was used. The resulting culture and antimicrobial susceptibility testing (AST) data could then be used to inform pen-level management decisions and antimicrobial treatment protocols for animals that become sick after testing. Despite limitations in laboratory testing data and cut points with which to interpret them, antimicrobial stewardship can be promoted by reducing the use of antimicrobials likely to be ineffective based on their resistance profiles. Providing laboratory data that AMU is evidence-based can also address the demands of stakeholders, trade partners, and consumers who are increasingly concerned with the use of antimicrobials in animal agriculture [35,36].

Identifying optimal sampling times used to inform AMU decisions requires a comprehensive understanding of bacterial dynamics and antimicrobial susceptibility changes within feedlot pens during the early feeding period, when calves are at increased risk of BRD. This study therefore focused on sampling cattle at arrival, and prior to metaphylaxis treatment and prolonged comingling with pen mates, then again two weeks after metaphylaxis. As such, this study provides foundational knowledge needed for subsequent studies to build upon.

The design of this study was distinct from and built on findings from others evaluating changes in the early feeding period in several ways. First, multiple samples were collected from almost 800 calves per year and the study was repeated over the course of two years and with different metaphylaxis protocols to provide robust data for description and comparison. Second, animals in this study were sourced from mixed farms of origin (30–81 different farms per pen in 2020 and 12–38 farms per pen in 2021) and housed in pens with calf numbers that more closely approach those observed in commercial feedlots than pen sizes typically reported in research studies. As a result, this information builds on the work of other longitudinal research investigations that have used smaller pen numbers (20–30 calves/pen) [18,19] or enrolled cattle with limited diversity in herd of origin [19,24]. Perhaps most distinctively, this study maintained a consistent sampling approach, collecting a sample from each animal at two specific time points, with all samples submitted for AST. This contrasts with other studies that only sampled a subset of animals from commercial feedlot pens [15] in which not all recovered bacteria underwent AST [37] or the second sampling times occurred at varying DOF [15,38,39]. Our study's second time point of 13DOF provided further data on bacterial changes during the first two weeks on feed as opposed to others focusing on the evolution of bacteria from arrival to ≥40DOF [20,24,40,41] or differences between bacterial isolation in auction-derived and ranch-direct calves [18].

While sampling cattle at 1DOF captures baseline pathogen and AMR statuses at feedlot entry, the 13DOF time point is significant for its insight into shifts in bacterial prevalence and antimicrobial susceptibility following feedlot placement. The first few weeks on feed are an important time in the feedlot for the development of clinical BRD and the transmission and dissemination of BRD pathogens and AMR within calves in pens and between pens [31,42]. Commingling of animals from different sources, stress, potential environmental contamination, exposure to fomites, changes in diet, and metaphylactic antimicrobial administration can each affect respiratory microflora [31,43]. In this study, the choice of 13DOF sampling coincides with the maximum PMI for tulathromycin [44] and the period when pathogen numbers might rebound, and any AMR-associated selection from AMU at arrival or transmission might emerge.

Although national surveillance programs support efforts to monitor BRD pathogens and AMR trends over time, allowing for prevalence comparisons between geographies, these programs generally collect samples from a subset of animals from a subset of pens to estimate commodity- or feedlot-level prevalence [45,46]. To enhance compliance, the samples for the surveillance program are collected when it is most convenient for feedlot staff, at arrival and reprocessing. In contrast, the purposeful design of the present study repeatedly sampled all animals within each pen across the two time points early in the feeding period, which is of great interest for potentially informing the treatment of BRD.

In the present study, *P. multocida* was the most commonly recovered bacteria at arrival in 2020, while *M. haemolytica* was more common at arrival in 2021 and remained the prominent bacteria recovered at 13DOF for both years. This contrasts with the recent studies by Nobrega et al. [22] and Guo et al. [40], where *P. multocida* was reported as the most commonly recovered bacteria at arrival and throughout most subsequent time points. Although both studies noted above agreed in describing changes in pathogen recovery within calf groups over time [22,40], the present study was unique in further evaluating the specific effect of the group-level clustering of outcomes.

While no formal assessments of group effects could be made due to the size of the study, a previous report did describe differences in AMR across groups [18]. Hirsch et al. [18] compared the presence of bacterial pathogens and AMR from two groups of 30 cattle either directly transported to a feedlot or first transported to an auction market prior to feedlot placement. Deep nasal swabs were collected at feedlot processing (e.g., on arrival), 2DOF, and 9DOF [18]. *P. multocida* was also the most frequently isolated bacteria at the time of feedlot processing in their study, regardless of transport group. While their objective was to compare sampling times between auction market and ranch-direct calves, differences were noted in the prevalence of bacteria over time between the two feedlot groups, with one group experiencing the spread of a multi-drug-resistant strain of *P. multocida* while the second group observed no recovery of *P. multocida* at 9DOF or 30DOF [18]. Together, these results demonstrate the potential variation in bacterial behavior over time observed across different populations of animals enrolled within the same study.

The percentage of calves from which the organisms of interest were recovered was generally higher in this study compared to others, particularly for *M. haemolytica* [15,34,38]. The *H. somni* trends were similar to those observed by Erickson et al. [38] during earlier time points, but prevalence was higher at later sampling times. The on-arrival prevalence of bacteria was also higher than the first year of national surveillance averages reported by the Canadian Integrated Program for Antimicrobial Resistance Surveillance (CIPARS) [47]. The CIPARS project observed a 2020 national on-arrival isolate recovery rate of 8.9% (34/384 samples) for *M. haemolytica*, 28% (108/384 samples) for *P. multocida*, and 2.6% (10/384 samples) for *H. somni* (Sheryl Gow, personal communication). In comparison, from all years combined, the present study observed a crude on-arrival recovery of 41% of calves with *M. haemolytica*, 49% with *P. multocida*, and 8% with *H. somni*.

The current study adhered to the identical protocol for bacterial culture and AST performed at the same diagnostic laboratory utilized by CIPARS. One possible reason for the discrepancies in recovery rates between this study and CIPARS is the differences in the risk levels of the cattle sampled. Surveillance by CIPARS collected samples from a range of risk categories and was not restricted to the fall season, whereas the present study focused on fall-placed auction-sourced calves. Moreover, the samples in the current study had the advantage of a fast turn-around time from sample collection to arrival and processing at the laboratory. This delay was a few hours for the present study compared to days for CIPARS samples collected and shipped by courier from commercial feedlots.

The present study reports a higher proportion of calves with *M. haemolytica* with resistance to gamithromycin and tulathromycin than to tilmicosin or tildipirosin in 2020, but the results from 2021 are more consistent with the CIPARS surveillance report for 2019 [47]. Both our study and the 2019 CIPARS report identified tetracycline as the antibiotic with the highest resistance prevalence for *P. multocida*, while *H. somni* exhibited the least amount of AMR among the three bacteria [47]. Overall, the present study agreed with CIPARS in observing low AMR on feedlot arrival, with subsequent increases noted at second sampling times. A direct comparison of resistance prevalence between the two studies should be conducted cautiously due to differences in reporting: CIPARS presents resistance prevalence at the per-isolate level, whereas our study reports at the per-calf level in the primary text. However, isolate-level information in the present study can be derived from the MIC tables available in the Supplementary Materials S1.

In the present study, the on-arrival resistance of *M. haemolytica* isolates was lower than that reported by Andrés-Lasheras et al. [34] in their longitudinal study using 10 commercial Alberta feedlots. Andrés-Lasheras et al. [34] reported that their most common on-arrival AMR resistance in *M. haemolytica* isolates collected from beef-type cattle included oxytetracycline (10%), tilmicosin (6.4%), and ampicillin (4.6%). In comparison, in the present study, only one oxytetracycline- (0.2%), two tilmicosin- (0.3%), and three ampicillin-resistant *M. haemolytica* (0.5%) were recovered from cattle on arrival. The resistance of *P. multocida* isolates was also slightly higher in the Andrés-Lasheras et al. study [34], with their most frequent resistance profiles being for tetracycline (8.4%), ampicillin (7.7%), and spectinomycin (8.1%), compared to recoveries in the present study for tetracycline (5.1%), ampicillin (4.1%), and spectinomycin (6.0%). *H. somni* consistently exhibited minimal AMR on arrival and was observed at similar levels in the present study, the study by Andrés-Lasheras et al., and CIPARS [34].

Antimicrobial exposure contributes to changes in microbiota [23,31]. Metaphylaxis is an important strategy for BRD control in high-risk calves that results in an alteration in pathogen load [3,6]. However, a likely secondary consequence of blanket AMU is a change in the antimicrobial susceptibility patterns of bacteria, even if transient. This phenomenon was observed in the pens studied, where select antimicrobial MIC distributions trended up from 1DOF to 13DOF and generally decreased again by 36DOF first in 2020 and then again in 2021 [48]. This shift in susceptibility distribution differed between pens and could have been missed if pens had only been sampled at arrival and later in the feeding period (>40DOF). Other longitudinal studies have also observed an increase in MIC values in respiratory bacteria over the feeding period [10,15,22,26].

The results of this study are also consistent with evidence of the dissemination of strains of BRD pathogens containing AMR among feedlot calves [18,40,49,50]. The rapid increase in macrolide-resistant *M. haemolytica* in calves within pens and the consistency in this pattern across five of eight pens of the 2020 study population was unique compared to what was observed in either metaphylaxis treatment groups from 2021. A more thorough investigation of the changes in phenotypic susceptibility of *M. haemolytica* as demonstrated by MIC results observed during the early feeding period in the year 2020 has previously been described [48]. In agreement with the present study, Guo et al. [40] also found that while the respiratory microbiota of beef calves can increase in diversity from calf ranch to feedlot, the results also vary by calf ranch and feedlot.

Also in agreement, Hirsch et al. [18] also noted bacterial and antimicrobial susceptibility variations in feedlot calves. They compared two groups, one directly transported to a feedlot and the other to an auction market before feedlot placement. Deep nasal swabs were collected at feedlot processing (e.g., on-arrival), 2DOF, and 9DOF [18]. While their objective was to compare sampling times between auction market and ranch-direct calves, differences were noted in the prevalence of bacteria recovered over time between the two feedlot groups, with one group experiencing the spread of a multi-drug-resistant strain of *P. multocida* while the second group observed no recovery of *P. multocida* at 9DOF or 30DOF [18]. Together with our study, these outcomes demonstrate the potential variability in bacterial behavior across different animal populations, reinforcing the importance of pen-level sampling.

The proportion of calves with bacteria resistant to the antibiotic class used for metaphylaxis increased between arrival processing and the second sample at 13DOF. Additionally, macrolide-treated pens in 2020 showed evidence of the clonal spread of macrolide-resistant *M. haemolytica* [48]. Our findings correspond with those of other studies on feedlots linking antimicrobial use to the selection of oxytetracycline-resistant *P. multocida* [40] and antimicrobial-resistant *M. haemolytica* clones [49,50]. Studies by Woolums et al., Snyder et al., and Crosby et al. also revealed a high prevalence of macrolide-resistant and multi-drug-resistant (MDR) *M. haemolytica* in stocker cattle post metaphylaxis treatment with macrolides [27–29].

Furthermore, Holman et al. [31] described significant alterations in nasopharyngeal microbiota due to single administrations of either oxytetracycline or tulathromycin metaphylaxis treatments in commercial feedlots. An association between oxytetracycline metaphylaxis and AMR was also evident, specifically shown by a significant increase in tet(H), a gene responsible for tetracycline efflux, observed from entry to exit of the feedlot period.

While recent studies suggest a trend between antimicrobial classes used in metaphylaxis and subsequent AMR patterns, inconsistencies exist. For instance, a longitudinal study of feedlot cattle by Nobrega et al. [22] found no association between tetracycline metaphylaxis and tetracycline MICs in respiratory bacteria. However, they did note higher MICs for macrolides after parenteral metaphylaxis treatment.

Moreover, Woolums et al. [28] highlighted that, despite receiving metaphylaxis treatment with tildipirosin on arrival, all *M. haemolytica* isolates from their group of stocker cattle were resistant to enrofloxacin prior to exposure to the fluoroquinolone antimicrobial class. This suggests that antimicrobial exposure might provide advantages to different resistant strains beyond the administered antimicrobial class, raising concerns about the unpredictable effects of such exposure [28]. A better understanding of the drivers of AMR, whether from selective pressure or the dissemination of resistant clones, is an area deserving of future research.

In addition to the potential to initiate resistance to antimicrobials beyond those used in treatment, AMU can also affect non-target bacterial populations. This was demonstrated in a longitudinal study by Holman et al. [23] that evaluated the effects of oxytetracycline and tulathromycin on the fecal and nasopharyngeal microbiota of feedlot cattle. Both antibiotics altered fecal and nasopharyngeal microbiota, highlighting AMU's broader impact. While antimicrobials are vital for BRD treatment, the collateral effects of disrupting the antimicrobial susceptibility of several bacterial species underscores the need for precise drug selection. This is essential to mitigate potential cost–benefit implications of AMR emergence in non-target bacterial populations.

The differences in the prevalence of both BRD bacteria and AMR observed at 1DOF and 13DOF in this study raise questions regarding the most appropriate time point for sampling feedlot calves within pens. As is often the case with complex systems, there is no one-size-fits-all answer, and sampling times will depend on the reason for action. There are limited publications directly comparing the temporality of BRD in cattle receiving on-arrival antimicrobials compared to cattle that do not. Older studies performed prior to the consistent use of metaphylaxis that examined the timing of BRD have reported a high incidence of disease during the first weeks on feed [14,51]. Therefore, sampling cattle on arrival can provide data on the bacteria and antimicrobial susceptibility profiles of incoming cattle and could be useful in situations where metaphylaxis is not used and calves are at risk of BRD treatment within the first few DOF. On-arrival sampling strategies would also benefit AMR surveillance efforts in monitoring AMR risk based on exposure prior to feedlot entry and targeting critical intervention points. This, in turn, would contribute to proactive mitigation strategies for AMR surveillance that are important for human and animal health.

However, sampling cattle at a single time point only provides a snapshot in time and, while useful for directing immediate therapeutic decisions in animals that fall ill shortly after sampling, this is insufficient for predicting the development and directionality of future AMR dynamics. Further, the low frequency of AMR bacteria on arrival and rapid evolution of bacterial communities in the early feed period suggests that sampling after an antimicrobial's PMI might prove more useful for informing subsequent treatment decisions than samples collected at arrival.

Animals in food production systems are aggregated and managed in groups. This hierarchical structure results in cattle within the same feedlot pen being more like each other and experiencing similar exposures to infectious disease than cattle in different pens. To the authors' knowledge, no other studies have specifically reported this effect of the clustering of cattle within pens on outcomes of interest related to BRD pathogen recovery and subsequent AMR by calculating the ICC. An ICC value of zero would imply no correlation in observations for calves within their cluster (i.e., pen), while an ICC value of one would indicate identical observations for calves within the same cluster [52]. In this study, the ICC was low for all outcomes analyzed at 1DOF, which was to be expected as the animals had been aggregated into their respective groups for less than 24 h. However, by 13DOF, ICCs increased substantially for *M. haemolytica* as well as for tulathromycin- and tetracycline-resistant *M. haemolytica*, implying high variability in the prevalence of resistance between pens. ICC estimates can vary between studies and populations and are important for understanding the effect of pen-level clustering [53].

Overall, this study provides support for practical, pen-level sampling. The explorations made lay the foundation necessary to begin building strategies for informed and justifiable antimicrobial treatment choices in BRD calves. While unanswered questions remain, fundamental aspects needed to be addressed regarding the need for pen-level sampling and were necessary first steps for future advancements in this realm.

Study Limitations

All samples were collected by two previously trained individuals in a facility with a hydraulic chute and neck extender with good restraint. Further in this study, an aliquot from a pooled sample of three DNP swabs per animal was used for bacterial culture. This technique might have improved the likelihood of obtaining a sample that represented the colonization status of the nasopharynx of the animal and increased the likelihood of a positive culture result in this study compared to those from which only one DNP was utilized [54–56]. Additionally, sample handling and transport time can affect the viability of bacteria prior to arrival at the laboratory. Samples were delivered to the diagnostic laboratory in less than an hour following sample collection from the last calf. Thus, direct comparisons between recovery rates of studies could be impacted by transport times.

The use of DNP swabs is also limited with respect to reflecting pathogens of the lower respiratory tract responsible for BRD. However, existing studies suggest reasonable agreement between upper and lower respiratory tract sampling techniques [54,57–60], despite the biological variability [61]. Regardless, DNP swabs are more easily implemented into commercial feedlot settings compared to more time-consuming and technically challenging sampling options such as transtracheal wash or bronchoalveolar lavage [62].

Bacterial culture has limited sensitivity, despite its frequent use as a gold standard [63]. The present investigation selected one isolate of each organism for AST. While this is a conventional practice used in part to reduce the costs associated with laboratory testing, the method assumes that all colonies within a culture plate display a common AMR pattern [22,34,47]. The number of colonies needed to estimate the diversity of isolates present on a primary culture plate has been determined for other bacterial pathogens [64]. To investigate this question in reference to *M. haemolytica* and describe the potential phenotypic and genotypic diversity of *M. haemolytica* isolates from individual animals, Carter et al. collected DNP swabs from 28 cattle at risk of, or treated for, BRD [65]. Up to 20 *M. haemolytica* colonies were selected per plate (up to 100 colonies per nasopharyngeal swab) [65]. Using a previously established genotyping technique [66], the study found *M. haemolytica* isolates from individual calf samples to be uniform in both genotype and AMR phenotype and suggested that the selection of few colonies could sufficiently represent the relevant susceptibility pattern of the plate [65]. In contrast, the use of pulse field gel electrophoresis (PFGE) led to the finding that calves can have more than one cluster of *M. haemolytica* when multiple isolates are taken from a plate, even if most were identical [67]. Another study also identified that calves have the potential to simultaneously shed *P. multocida* isolates with differing plasmid profiles [30]. Together, these studies indicate that while a dominant strain of bacteria might exist within a sample, the present study could have missed resistant isolates if only one colony per plate was selected for testing, particularly in the case of *H. somni*, for which the diversity of isolates within individual plates has not been evaluated [65,67].

In addition to the bacteria belonging to the *Pasteurellaceae* family, *Mycoplasma bovis* is another bacterium implicated in BRD, particularly in chronically ill animals [68]. While *M. bovis* was investigated, the approach to culture and MIC data followed a distinct protocol that was not directly comparable to the methodology applied for the other three bacteria of interest. As such, *M. bovis* data will be examined in a subsequent report.

Commercial feedlots are described as having BRD morbidity rates of 10–30% in auction-derived calves, and mortality rates for animals treated for BRD are posited to be around 5–10% [69]. In contrast, the present study observed a relatively low number of calves receiving first treatment for BRD, with an average of 8.1%. The BRD mortality rate was also low at 0.7% and not all of these animals had received prior BRD treatment. The smaller pen size of 100 cattle in our study might have played a role in contributing to the comparatively lower rates when contrasted with commercial feedlots that accommodate 150–300 cattle per pen, with total holding capacities ranging from 15,000 to 25,000. A higher morbidity rate would likely have led to increased AMU, potentially leading to a greater prevalence of AMR over time or a greater diversity in resistance more comparable to commercial feedlots. Most studies of AMR in BRD pathogens in feedlot cattle have not been designed to assess the specific impact of phenotypic AMR on BRD outcomes [70], which is an important area for future research.

4. Materials and Methods

4.1. Ethical Statement

The research protocol was approved by the University of Saskatchewan Animal Care Committee (AUP 20190069).

4.2. Study Population

Recently weaned steers of various beef breeds were sourced from a regional auction market in Saskatchewan, typical of Western Canada. One hundred calves were purchased once a week for 8 weeks in 2020 and again in 2021. Placements occurred in the fall from 6 October to 1 December 2020 and 28 September to 16 November 2021. Each group of 100 calves was maintained as a single cohort and assigned to consecutive feedlot pens. Herd of origin was approximated using the first 12 digits of the calves' 15-digit RFID tags. Calves from the same herd of origin have unique tags typically consecutively placed from a sequentially numbered, commercially sourced package either at birth, spring processing, or shipping and, therefore, share the initial numeric sequence unique to a herd. The mean weight of calves from the 2020 study population was 253 kg (range 211–291 kg). Lighter-weight calves were targeted for 2021 and, hence, the mean weight was 225 kg (range 351–694 kg). On the day of purchase, calves were transported 51 km to a research feedlot at the Livestock and Forage Centre of Excellence (LFCE) in Clavet, Saskatchewan. Calves were rested in a holding pen overnight and processed the following morning.

4.3. Calf Processing Procedure

All animals ($n = 1600$) were processed at 1DOF following industry protocol that included the placement of a feedlot identification ear tag, verification of castration, and subcutaneous administration of *M. haemolytica* and a modified live viral vaccine (Pyramid® 5 + Presponse®, Boehringer Ingelheim Animal Health, Duluth, GA, USA) and a multivalent clostridial vaccine (Ultrachoice® 7, Zoetis Inc., Florham Park, NJ, USA). All calves received a growth implant (Ralgro®, Merck Animal Health, Rahway, NJ, USA) and a topical anthelmintic (Solmectin™, Solvet, Calgary, AB, Canada). In 2020, calves ($n = 800$) received metaphylactic tulathromycin as a single dose of 2.5 mg/kg of body weight (Draxxin®, Zoetis Inc., Florham Park, NJ, USA), administered subcutaneously based on the average weight of the cohort. In the fall of 2021, cattle in four pens ($n = 400$ calves) were administered metaphylactic tulathromycin and cattle in the other four pens ($n = 400$ calves) were administered oxytetracycline (Oxyvet®200 LA, Vetoquinol, Lavaltrie, QC, Canada) subcutaneously as a single dose of 20mg/kg of body weight. Following processing, each cohort of 100 calves was placed in their designated home pen, where they remained for the duration of the study.

4.4. Animal Housing and Management

Calves were housed in eight outdoor, dirt floor pens, designed as per the Canadian guidelines for feedlot cattle [71]. Each pen held 100 animals and pens were filled consecutively. The four adjacent pairs of pens each shared fence-line watering bowls. The first four and last four pens had contact through consecutive cross-fences. A building separated the first four pens from the last four pens.

On day 1, calves were fed a high-forage starter feed ration (34% barley silage, 15% barley, 44% hay, 7% canola meal) to encourage bunk eating. For the remainder of the feeding period, the diet consisted of 59% barley silage, 15% barley, 20% hay, and 6% canola meal. The calves were started at an estimated 15 lbs/head on an as-fed basis (10 lbs dry-matter intake (DMI)), which was steadily increased until the calves reached 30–33 lbs as-fed or 20–22 lbs of DMI. A supplement was provided consisting of salt (1500 mg/kg) and vitamins A, D, and E. Monensin (33mg/kg dry matter, concentrate) was included in this supplement and was the only in-feed antimicrobial administered.

4.5. Sampling Procedures

All calves were sampled at two time points: at the time of arrival and processing (1DOF) prior to metaphylaxis administration and again at 13DOF. A random subset of calves from each pen was sampled at 36DOF as determined by available resources and laboratory capacity: 10 calves/pen in 2020 and 30 calves/pen in 2021. A snowstorm in 2021 delayed the initial sampling of one pen (#16) by a day. Thus, sampling time points occurred on 12DOF and 35DOF, respectively. The results from these samplings were incorporated into the analysis along with samples from the regular 13DOF and 36DOF sampling performed for all other pens. In addition, above-average mortalities in pen 16 (the last pen filled in the year 2021) led to mass-treatment with oxytetracycline at 30DOF. As a result, 20 calves from pen 16 were sampled pre-treatment (30DOF) and post-treatment (35DOF).

At each sampling time point, calves were restrained in a hydraulic chute and sampled via three DNP swabs; a neck extender was used to stabilize the calves' heads during sampling. A single-use paper towel was used to wipe clean the external nares, and a double-guarded culture swab (Continental Plastic Corp., Delevan, WI, USA) was directed into the ventral meatus of the nostril. The polyester-tipped swab was advanced through the inner sheath and vigorously rotated against the nasopharyngeal mucosa for 5–6 rotations. The swab was withdrawn into the inner sheath and outer guard prior to removal from the nostril, and approximately 3 cm of swab tip was cut and placed in a 15 mL vial containing 3 mL of liquid Amies transport medium. Two additional samples were obtained from alternating nostrils using the same procedure and all three DNP swabs per calf were pooled in the same vial.

4.6. Bacteriology

Samples were transported to the University of Saskatchewan for same-day processing. The samples were vortexed for 1 min and a 300 uL aliquot was submitted to Prairie Diagnostic Services, Inc. (Saskatoon, SK, Canada, PDS). For *M. haemolytica*, *P. multocida*, and *H. somni* cultures, a 10 uL inoculation loop of sample was cultured on Columbia agar with 5% sheep blood (BA) and a second loop was cultured on chocolate agar (CHOC); plates were incubated at 35 °C for 18 h in 5% CO_2. Bacterial colonies were examined at 18 h and 42 h of incubation. By examining both BA and CHOC plates, one isolate exhibiting phenotypic morphologies for each bacterium of interest was selected and confirmed using MALDI-TOF MS (Bruker Daltonik, Bremen, Germany), according to manufacturer guidelines. MALDI-TOF MS Biotyper Microflex LT Compass version 1.4 software and the MSP library were used for direct testing. If visible characteristics suggested the presence of multiple isolates of interest from one sample, representative colonies of each unique colony morphology were selected for identification. Positive and negative controls were processed for each day of sample setup and for each new media lot using *Staphylococcus aureus* ATCC 29213, *Escherichia coli* ATCC 25922, and *Histophilus somni* ATCC 700025. Only MALDI-TOF identification scores of ≥ 2 indicating secure species-level identification were used for further analysis. A plain matrix spot was run with every MALDI run to ensure no contamination. The diagnostic laboratory used in this study also processed the respiratory samples for the national surveillance program, CIPARS, following the same methods for bacterial isolation and identification.

4.7. Antimicrobial Susceptibility Testing

The AST procedures were also consistent with those used in CIPARS. This included utilizing the same AST microdilution panel and adhering to identical reporting standards for distinguishing between susceptible and resistant isolates.

For AST, all isolates that showed positive MALDI-TOF MS identification were streaked from the inoculated Todd Hewitt broth onto purity plates specific to each bacteria type: BA for *M. haemolytica* and *P. multocida* and CHOC for *H. somni*. Each colony of interest from the purity plates underwent AST using a commercially available bovine serial broth microdilution panel (Thermo Fisher Scientific™, Waltham, MA, USA, Bovine AST BOPO7F Plate) on the Sensititre™ platform. *Escherichia coli* ATCC 25922, *Staphylococcus aureus* ATCC 29213, and *Histophilus somni* ATCC 700025 were used as positive controls. The minimum inhibitory concentrations (MICs) plate was placed and read on the BIOMIC® V3 microplate reader. The MIC value was considered equal to the lowest concentration of antimicrobial that inhibited visible growth. The MIC for each antimicrobial was compared to Clinical and Laboratory Standards Institute (CLSI) breakpoints, where available [72]. Isolates with "intermediate" MICs were categorized as "susceptible".

MIC 50 and MIC 90 were defined as the MIC values at which $\geq$50% or $\geq$90% of the isolates were inhibited [73]. Results were summarized in distribution tables by pathogen, antimicrobial, and sampling time point (1DOF, 13DOF, and 36DOF). The MIC results are provided in the Supplementary Materials S1 (Tables S1–S9) for all antimicrobials tested, regardless of whether a CLSI breakpoint was available.

4.8. Treatment of Calves with BRD

Experienced feedlot personnel monitored the animals daily for signs of illness. Calves exhibiting signs of respiratory disease were identified using a DART (depression, appetite, respiratory system, temperature) BRD clinical scoring system [74]. The severity of clinical signs was graded using a standardized numerical scale ranging from 0 (clinically normal) to 4 (moribund). To meet the BRD case definition and receive treatment, calves needed to have a score of 1 or 2 with a rectal temperature $\geq$ 40 °C or a score of 3 or 4 regardless of temperature (and with no other obvious causes of illness).

For calves that received metaphylactic tulathromycin, a PMI (waiting period before eligibility for retreatment) of 7 d was observed. Calves developing BRD after the metaphylaxis were administered florfenicol 40 mg/kg BW and flunixin 2.2 mg flunixin/kg BW (Resflor Gold®, Merck Animal Health, Rahway, NJ, USA,) subcutaneously. For calves that received metaphylactic oxytetracycline, a 5 d PMI was observed. The treatment regimen for those calves was tulathromycin (Draxxin®, Zoetis Inc., Florham, NJ, USA) administered subcutaneously at a dose of 2.5 mg/kg of body weight. Calves were returned to their home pen following treatment.

Morbidities and mortalities associated with *H. somni* in pen 16 during the year 2021 were mass-medicated, resulting in the cooperative decision between the feedlot manager, feedlot veterinarians, and the research team to mass-treat the pen cohort at 30DOF. Animals in this pen were administered oxytetracycline (Vetoquinol, Oxyvet® 200 LA, Lavaltrie, QC, Canada) subcutaneously at a dose of 20 mg/kg of BW.

4.9. Statistical Analysis

Data were entered and managed in a spreadsheet (Microsoft Excel, version 2401, Microsoft Corporation, Redmond, Washington, DC, USA), and analyses were completed using a commercial statistical software package (Stata/IC, version 16.1, StataCorp LLC, College Station, TX, USA). Bacterial recovery was summarized at the calf level; antimicrobial susceptibility was also summarized by reporting prevalence at the calf level rather than at the recovered isolate level (i.e., the denominators for all antimicrobial susceptibility prevalence calculations were the total number of calves, not the total number of recovered isolates). Data from this study were used to inform simulation-based sample size calculations for pen-level sampling (Supplementary Materials S1).

The total number of calves from each pen was summarized for the positive recovery of *M. haemolytica*, *P. multocida*, and *H. somni*; bacterial co-isolation; and bacteria of interest classified as resistant to antimicrobials, with breakpoints established by the CLSI [72].

The recovery of *M. haemolytica*, *P. multocida*, and *H. somni* at the calf level at 1DOF and 13DOF were each compared between years 2020 and 2021 using mixed-effects logistic regression models [52]. Differences between years in the frequency of calves from which AMR pathogens were recovered were also examined when the crude prevalence of AMR was $\geq$5%. Pen-level clustering was accounted for as a random intercept in all models. Models for differences between years in the recovery of bacteria of interest and bacteria with AMR at 1DOF included year as a fixed effect. Differences at 13DOF also accounted for the choice of metaphylaxis. A single fixed effect accounted for both study population year and the metaphylactic antimicrobial used: year 2020 calves treated with metaphylactic tulathromycin, year 2021 calves treated with metaphylactic tulathromycin, and year 2021 calves treated with metaphylactic oxytetracycline.

Population-averaged prevalence and 95% confidence intervals (CIs) were estimated for calves with BRD pathogens of interest, as well as for calves from which pathogens with resistance patterns of interest were recovered from mixed models for each time point, as described (1DOF, 13DOF, and 36DOF) using the variance of the random pen effects estimated from the models as follows: $\beta^{PA} = \beta^{SS}/(1 + 0.346\,\sigma^2_h)^{0.5}$ [52]. Similarly, population-averaged odds ratios (ORs) were determined when summarizing relative differences among groups. Intraclass correlation coefficients (ICC values) were reported to estimate the extent of the clustering of outcomes within pens (h), as followed from the variance from the random effects for pen: $\rho = \sigma^2_h/(\sigma^2_h + \pi^2/3)$ [52].

Differences in pathogen recovery between sampling time points (1DOF, 13DOF, and 36DOF) accounting for repeated measures on individual calves within pens were examined with three-level mixed-effects logistic regression models for each combination of year and metaphylactic treatment. For each model, year, metaphylaxis, sampling time point, and an interaction term between the year/metaphylaxis and sampling time point variables were included as fixed effects. Individual calves and calves nested within pens were set as random intercepts. Post-hoc Wald tests were used to test the significance ($p < 0.05$) of the coefficients of the interaction term for year/metaphylaxis and sampling time for each model. The likelihood of pathogen recovery at 13DOF vs. 1DOF, 36DOF vs. 1DOF, and 36DOF vs. 13DOF were then compared for calves in year 2020 treated with metaphylactic tulathromycin, calves in year 2021 treated with metaphylactic tulathromycin, and calves in year 2021 treated with metaphylactic oxytetracycline.

Similar models were repeated for AMR patterns where the crude prevalence was $\geq 5\%$ for at least one sampling time to estimate differences in recovery for each antimicrobial-resistant pathogen of interest between sampling points (1DOF, 13DOF, and 36DOF). When mixed-effects models failed to converge due to sampling time points with few or zero calves, exact logistic regression models (SAS for Windows, version 9.4, Cary, NC, USA) were used to generate estimates.

Simulation models were also developed to examine the effectiveness of different sample sizes to support pen-level testing (Supplementary Materials S2, Figure S1).

5. Conclusions

This study highlights substantial variability in the prevalence of target BRD bacteria and antimicrobial susceptibility profiles among pens of fall-placed calves at higher risk of BRD, observed both at feedlot arrival and again at 13DOF. The culture and AST demonstrated considerable pen-level variability over time within individual cohorts of fall-placed high-risk calves as well as between years and across different metaphylaxis protocols.

These findings emphasize the importance of pen-level management and the challenge of making antimicrobial drug choices without laboratory guidance. Sampling a subset of 20–30 calves in a feedlot pen, either near arrival in pens where cattle did not receive metaphylaxis or after a PMI in treated cattle, would allow feedlot veterinarians and managers to make more informed AMU decisions based on the risk assessment of individual pens. The use of laboratory-based results to target antimicrobial drug selection in feedlot pens will allow the industry to remain aligned with WHO recommendations [7] and establish practical antimicrobial stewardship recommendations. These are important first steps toward improving prudent AMU in feedlots. However, considering the demonstrated dynamics of bacterial populations, exploring technologies that improve turn-around times from sampling to results and provide comprehensive data on both organisms of interest and AMR are warranted. Such advancements could better support timely antimicrobial decision-making in commercial settings.

Supplementary Materials: The following supporting information can be downloaded at https://www.mdpi.com/article/10.3390/antibiotics13040322/s1: Supplementary Materials S1—Tables S1–S9: Supplemental MIC tables. Supplementary Materials S2—Figure S1: Simulation model to evaluate sample size estimates in support of pen-level testing.

Author Contributions: Conceptualization, J.R.C. and C.L.W.; methodology, C.L.W., J.N.A.Y. and S.L.; sample curation, J.N.A.Y. and S.L.; project administration, C.L.W., J.N.A.Y. and S.L.; funding acquisition, J.R.C. and C.L.W.; data analysis, J.N.A.Y. and C.L.W.; writing—original draft preparation, J.N.A.Y. and C.L.W.; writing—editing and review, C.L.W., S.J.G.O., M.J., S.P.G., A.R.W., J.R.C. and J.N.A.Y. All authors have read and agreed to the published version of the manuscript.

Funding: This research was funded by the Saskatchewan Agriculture Development Fund, grant number 20180129.

Institutional Review Board Statement: This study was conducted in accordance with the Declaration of Helsinki and approved by the University of Saskatchewan Animal Care Committee (AUP 20190069, last renewed on 25 May 2022).

Informed Consent Statement: Not applicable.

Data Availability Statement: The original contributions presented in this study are included in the article/Supplementary Material; further inquiries can be directed to the corresponding author.

Acknowledgments: We extend our sincere gratitude to the entire laboratory team for their expertise and contributions to the success of this project. Special thanks to Jayce Fossen for his diligent work in animal sampling and to Morgan Lehmann and Sharlene April for their assistance in sample handling. We are also grateful to Dana Ramsay for her support in coding for the development of the MIC tables and to the dedicated crew of the Livestock and Forage Center of Excellence for their cooperation and animal handling skills.

Conflicts of Interest: The authors declare no conflicts of interest.

References

1. Ives, S.E.; Richeson, J.T. Use of Antimicrobial Metaphylaxis for the Control of Bovine Respiratory Disease in High-Risk Cattle. *Vet. Clin. N. Am. Food Anim. Pract.* **2015**, *31*, 341–350. [CrossRef] [PubMed]
2. Brault, S.A.; Hannon, S.J.; Gow, S.P.; Warr, B.N.; Withell, J.; Song, J.; Williams, C.M.; Otto, S.J.G.; Booker, C.W.; Morley, P.S. Antimicrobial Use on 36 Beef Feedlots in Western Canada: 2008–2012. *Front. Vet. Sci.* **2019**, *6*, 329. [CrossRef] [PubMed]
3. Edwards, T.A. Control Methods for Bovine Respiratory Disease for Feedlot Cattle. *Vet. Clin. N. Am. Food Anim. Pract.* **2010**, *26*, 273–284. [CrossRef] [PubMed]
4. O'Connor, A.M.; Hu, D.; Totton, S.C.; Scott, N.; Winder, C.B.; Wang, B.; Wang, C.; Glanville, J.; Wood, H.; White, B.; et al. A systematic review and network meta-analysis of bacterial and viral vaccines, administered at or near arrival at the feedlot, for control of bovine respiratory disease in beef cattle. *Anim. Health Res. Rev.* **2019**, *20*, 143–162. [CrossRef] [PubMed]
5. Smith, R.A.; Step, D.L.; Woolums, A.R. Bovine Respiratory Disease: Looking Back and Looking Forward, What Do We See? *Vet. Clin. N. Am. Food Anim. Pract.* **2020**, *36*, 239–251. [CrossRef] [PubMed]
6. Nickell, J.S.; White, B.J. Metaphylactic Antimicrobial Therapy for Bovine Respiratory Disease in Stocker and Feedlot Cattle. *Vet. Clin. N. Am. Food Anim. Pract.* **2010**, *26*, 285–301. [CrossRef] [PubMed]
7. Aidara-Kane, A.; Angulo, F.J.; Conly, J.M.; Minato, Y.; Silbergeld, E.K.; McEwen, S.A.; Collignon, P.J.; Balkhy, H.; Collignon, P.; Conly, J.; et al. World Health Organization (WHO) guidelines on use of medically important antimicrobials in food-producing animals. *Antimicrob. Resist. Infect. Control* **2018**, *7*, 7. [CrossRef] [PubMed]
8. Ryan, M. *Assessing National Action Plans on Antimicrobial Resistance in Animal Production. What Lessons Can Be Drawn?* OECD Food, Agriculture and Fisheries Papers, No. 153; OECD Publishing: Paris, France, 2021. [CrossRef]
9. Klima, C.L.; Holman, D.B.; Cook, S.R.; Conrad, C.C.; Ralston, B.J.; Allan, N.; Anholt, R.M.; Niu, Y.D.; Stanford, K.; Hannon, S.J.; et al. Multidrug Resistance in *Pasteurellaceae* Associated With Bovine Respiratory Disease Mortalities in North America From 2011 to 2016. *Front. Microbiol.* **2020**, *11*, 606438. [CrossRef] [PubMed]
10. Stanford, K.; Zaheer, R.; Klima, C.; McAllister, T.; Peters, D.; Niu, Y.D.; Ralston, B. Antimicrobial Resistance in Members of the Bacterial Bovine Respiratory Disease Complex Isolated from Lung Tissue of Cattle Mortalities Managed with or without the Use of Antimicrobials. *Microorganisms* **2020**, *8*, 288. [CrossRef] [PubMed]
11. Watts, J.L.; Sweeney, M.T. Antimicrobial Resistance in Bovine Respiratory Disease Pathogens: Measures, Trends, and Impact on Efficacy. *Vet. Clin. N. Am. Food Anim. Pract.* **2010**, *26*, 79–88. [CrossRef] [PubMed]
12. Taylor, J.D.; Fulton, R.W.; Lehenbauer, T.W.; Step, D.L.; Confer, A.W. epidemiology of bovine respiratory disease: What is the evidence for predisposing factors? *Can. Vet. J. = La Rev. Vet. Can.* **2010**, *51*, 1095–1102.
13. Snowder, G.D.; Van Vleck, L.D.; Cundiff, L.V.; Bennett, G.L. Bovine respiratory disease in feedlot cattle: Environmental, genetic, and economic factors. *J. Anim. Sci.* **2006**, *84*, 1999–2008. [CrossRef] [PubMed]
14. Sanderson, M.W.; Dargatz, D.A.; Wagner, B.A. Risk factors for initial respiratory disease in United States' feedlots based on producer-collected daily morbidity counts. *Can. Vet. J.* **2008**, *49*, 373–378. [PubMed]

15. Noyes, N.R.; Benedict, K.M.; Gow, S.P.; Booker, C.W.; Hannon, S.J.; McAllister, T.A.; Morley, P.S. *Mannheimia haemolytica* in feedlot cattle: Prevalence of recovery and associations with antimicrobial use, resistance, and health outcomes. *J. Vet. Intern. Med.* **2015**, *29*, 705–713. [CrossRef] [PubMed]
16. Smith, K.J.; White, B.J.; Amrine, D.E.; Larson, R.L.; Theurer, M.E.; Szasz, J.I.; Bryant, T.C.; Waggoner, J.W. Evaluation of First Treatment Timing, Fatal Disease Onset, and Days from First Treatment to Death Associated with Bovine Respiratory Disease in Feedlot Cattle. *Vet. Sci.* **2023**, *10*, 204. [CrossRef] [PubMed]
17. Smith, R.A. Impact of disease on feedlot performance: A review. *J. Anim. Sci.* **1998**, *76*, 272–274. [CrossRef] [PubMed]
18. Hirsch, C.; Timsit, E.; Uddin, M.S.; Guan, L.L.; Alexander, T.W. Comparison of pathogenic bacteria in the upper and lower respiratory tracts of cattle either directly transported to a feedlot or co-mingled at auction markets prior to feedlot placement. *Front. Vet. Sci.* **2023**, *9*, 1026470. [CrossRef] [PubMed]
19. McMullen, C.; Alexander, T.W.; Orsel, K.; Timsit, E. Progression of nasopharyngeal and tracheal bacterial microbiotas of feedlot cattle during development of bovine respiratory disease. *Vet. Microbiol.* **2020**, *248*, 108826. [CrossRef] [PubMed]
20. McMullen, C.; Orsel, K.; Alexander, T.W.; van der Meer, F.; Plastow, G.; Timsit, E. Evolution of the nasopharyngeal bacterial microbiota of beef calves from spring processing to 40 days after feedlot arrival. *Vet. Microbiol.* **2018**, *225*, 139–148. [CrossRef] [PubMed]
21. O'Connor, A.M.; Hu, D.; Totton, S.C.; Scott, N.; Winder, C.B.; Wang, B.; Wang, C.; Glanville, J.; Wood, H.; White, B.; et al. A systematic review and network meta-analysis of injectable antibiotic options for the control of bovine respiratory disease in the first 45 days post arrival at the feedlot. *Anim. Health Res. Rev.* **2019**, *20*, 163–181. [CrossRef] [PubMed]
22. Nobrega, D.; Andres-Lasheras, S.; Zaheer, R.; McAllister, T.; Homerosky, E.; Anholt, R.M.; Dorin, C. Prevalence, Risk Factors, and Antimicrobial Resistance Profile of Respiratory Pathogens Isolated From Suckling Beef Calves to Reprocessing at the Feedlot: A Longitudinal Study. *Front. Vet. Sci.* **2021**, *8*, 764701. [CrossRef] [PubMed]
23. Holman, D.B.; Yang, W.; Alexander, T.W. Antibiotic treatment in feedlot cattle: A longitudinal study of the effect of oxytetracycline and tulathromycin on the fecal and nasopharyngeal microbiota. *Microbiome* **2019**, *7*, 86. [CrossRef] [PubMed]
24. Timsit, E.; Workentine, M.; Schryvers, A.B.; Holman, D.B.; van Der Meer, F.; Alexander, T.W. Evolution of the nasopharyngeal microbiota of beef cattle from weaning to 40days after arrival at a feedlot. *Vet. Microbiol.* **2016**, *187*, 75–81. [CrossRef] [PubMed]
25. Loerch, S.C.; Fluharty, F.L. Physiological changes and digestive capabilities of newly received feedlot cattle. *J. Anim. Sci.* **1999**, *77*, 1113–1119. [CrossRef] [PubMed]
26. Anholt, R.M.; Klima, C.; Allan, N.; Matheson-Bird, H.; Schatz, C.; Ajitkumar, P.; Otto, S.J.; Peters, D.; Schmid, K.; Olson, M.; et al. Antimicrobial Susceptibility of Bacteria That Cause Bovine Respiratory Disease Complex in Alberta, Canada. *Front. Vet. Sci.* **2017**, *4*, 207. [CrossRef] [PubMed]
27. Crosby, W.B.; Karisch, B.B.; Hiott, L.M.; Pinnell, L.J.; Pittman, A.; Frye, J.G.; Jackson, C.R.; Loy, J.D.; Epperson, W.B.; Blanton, J.; et al. Tulathromycin metaphylaxis increases nasopharyngeal isolation of multidrug resistant *Mannheimia haemolytica* in stocker heifers. *Front. Vet. Sci.* **2023**, *10*, 1256997. [CrossRef] [PubMed]
28. Woolums, A.; Karisch, B.; Frye, J.; Epperson, W.; Smith, D.; Blanton, J.; Austin, F.; Kaplan, R.; Hiott, L.; Woodley, T.; et al. Multidrug resistant *Mannheimia haemolytica* isolated from high-risk beef stocker cattle after antimicrobial metaphylaxis and treatment for bovine respiratory disease. *Vet. Microbiol.* **2018**, *221*, 143. [CrossRef] [PubMed]
29. Snyder, E.; Credille, B.; Berghaus, R.; Giguère, S. Prevalence of multi drug antimicrobial resistance in isolated from high-risk stocker cattle at arrival and two weeks after processing. *J. Anim. Sci.* **2017**, *95*, 1124–1131. [CrossRef] [PubMed]
30. Briggs, R.E.; Frank, G.H.; Purdy, C.W.; Zehr, E.S.; Loan, R.W. Rapid spread of a unique strain of *Pasteurella haemolytica* serotype 1 among transported calves. *Am. J. Vet. Res.* **1998**, *59*, 401–405. [CrossRef] [PubMed]
31. Holman, D.B.; Timsit, E.; Booker, C.W.; Alexander, T.W. Injectable antimicrobials in commercial feedlot cattle and their effect on the nasopharyngeal microbiota and antimicrobial resistance. *Vet. Microbiol.* **2018**, *214*, 140–147. [CrossRef] [PubMed]
32. Harner, J.; Boyer, W.; Davidson, J.; George, H.; Graber, R.; Murphy, J.P. Guidelines for Planning Cattle Feedlots. 2017. Available online: https://www.kcare.k-state.edu/pubs/livestock_management/Guidelines%20for%20Planning%20Cattle%20Feedlots%20MF3392.pdf (accessed on 5 August 2023).
33. Pound-Maker. Feedlot Virtual Tour. Available online: https://pound-maker.ca/pages/feedlot-virtual-tour (accessed on 30 January 2024).
34. Andrés-Lasheras, S.; Ha, R.; Zaheer, R.; Lee, C.; Booker, C.W.; Dorin, C.; Van Donkersgoed, J.; Deardon, R.; Gow, S.; Hannon, S.J.; et al. Prevalence and Risk Factors Associated With Antimicrobial Resistance in Bacteria Related to Bovine Respiratory Disease—A Broad Cross-Sectional Study of Beef Cattle at Entry Into Canadian Feedlots. *Front. Vet. Sci.* **2021**, *8*, 692646. [CrossRef] [PubMed]
35. European Medicines Agency. *Categorisation of Antibiotics in the European Union*; EMA/CVMP/CHMP/682198/2017; European Medicines Agency: Amsterdam, The Netherlands, 2019. Available online: https://www.ema.europa.eu/en/documents/report/categorisation-antibiotics-european-union-answer-request-european-commission-updating-scientific-advice-impact-public-health-and-animal-health-use-antibiotics-animals_en.pdf (accessed on 12 June 2022).
36. Barrett, J.R.; Innes, G.K.; Johnson, K.A.; Lhermie, G.; Ivanek, R.; Greiner Safi, A.; Lansing, D. Consumer perceptions of antimicrobial use in animal husbandry: A scoping review. *PLoS ONE* **2021**, *16*, e0261010. [CrossRef] [PubMed]
37. Andrés-Lasheras, S.; Jelinski, M.; Zaheer, R.; McAllister, T.A. Bovine Respiratory Disease: Conventional to Culture-Independent Approaches to Studying Antimicrobial Resistance in North America. *Antibiotics* **2022**, *11*, 487. [CrossRef] [PubMed]

38. Erickson, N.E.; Ngeleka, M.G.; Lubbers, B.V.; Trokhymchuk, A. Changes in the rates of field isolation and antimicrobial susceptibility of bacterial pathogens collected from fall-placed feedlot steers between arrival at the feedlot and 90 to 120 days on feed. *Bov. Pract.* **2017**, *51*, 165–173. [CrossRef]

39. Wennekamp, T.R.; Waldner, C.L.; Windeyer, M.C.; Larson, K.; Trokhymchuk, A.; Campbell, J.R. Antimicrobial resistance in bovine respiratory disease: Auction market- and ranch-raised calves. *Can. Vet. J.* **2022**, *63*, 47–54. [CrossRef] [PubMed]

40. Guo, Y.; McMullen, C.; Timsit, E.; Hallewell, J.; Orsel, K.; van der Meer, F.; Yan, S.; Alexander, T.W. Genetic relatedness and antimicrobial resistance in respiratory bacteria from beef calves sampled from spring processing to 40 days after feedlot entry. *Vet. Microbiol.* **2020**, *240*, 108478. [CrossRef] [PubMed]

41. Holman, D.B.; Timsit, E.; Alexander, T.W. The nasopharyngeal microbiota of feedlot cattle. *Sci. Rep.* **2015**, *5*, 15557. [CrossRef] [PubMed]

42. Ribble, C.S.; Meek, A.H.; Janzen, E.D.; Guichon, P.T.; Jim, G.K. Effect of time of year, weather, and the pattern of auction market sales on fatal fibrinous pneumonia (shipping fever) in calves in a large feedlot in Alberta (1985–1988). *Can. J. Vet. Res.* **1995**, *59*, 167–172. [PubMed]

43. Timsit, E.; Workentine, M.; van der Meer, F.; Alexander, T. Distinct bacterial metacommunities inhabit the upper and lower respiratory tracts of healthy feedlot cattle and those diagnosed with bronchopneumonia. *Vet. Microbiol.* **2018**, *221*, 105–113. [CrossRef] [PubMed]

44. Apley, M.D. Treatment of Calves with Bovine Respiratory Disease: Duration of Therapy and Posttreatment Intervals. *Vet. Clin. N. Am. Food Anim. Pract.* **2015**, *31*, 441–453. [CrossRef] [PubMed]

45. Hannon, S.J.; Brault, S.A.; Otto, S.J.G.; Morley, P.S.; McAllister, T.A.; Booker, C.W.; Gow, S.P. Feedlot Cattle Antimicrobial Use Surveillance Network: A Canadian Journey. *Front. Vet. Sci.* **2020**, *7*, 596042. [CrossRef] [PubMed]

46. Food and Drug Administration (FDA). NARMS Now. Rockville, MD: U.S. Department of Health and Human Services. Available online: https://www.fda.gov/animal-veterinary/national-antimicrobial-resistance-monitoring-system/narms-now-integrated-data (accessed on 19 February 2022).

47. Gow, S.; Bergen, R.; Booker, C.; Butters, A.; Dorin, C.; Dimmers, G.; Erickson, N.; Hannon, S.; Hendrick, S.; Ramsay, D.; et al. National surveillance of antimicrobial use and antimicrobial resistance in Canadian feedlots. In Proceedings of the Fifty-Fourth Annual Conference, American Association of Bovine Practitioners, Salt Lake City, UT, USA, 7–9 October 2021.

48. Younes, J.A.; Ramsay, D.E.; Lacoste, S.; Deschner, D.; Hill, J.E.; Campbell, J.; Waldner, C.L. Changes in the phenotypic susceptibility of *Mannheimia haemolytica* isolates to macrolide antimicrobials during the early feeding period following metaphylactic tulathromycin use in western Canadian feedlot calves. *Can. Vet. J.* **2022**, *63*, 920–928. [PubMed]

49. Timsit, E.; Christensen, H.; Bareille, N.; Seegers, H.; Bisgaard, M.; Assie, S. Transmission dynamics of *Mannheimia haemolytica* in newly-received beef bulls at fattening operations. *Vet. Microbiol.* **2013**, *161*, 295–304. [CrossRef] [PubMed]

50. Klima, C.L.; Alexander, T.W.; Hendrick, S.; McAllister, T.A. Characterization of *Mannheimia haemolytica* isolated from feedlot cattle that were healthy or treated for bovine respiratory disease. *Can. J. Vet. Res. = Rev. Can. De Rech. Vet.* **2014**, *78*, 38.

51. Babcock, A.H.; White, B.J.; Dritz, S.S.; Thomson, D.U.; Renter, D.G. Feedlot health and performance effects associated with the timing of respiratory disease treatment1. *J. Anim. Sci.* **2009**, *87*, 314–327. [CrossRef] [PubMed]

52. Dohoo, I.R.; Martin, W.; Stryhn, H. *Veterinary Epidemiologic Research*, 2nd ed.; Charlotte, P.E.I., Ed.; VER, Inc.: Charlottetown, PE, Canada, 2014.

53. Stevenson, M.A. Sample Size Estimation in Veterinary Epidemiologic Research. *Front. Vet. Sci.* **2021**, *7*, 539573. [CrossRef] [PubMed]

54. Capik, S.F.; White, B.J.; Lubbers, B.V.; Apley, M.D.; DeDonder, K.D.; Larson, R.L.; Harhay, G.P.; Chitko-McKown, C.G.; Harhay, D.M.; Kalbfleisch, T.S.; et al. Comparison of the diagnostic performance of bacterial culture of nasopharyngeal swab and bronchoalveolar lavage fluid samples obtained from calves with bovine respiratory disease. *Am. J. Vet. Res.* **2017**, *78*, 350–358. [CrossRef] [PubMed]

55. Pass, D.A.; Thompson, R.G. Wide distribution of *Pasteurella haemolytica* type 1 over the nasal mucosa of cattle. *Can. J. Comp. Med.* **1971**, *35*, 181–186. [PubMed]

56. Magwood, S.E.; Barnum, D.A.; Thomson, R.G. Nasal bacterial flora of calves in healthy and in pneumonia-prone herds. *Can. J. Comp. Med.* **1969**, *33*, 237–243. [PubMed]

57. Doyle, D.; Credille, B.; Lehenbauer, T.W.; Berghaus, R.; Aly, S.S.; Champagne, J.; Blanchard, P.; Crossley, B.; Berghaus, L.; Cochran, S.; et al. Agreement Among 4 Sampling Methods to Identify Respiratory Pathogens in Dairy Calves with Acute Bovine Respiratory Disease. *J. Vet. Intern. Med.* **2017**, *31*, 954–959. [CrossRef] [PubMed]

58. DeRosa, D.C.; Mechor, G.D.; Staats, J.J.; Chengappa, M.M.; Shryock, T.R. Comparison of *Pasteurella* spp. simultaneously isolated from nasal and transtracheal swabs from cattle with clinical signs of bovine respiratory disease. *J. Clin. Microbiol.* **2000**, *38*, 327–332. [CrossRef] [PubMed]

59. Godinho, K.S.; Sarasola, P.; Renoult, E.; Tilt, N.; Keane, S.; Windsor, G.D.; Rowan, T.G.; Sunderland, S.J. Use of deep nasopharyngeal swabs as a predictive diagnostic method for natural respiratory infections in calves. *Vet. Rec.* **2007**, *160*, 22–25. [CrossRef] [PubMed]

60. McDaneld, T.G.; Kuehn, L.A.; Keele, J.W. Evaluating the microbiome of two sampling locations in the nasal cavity of cattle with bovine respiratory disease complex (BRDC). *J. Anim. Sci.* **2018**, *96*, 1281–1287. [CrossRef] [PubMed]

61. McMullen, C.; Alexander, T.W.; Léguillette, R.; Workentine, M.; Timsit, E. Topography of the respiratory tract bacterial microbiota in cattle. *Microbiome* **2020**, *8*, 91. [CrossRef] [PubMed]

62. Pardon, B.; Buczinski, S. Bovine Respiratory Disease Diagnosis: What Progress Has Been Made in Infectious Diagnosis? *Vet. Clin. N. Am. Food Anim. Pract.* **2020**, *36*, 425–444. [CrossRef] [PubMed]

63. Limmathurotsakul, D.; Jamsen, K.; Arayawichanont, A.; Simpson, J.A.; White, L.J.; Lee, S.J.; Wuthiekanun, V.; Chantratita, N.; Cheng, A.; Day, N.P.J.; et al. Defining the true sensitivity of culture for the diagnosis of melioidosis using Bayesian latent class models. *PLoS ONE* **2010**, *5*, e12485. [CrossRef] [PubMed]

64. Singer, R.S.; Johnson, W.O.; Jeffrey, J.S.; Chin, R.P.; Carpenter, T.E.; Atwill, E.R.; Hirsh, D.C. A statistical model for assessing sample size for bacterial colony selection: A case study of *Escherichia coli* and avian cellulitis. *J. Vet. Diagn. Investig.* **2000**, *12*, 118–125. [CrossRef] [PubMed]

65. Carter, H.F.; Wills, R.W.; Scott, M.A.; Thompson, A.C.; Singer, R.S.; Loy, J.D.; Karisch, B.B.; Epperson, W.B.; Woolums, A.R. Assessment of Diversity of Antimicrobial Resistance Phenotypes and Genotypes of *Mannheimia haemolytica* Isolates From Bovine Nasopharyngeal Swabs. *Front. Vet. Sci.* **2022**, *9*, 883389. [CrossRef] [PubMed]

66. Loy, J.D.; Clawson, M.L. Rapid typing of *Mannheimia haemolytica* major genotypes 1 and 2 using MALDI-TOF mass spectrometry. *J. Microbiol. Methods* **2017**, *136*, 30–33. [CrossRef] [PubMed]

67. Capik, S.F.; White, B.J.; Lubbers, B.V.; Apley, M.D.; Mosier, D.A.; Larson, R.L.; Murray, R.W. Characterization of *Mannheimia haemolytica* in beef calves via nasopharyngeal culture and pulsed-field gel electrophoresis. *J. Vet. Diagn. Investig.* **2015**, *27*, 568–575. [CrossRef] [PubMed]

68. Maunsell, F.P.; Woolums, A.R.; Francoz, D.; Rosenbusch, R.F.; Step, D.L.; Wilson, D.J.; Janzen, E.D. Mycoplasma bovis Infections in Cattle. *J. Vet. Intern. Med.* **2011**, *25*, 772–783. [CrossRef] [PubMed]

69. Booker, C.W.; Abutarbush, S.M.; Morley, P.S.; Jim, G.K.; Pittman, T.J.; Schunicht, O.C.; Perrett, T.; Wildman, B.K.; Fenton, R.K.; Guichon, P.T.; et al. Microbiological and histopathological findings in cases of fatal bovine respiratory disease of feedlot cattle in Western Canada. *Can. Vet. J.* **2008**, *49*, 473–481. [PubMed]

70. Sarchet, J.J.; Pollreisz, J.P.; Bechtol, D.T.; Blanding, M.R.; Saltman, R.L.; Taube, P.C. Limitations of bacterial culture, viral PCR, and tulathromycin susceptibility from upper respiratory tract samples in predicting clinical outcome of tulathromycin control or treatment of bovine respiratory disease in high-risk feeder heifers. *PLoS ONE* **2022**, *17*, e0247213. [CrossRef] [PubMed]

71. National Farm Animal Council. *Code of Practice for the Care and Handling of Beef Cattle*; National Farm Animal Council: Lacombe, AB, Canada, 2013.

72. Laboratory Standards Insitute. *Performance Standards for Antimicrobial Disk and Dilution Susceptibility Tests for Bacterial Isolated from Animals*, 6th ed.; Clinical and CLSI Supplement VET01S; Laboratory Standards Insitute: Wayne, PA, USA, 2023.

73. Schwarz, S.; Silley, P.; Simjee, S.; Woodford, N.; van Duijkeren, E.; Johnson, A.P.; Gaastra, W. Editorial: Assessing the antimicrobial susceptibility of bacteria obtained from animals. *J. Antimicrob. Chemother.* **2010**, *65*, 601–604. [CrossRef] [PubMed]

74. Step, D.L.; Krehbiel, C.R.; DePra, H.A.; Cranston, J.J.; Fulton, R.W.; Kirkpatrick, J.G.; Gill, D.R.; Payton, M.E.; Montelongo, M.A.; Confer, A.W. Effects of commingling beef calves from different sources and weaning protocols during a forty-two-day receiving period on performance and bovine respiratory disease. *J. Anim. Sci.* **2008**, *86*, 3146–3158. [CrossRef] [PubMed]

Article

Multidrug-Resistant Extended-Spectrum Beta-Lactamase (ESBL)-Producing *Escherichia coli* in a Dairy Herd: Distribution and Antimicrobial Resistance Profiles

Martina Penati [1], Laura Musa [1], Laura Filippone Pavesi [1], Alessandro Guaraglia [2], Fernando Ulloa [3], Paolo Moroni [1,4], Renata Piccinini [1,4,†] and Maria Filippa Addis [1,4,*,†]

1 Department of Veterinary Medicine and Animal Sciences, University of Milan, 26900 Lodi, Italy; martina.penati@unimi.it (M.P.); laura.musa@unimi.it (L.M.); laura.filippone@unimi.it (L.F.P.); paolo.moroni@unimi.it (P.M.); renata.piccinini@unimi.it (R.P.)
2 Department of Humanities and Social Sciences, University of Sassari, 07100 Sassari, Italy; aguaraglia@uniss.it
3 Escuela de Graduados, Facultad de Ciencias Veterinarias, Universidad Austral de Chile, Valdivia 5090000, Chile; fernando.ulloa@uach.cl
4 Laboratory of Infectious Diseases of Domestic Animals (MiLab), University of Milan, 26900 Lodi, Italy
* Correspondence: filippa.addis@unimi.it
† These authors contributed equally to this work.

Abstract: This study investigated the presence, distribution, and antimicrobial resistance profiles of extended-spectrum beta-lactamase (ESBL)-producing *Escherichia coli* in a dairy herd located in Northern Italy. The feces of clinically healthy calves, their mothers, and the cows treated for mastitis, as well as water, environmental samples, and waste milk were collected and subjected to bacteriological culture on CHROMagar™ ESBL plates. A questionnaire was administered to identify risk factors. The isolates were identified as *E. coli* by MALDI-TOF MS and subjected to the double-disk synergy test (DDST) and minimal inhibitory concentration (MIC) assay. As a result, ESBL *E. coli* was isolated from the feces of 28 of 37 (75.67%) calves, the feces of 2 of 3 (66.67%) treated cows, 8 of 14 (57.15%) environmental samples, and waste milk. All ESBL isolates showed multiple resistances and were categorized as multidrug-resistant (MDR). Several risk factors for ESBL *E. coli* selection and diffusion were identified, including lack of routine cleaning of calf feeding and housing equipment, administration of waste milk to male calves, and blanket dry cow therapy. In conclusion, this study highlighted the presence of MDR, ESBL *E. coli* in the feces of most dairy calves, and their association with different sample sources. Accordingly, adding to the prudent use of antibiotics, the adoption of adequate farm hygiene and biosecurity measures might also help prevent the spread and transmission of ESBL *E. coli* within the herd.

Keywords: antimicrobial resistance; biosecurity; dairy calves; ESBL *E. coli*; multidrug-resistant; waste milk

Citation: Penati, M.; Musa, L.; Filippone Pavesi, L.; Guaraglia, A.; Ulloa, F.; Moroni, P.; Piccinini, R.; Addis, M.F. Multidrug-Resistant Extended-Spectrum Beta-Lactamase (ESBL)-Producing *Escherichia coli* in a Dairy Herd: Distribution and Antimicrobial Resistance Profiles. *Antibiotics* **2024**, *13*, 241. https://doi.org/10.3390/antibiotics13030241

Academic Editor: Jonathan Frye

Received: 30 January 2024
Revised: 26 February 2024
Accepted: 2 March 2024
Published: 5 March 2024

1. Introduction

Antimicrobial resistance (AMR) is a problem of global concern [1]. The rise in antibiotic-resistant bacteria, as stated by the World Health Organization (WHO), can render many drugs previously essential for treating infections in humans and animals ineffective [2]. About 50–80% of the total antibiotic use in developed countries has been attributed to livestock [3], although significant efforts are being made to reduce antibiotic use. According to a report published by the European Food Safety Authority [4], in Europe antibiotic use has decreased to become lower in food-producing animals than in humans. Between 2016 and 2018, animals used less antimicrobials than humans overall. In 2017, for example, animals averaged 108.3 mg/kg (range 3.1–423.1) compared to 130.0 mg/kg (range 52.8–212.6) for humans [4]. Nevertheless, the animal industry still plays a crucial role in the occurrence and transmission of AMR [5].

In recent years, the prevalence of bacterial strains producing extended-spectrum beta-lactamases (ESBLs) has increased worldwide [6,7]. These enzymes are able to hydrolyze third-generation cephalosporins and aztreonam but are inhibited by clavulanic acid. ESBL-producing organisms often show co-resistance to many other classes of antibiotics. Adding to problems in veterinary medicine, this group of plasmid-mediated, rapidly evolving, and diverse enzymes poses major therapeutic challenges in human medicine, especially concerning the treatment of hospitalized and community-based patients [7].

ESBL-producing *Escherichia coli* is a common occurrence in dairy cattle, with the highest incidence observed in calves [8,9]. Food-producing animals acquire AMR microorganisms due to several factors such as antibiotic use, forage, soil, water [3], interaction with wildlife [10,11], and by contact with humans [12,13]. AMR bacteria colonizing the animal gastrointestinal tract are shed in feces, thus favoring intra-farm spread and maintenance as well as environmental contamination via farm waste products, including untreated wastewater, sewage sludge, and organic fertilizers such as manure [3,14,15]. Adding to the control of pathogens from outside and inside the farm, biosecurity measures can therefore play a crucial role also in avoiding the selection, maintenance, and spread of AMR microbes within and outside the farm. Accordingly, although the prudent use of antibiotics is key for reducing AMR, adequate farm management practices can also play a fundamental role in containing the AMR burden [4].

With these premises, we investigated a medium-to-large herd where a very high prevalence of ESBL *E. coli* in the feces of calves had been identified in previous unpublished observations. To understand their prevalence, distribution, and antimicrobial resistance traits, as well as to identify possible biosecurity issues, calf feces, cow feces, and the farm and animal environment were investigated by bacteriological culture and microbial sensitivity assays, and a detailed questionnaire was administered to evaluate farming practices.

2. Results

2.1. Bacteriological Culture Results

The results obtained for all samples collected in the farm are summarized in Table 1.

Table 1. Summary of analyzed samples and respective bacteriological results.

Sample Type	*n*	ESBL *E. coli* (%)
Female calf feces	18	15 (83.3%)
Male calf feces	19	13 (68.4%)
Treated cow feces	3	2 (66%)
Dam feces	26	0
Waste milk	1	1 (100%)
Male calf pens	1	1 (100%)
Female calf pens	1	1 (100%)
Mixed-use calf pens	1	0
Calf feeding bucket	2	1 (50%)
Calf drinking water	2	1 (50%)
Cow alleys	3	3 (100%)
Cow's berth tube	1	0
Cow water trough	1	0
Cow feeding rack	2	1 (50%)

ESBL *E. coli* was isolated from the feces of 15 of 18 (83.3%) female calves, 13 of 19 (68.4%) male calves, 2 of 3 (66.7%) cows treated for IMI, and waste milk. The feces of the dams of the enrolled calves were all negative. ESBL *E. coli* was also isolated from 8 of 14 (57.14%) environmental samples, including male and female calf pens, the cow feeding rack, the alley floors, the calf drinking water, and the calf feeding buckets.

2.2. Antimicrobial Susceptibility Testing of ESBL E. coli

All *E. coli* isolates were phenotypically positive for the double-disk synergy test (DDST), confirming the production of ESBL. However, all of them were susceptible to the carbapenem class. Based on the MIC results (Figure 1), the highest level of resistance was observed for β-lactams, with all isolates being resistant to ampicillin, cefazolin, and cefotaxime (100%), while only 2.6% of isolates were resistant to amoxicillin/clavulanic acid. Concerning aminoglycosides, 97.4% of isolates were resistant to kanamycin, 97.4% to aminosidine, and 15.4% to gentamicin. Concerning fluoroquinolones, 12.8% were resistant to enrofloxacin and 17.9% to flumequine. Resistance to florfenicol was 46.2%. Concerning sulfonamides, 84.6% were resistant to sulfisoxazole, and 48.7% to trimethoprim/sulfamethoxazole. For the tetracycline class, 89.7% of isolates were resistant. All isolates were susceptible to colistin (100%). Notably, all ESBL *E. coli* isolates were MDR, being resistant to at least three classes of antibiotics. The MIC results are detailed in Supplementary Materials File S1.

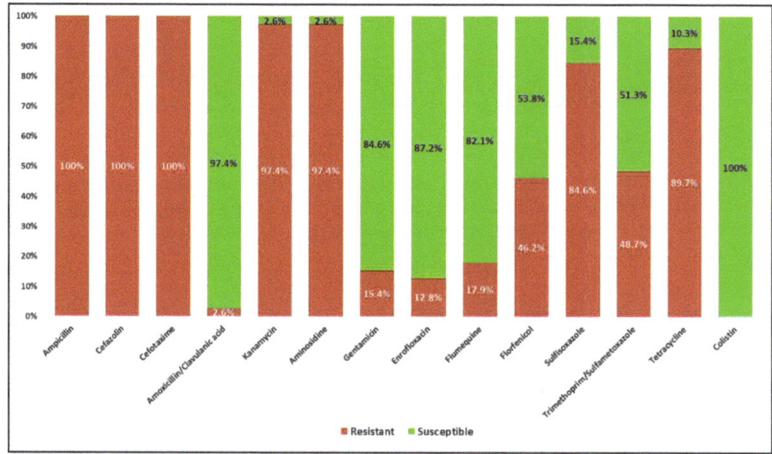

Figure 1. Distribution of resistance and susceptibility of the ESBL *E. coli* isolates to the different antimicrobials according to the plate MIC test.

Table 2 reports the 12 different ESBL *E. coli* resistance profiles observed in this study. Profile 1 was the most frequent (10 out of 39) and was found in 9 calves (3 male and 6 female calves) and the calf feeding bucket. Profile 2 (8 of 39) was found in 7 calves (4 male and 3 female calves) and waste milk. Profile 3 (6 of 39) was found in 2 out of 3 cows treated for IMI, 3 alley floor samples, and 1 male calf. Profile 4 (4 of 39) was found in 3 female calves and 1 male calf. Profile 5 (3 of 39) was found in 2 female calves and 1 female calf pen. Profile 6 (2 of 39) was found in a male pen and the cow feeding rack. Profiles from 7 to 12 were found only once and in 4 male calves, the calf drinking water, and 1 female calf, respectively.

2.3. Hierarchical Clustering of E. coli Isolates Based on the MIC Results

Figure 2 illustrates the hierarchical clustering of ESBL *E. coli* isolates based on the MIC test results. Two main branches were observed. One included the isolates from 12 calves, waste milk, treated cows, and alley floors. Within this branch, all the isolates from cows and alley floors and one calf isolate were separated from the waste milk isolate and eleven calf isolates with similar AMR profiles. Another branch included 16 calf isolates and the isolates from calf pens, water and feeding buckets, and the cow feeding rack. Within this branch, one subgroup included the isolates from four calves, male pens, and the cow feeding rack, separated from the isolates from nine calves and the water and feeding buckets, while another subgroup included three calves and the female pen.

Table 2. Antimicrobial resistance profiles of the ESBL *E. coli* isolates. R, resistant; S, sensitive; MDR, multidrug-resistant. The number of antimicrobial classes is reported in parentheses.

Resistance Profile'	Number of Isolates	Aminosidine	Amoxicillin/Clavulanic Acid	Ampicillin	Cefazolin	Cefotaxime	Colistin	Enrofloxacin	Florfenicol	Flumequine	Gentamicin	Kanamycin	Sulfisoxazole	Tetracycline	Trimethoprim/Sulfamethoxazole	MDR (Class Number)
1	10	R	S	R	R	R	S	S	R	S	S	R	R	R	R	(5)
2	8	R	S	R	R	R	S	S	S	S	S	R	R	R	S	(4)
3	6	R	S	R	R	R	S	S	S	S	S	R	S	R	S	(3)
4	4	R	S	R	R	R	S	S	R	S	S	R	R	R	S	(5)
5	3	R	S	R	R	R	S	R	S	R	S	R	R	S	R	(4)
6	2	R	S	R	R	R	S	S	R	S	R	R	R	R	R	(5)
7	1	R	S	R	R	R	S	S	S	R	R	R	R	R	S	(5)
8	1	S	S	R	R	R	S	S	S	S	S	S	R	R	R	(3)
9	1	R	S	R	R	R	S	S	S	R	R	R	R	S	R	(4)
10	1	R	S	R	R	R	S	R	S	R	S	R	R	R	R	(5)
11	1	R	S	R	R	R	S	S	R	S	R	R	R	R	S	(5)
12	1	R	R	R	R	R	S	R	R	R	R	R	R	R	R	(6)

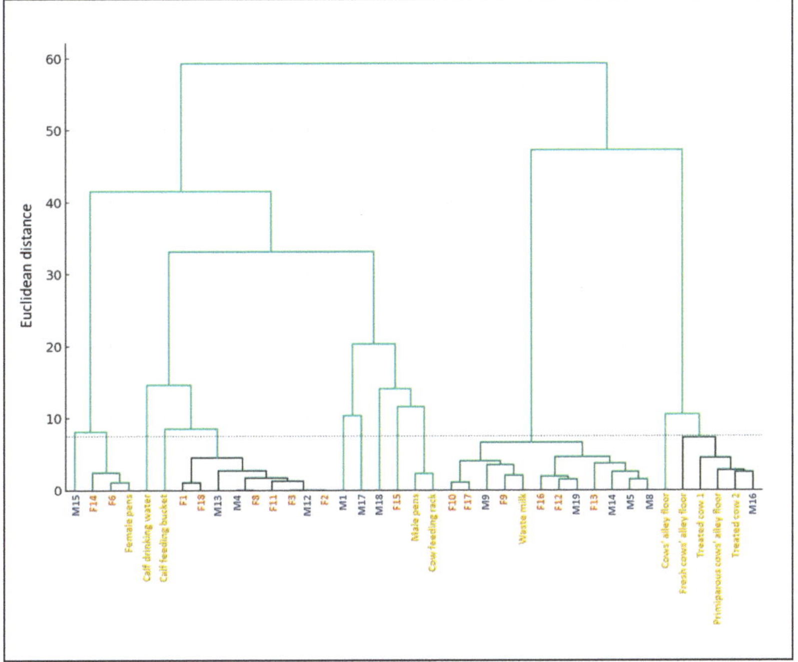

Figure 2. Hierarchical clustering of the ESBL *E. coli* isolates' MIC profiles based on Ward's method. Female calves (F) are illustrated in red. Male calves (M) are illustrated in blue. Cows, environment, and water samples are illustrated in gold. The dotted line indicates the significant Euclidean distance cut-off point at 7.5.

Based on the threshold Euclidean distance of 7.5, several statistically significant clusters grouped more than one sample. The largest cluster included 11 calf isolates and the waste milk isolate. The second largest cluster included the isolates from nine calves, closely related to the common feeding bucket and the calf drinking water. The third largest cluster grouped the isolates from one calf, treated cows, and the primiparous and fresh cow alleys, indicating fecal shedding from treated cows as another diffusion route. The fourth one included two female calf isolates and the female calf pens. The other four calf isolates, the male pens, and the cow feeding rack also had similar AMR profiles.

2.4. Results of the Biosecurity Questionnaire

The BioCheck.UGent biosecurity questionnaire assessed various aspects of farm management, including general farm organization, sick pen health management and outbreak management, reproduction management, calf pen management and hygiene, calf rearing, herd health management, and milking management. We identified several potential biosecurity issues that might impact ESBL maintenance and diffusion. Concerning calf management and hygiene, cleaning of teat buckets with water without detergents or disinfectants, feeding waste milk to male calves, then sharing buckets between male and female calves, mixed use of calf pens, and occasional cleaning of individual calf pens during the winter season were identified as risk factors. Regarding hygiene issues, lack of pasteurizer cleaning and occasional cleaning of troughs were identified as additional potential risk factors. Finally, the absence of a sick pen was identified as a lack of biosecurity, and the application of dry cow therapy with a first-generation cephalosporin (dihydrated cefalonium) was identified as a factor that could establish a relationship between farm practices and the occurrence and spread of AMR (Supplementary File S2).

3. Discussion

This study assessed the presence, distribution, and antimicrobial resistance profiles of ESBL-producing *E. coli* in a medium-sized dairy herd in Northern Italy, hosting nearly 1000 animals, including calves, heifers, and lactating and dry cows. We collected calf and cow feces, waste milk, environmental samples, and water, and we administered a questionnaire to assess the associated risk factors. As a result, most pre-weaned calves, including males and females, carried ESBL *E. coli* in their intestines, and ESBL *E. coli* was also present in the environment and farm equipment in contact with them. Not much is known about the transmission of ESBL *E. coli* among calves and cows, and how it is affected by environmental factors [16]. The hierarchical clustering of ESBL *E. coli* isolates based on the MIC results suggested that multiple MDR strains with different resistance characteristics were circulating in the farm and were found throughout sample types. Different sources and routes could therefore be involved in their dissemination and maintenance, facilitated by incorrect or inadequate management, biosecurity, and hygiene practices.

The herd management interview enabled gathering of information on potential risk factors for the distribution of ESBL microorganisms on the farm. The farmer used waste milk with antibiotic residues for feeding male calves. According to hierarchical clustering based on the isolate MIC profiles, the largest statistically significant cluster included about 40% of all calves' fecal isolates, the feces of two out of three cows treated for mastitis that contributed to the waste milk, and the waste milk isolate. Among the risk factors associated with the spread of ESBL *E. coli* on cattle farms, the use of waste milk containing antibiotic residues as calf feed appears to play an important role [3,17].

Based on hierarchical clustering, the MIC profiles of the isolates from the calf feeding bucket and drinking water clustered with those of the fecal isolates of over 30% of the calves. Incorrect management practices such as shared or improperly cleaned feeding equipment [3] can favor the diffusion of AMR-carrying bacteria on the farm. Notably, the feeding buckets and calf water buckets were not cleaned with detergents or disinfectants, and sometimes not even rinsed with water between feedings; furthermore, the number of buckets was not adequate for the number of animals on the farm. ESBL *E. coli* was isolated from all these pieces of equipment as well as from the calf drinking water. Moreover, the pasteurizer used to reduce bacterial contamination of milk was not cleaned between cycles. Poor cleaning is one of the factors favoring bacterial contamination and multiplication, leading to higher microbial loads [18]. Indeed, we also isolated ESBL *E. coli* from the farm's pasteurized waste milk. Moreover, many positive calves were females, which should not have received waste milk. Shared feeding buckets, as well as their unproper cleaning, might also facilitate the transmission of ESBL *E. coli* between male and female calves.

Shared calf pens and their poor hygiene may also play a role in promoting the diffusion of AMR bacteria. We isolated ESBL *E. coli* from both male and female calf pens, and we observed a relationship between the MIC profiles of these isolates and those from calf feces based on hierarchical clustering. As highlighted in an EFSA scientific opinion paper on calf welfare, the level of cleanliness of the areas used for housing calves is a major determinant of their health [19]. Inaccurate cleaning procedures of the single pens or calf hutches may not adequately remove fecal contamination from the walls, leading them to serve as a reservoir [16].

The cows underwent blanket dry cow therapy (BDCT) with a β-lactam, specifically dihydrated cefalonium. Although this practice is not allowed in Italy, some farms are still using it. BDCT has been reported to be linked to a significant increase in ESBL *E. coli* in calf feces during the colostral phase [8].

The MIC profiles of the ESBL *E. coli* isolated from the feces of cows treated for mastitis were similar to those of the isolates from the cows' alley floors, suggesting fecal shedding. *E. coli* ESBL shedding can vary greatly among individuals [20], and antibiotic treatment for mastitis could play a role in increasing animal colonization, shedding, and subsequent environmental contamination by AMR-carrying bacteria [9]. The farm evaluated in this study did not have a sick pen, and this represents a lack of biosecurity. Early isolation of sick animals is a crucial practice for preventing the spread of pathogenic bacteria and maintaining herd health [21].

ESBL *E. coli* often carries multiple resistance genes for other antimicrobial drugs than β-lactams, leading to MDR [22]. All the isolates obtained in our study, from the calves, their equipment, and the farm environment, were MDR. On the other hand, resistance to colistin, an antibiotic of last resort for humans [23], was not detected, probably because many developed countries, including Italy, have prohibited its usage in food-producing animals. All ESBL *E. coli* isolates were also carbapenem-sensitive. This is also a positive finding, as carbapenemase-producing *E. coli* causes serious human infections. The study by Waade et al. conducted in Germany in 2021 reported similar results since the ESBL-producing isolates were 92.9% *E. coli*, and 60.6% of ESBL-producing isolates were resistant to one or more classes of antibiotics including penicillins and cephalosporins but were sensitive to carbapenems [24].

4. Materials and Methods

4.1. Farm Description and Ethics Statement

The farm was located in Northern Italy and consisted of 1000 animals of which 450 were lactating Italian Friesian cows. It is accredited free from infectious bovine rhinotracheitis (IBR) and vaccinated for neonatal diarrhea agents and type-1 and type-2 bovine viral diarrhea virus (BVDV). The farm does not use an in-house colostrum bank, but the colostrum is taken by the calves directly from the dam. The farmer used pasteurized waste milk to feed male calves. Based on the questionnaire, waste milk given to the calves was mainly represented by milk with high somatic cell count (SCC) and milk from cows treated with antibiotics. The waste milk produced on the farm at the time of the visit came from animals treated for mastitis with beta-lactam antibiotics (amoxicillin/clavulanic acid). The study was conducted in accordance with the Declaration of Helsinki and approved by the Institutional Committee for Animal Welfare of the University of Milan (protocol number 99_2023).

4.2. Questionnaire

A questionnaire was completed together with the herd manager (Supplementary File S2). The questionnaire follows the Biocheck.UGent checklist [25] and was used to assess different aspects of herd management, the use of antibiotics, and farm biosecurity. The questionnaire was also integrated with further aspects based on previous studies on ESBL *E. coli* risk factors [9,17,26]. The form was divided into several sections: I. General questions about farm organization: how many animals are present in the different categories, who works with the animals; II. Health management in the sick pen and management of outbreaks; III. Reproduction management; IV. Calving pen management and hygiene questions; V. Calf rearing: colostrum feeding management, milk feeding management, calf housing, vaccinations, and treatments; VI. Health management of the herd; VII. Milking management.

4.3. Animals and Sample Collection

We collected fecal samples from 37 healthy dairy calves (19 males, 18 females) aged 7–21 days. All calves were free from diarrhea and had not been treated with antibiotics. Males were fed waste milk, while females received commercial milk replacer. We sampled the feces of the 26 dams present on the farm (the 9 missing dams had been sold or sent to the slaughterhouse) and of 3 cows treated for intramammary infection (IMI) with amoxicillin/clavulanic acid that contributed to the waste milk. All the fecal samples were collected from the rectal ampoule using gloves, transported to the laboratory in refrigerated conditions, and frozen at $-20\ ^{\circ}$C for two to five days until analysis [27]. Waste milk was collected directly from the pasteurizer and kept refrigerated until arrival at the laboratory. Three calf pens were sampled by rubbing sterile gauzes against the inner wall of the pens over an area of about $150 \times 30\ \mathrm{cm}^2$ at the height of the calves' noses, avoiding obvious fecal smears, then stored in sterile 50 mL Falcon® tubes. During the sampling process, two separate sterile swabs were used. One swab was rubbed thoroughly against the bottom and inner wall of a calf feeding bucket, while the other swab was used to collect samples from the inside of the nipple. Two water samples of 150 mL were collected into sterile containers from the calf watering buckets and one from the cow watering trough, respectively. Two environmental samples were also taken with gauzes from the cow feeding rack and one from the cow's berth tube. Disposable fabric socks were used to collect three samples from the barn floors by walking down the alleys one time, and then inserted in sterile plastic bags: one from the cow alley, one from the primiparous cow alley, and one from the fresh cow alley, respectively. All environmental and water samples were stored at refrigerated temperature until arrival at the laboratory.

4.4. Isolation and Characterization of ESBL-Producing E. coli

Environmental swabs and feces (0.1 g) were enriched in 5 mL of Müeller Hinton broth (Microbiol, Cagliari, Italy) and incubated at 37 °C under aerobic conditions for 18–24 h. For environmental samples, 30 mL of Müeller Hinton broth was added to the Falcon tubes and plastic bags containing the samples and incubated at 37 °C for 18–24 h. One-hundred milliliters of water was added to an equal amount of double-strength enrichment broth and incubated at 37 °C for 48 h. All the feces and environmental samples were cultured on CHROMagar™ ESBL agar plates (CHROMagar, Paris, France) and MacConkey agar as a control medium (Oxoid Ltd., Basingstoke, UK), and incubated at 37 °C for 18–24 h. Pasteurized waste milk was seeded on blood agar plates (Microbiol, Cagliari, Italy) and CHROMagar™ ESBL agar plates in amounts of 100 µL and incubated at 37 °C for 24 h. Colonies indicating ESBL bacteria grown on CHROMagar™ ESBL agar plates were picked and submitted to species identification with the MBT Microflex LT/SH MALDI-TOF mass spectrometer (Bruker Daltonik GmbH, Bremen, Germany) as described previously [28]. After species identification, the colonies recovered from CHROMagar™ ESBL agar plates were sub-cultured on blood agar plates (Microbiol, Cagliari, Italy) and subjected to ESBL phenotyping assessment using the double-disk synergy test (DDST) to assess carbapenemase production according to the EUCAST guidelines [29].

4.5. Antimicrobial Susceptibility Testing

A Sensititre™ ITISVE1 plate (Thermo Fisher Scientific®, Waltham, MA, USA) was used to determine the MIC of the antimicrobials commonly used in dairy herds against the ESBL *E. coli* isolates. The plate contained the following antibiotics: flumequine (range 1–16 µg/mL); amoxicillin/clavulanic acid (0.25–32 µg/mL); ampicillin (0.25–32 µg/mL); cefazolin (0.5–8 µg/mL); cefotaxime (0.5–4 µg/mL); sulfisoxazole (128–512 µg/mL); colistin (0.03–8 µg/mL); enrofloxacin (0.02–32 µg/mL); florfenicol (1–64 µg/mL); gentamicin (0.25–32 µg/mL); tetracycline (0.5–16 µg/mL); trimethoprim/sulfamethoxazole (0.06–16 µg/mL); aminosidine (1–32 µg/mL); kanamycin (2–32 µg/mL). Quality control for Sensititre plates was performed using *E. coli* strain ATCC 25922 and the Sensititre™ SWIN™ Software System V. 3.4 (3.4.6.2) (Sensititre™, Thermo Fisher Scientific®, Waltham, MA, USA). The MIC results were interpreted according to the manufacturer's instructions using CLSI VET08 4th edition [30] (V = Vet), CLSI VET06 1st edition [31] (V = Vet), CLSI M100 29th edition (H = Human) [32], EUCAST v.11.0 [33], CASFM 2019 [34]. ESBL *E. coli* isolates resistant to at least 3 classes of antibiotics were classified as MDR [35], and intermediate isolates were classified as susceptible.

4.6. Hierarchical Clustering

Non-supervised hierarchical cluster analysis using Ward's method was performed based on the MIC values [36,37]. A total of 21 parameters were obtained by assigning to each MIC value ranging from >512 µg/mL to ≤0.015625 a number from 1 to 21 according to decreasing antibiotic concentrations. The profiles obtained for each sample after the conversion were used to construct a dendrogram. This technique was chosen for its effectiveness in minimizing variance within the clusters, allowing us to identify groups of isolates with similar resistance patterns. The resulting dendrogram provides a visual representation of the progressive merging of the clusters based on the Euclidean distance. The dendrogram was cut (maximum distance for clustering) at a height of 7.5. This cut-off point was chosen based on statistical significance, ensuring that each cluster represented a distinctive group of isolates with similar characteristics. The analysis was conducted using the SciPy library (version 1.11.4, https://scipy.org/, accessed on 18 January 2024) within the Python environment (version 3.10.12, https://www.python.org/, accessed on 18 January 2024).

5. Conclusions

This study highlighted the widespread presence of ESBL *E. coli* in the dairy farm and the relevant presence and circulation of MDR strains in association with different sources and sample types. Prudent antibiotic use remains the most relevant driver enabling the reduction in and control of AMR bacteria. Nevertheless, adherence to good internal and external biosecurity practices, hygiene of facilities and equipment, correct feeding procedures, and correct animal management might also significantly contribute to reducing and controlling AMR bacteria.

Supplementary Materials: The following supporting information can be downloaded at: https://www.mdpi.com/article/10.3390/antibiotics13030241/s1, Supplementary File S1: MIC values expressed in µg/mL for the ESBL *E. coli* isolates assessed in this study with the plate assay. Supplementary File S2: Questionnaire of the study Multidrug-resistant extended-spectrum beta-lactamase (ESBL)-producing *Escherichia coli* in a dairy herd: distribution and antimicrobial resistance profiles.

Author Contributions: Conceptualization: M.P., R.P. and M.F.A.; methodology, M.P., R.P., M.F.A., A.G., F.U. and P.M.; farm activities and sample collection: M.P. and L.F.P.; formal analysis: M.P., L.F.P. and L.M.; resources: R.P. and M.F.A.; data curation: M.P., A.G. and M.F.A.; writing, original draft: M.P. and M.F.A.; writing, review and editing, all authors; supervision: R.P. and M.F.A. All authors have read and agreed to the published version of the manuscript.

Funding: This research received no external funding. All funds were from the University of Milan.

Institutional Review Board Statement: The study was conducted in accordance with the Declaration of Helsinki and approved by the Institutional Committee for Animal Welfare of the University of Milan (protocol number 99_2023).

Informed Consent Statement: Written informed consent has been obtained from the farmer to publish this paper.

Data Availability Statement: All pertinent data generated in this study are provided in the Supplementary Materials.

Acknowledgments: Fernando Ulloa was supported by a fellowship from the Universidad Austral de Chile.

Conflicts of Interest: The authors declare no conflicts of interest.

References

1. Antimicrobial Resistance Surveillance in Europe 2022—2020 Data. Available online: https://www.ecdc.europa.eu/en/publications-data/antimicrobial-resistance-surveillance-europe-2022-2020-data (accessed on 29 January 2024).
2. Global Antimicrobial Resistance and Use Surveillance System (GLASS) Report: 2022. Available online: https://www.who.int/publications-detail-redirect/9789240062702 (accessed on 29 January 2024).
3. Ma, Z.; Lee, S.; Jeong, K.C. Mitigating Antibiotic Resistance at the Livestock-Environment Interface: A Review. *J. Microbiol. Biotechnol.* **2019**, *29*, 1683–1692. [CrossRef]
4. European Centre for Disease Prevention and Control; European Food Safety Authority; European Medicines Agency. Third Joint Inter-Agency Report on Integrated Analysis of Consumption of Antimicrobial Agents and Occurrence of Antimicrobial Resistance in Bacteria from Humans and Food-Producing Animals in the EU/EEA. *EFSA J.* **2021**, *19*, e06712. [CrossRef]
5. Cuong, N.V.; Padungtod, P.; Thwaites, G.; Carrique-Mas, J.J. Antimicrobial Usage in Animal Production: A Review of the Literature with a Focus on Low- and Middle-Income Countries. *Antibiotics* **2018**, *7*, 75. [CrossRef] [PubMed]
6. Reygaert, W.C. An Overview of the Antimicrobial Resistance Mechanisms of Bacteria. *AIMS Microbiol.* **2018**, *4*, 482–501. [CrossRef] [PubMed]
7. Rawat, D.; Nair, D. Extended-Spectrum β-Lactamases in Gram Negative Bacteria. *J. Glob. Infect. Dis.* **2010**, *2*, 263–274. [CrossRef]
8. Tetens, J.L.; Billerbeck, S.; Schwenker, J.A.; Hölzel, C.S. Short Communication: Selection of Extended-Spectrum β-Lactamase-Producing *Escherichia coli* in Dairy Calves Associated with Antibiotic Dry Cow Therapy—A Cohort Study. *J. Dairy. Sci.* **2019**, *102*, 11449–11452. [CrossRef]
9. Gonggrijp, M.A.; Santman-Berends, I.M.G.A.; Heuvelink, A.E.; Buter, G.J.; van Schaik, G.; Hage, J.J.; Lam, T.J.G.M. Prevalence and Risk Factors for Extended-Spectrum β-Lactamase- and AmpC-Producing *Escherichia coli* in Dairy Farms. *J. Dairy. Sci.* **2016**, *99*, 9001–9013. [CrossRef]
10. Prandi, I.; Bellato, A.; Nebbia, P.; Stella, M.C.; Ala, U.; von Degerfeld, M.M.; Quaranta, G.; Robino, P. Antibiotic Resistant *Escherichia coli* in Wild Birds Hospitalised in a Wildlife Rescue Centre. *Comp. Immunol. Microbiol. Infect. Dis.* **2023**, *93*, 101945. [CrossRef] [PubMed]
11. Musa, L.; Stefanetti, V.; Casagrande Proietti, P.; Grilli, G.; Gobbi, M.; Toppi, V.; Brustenga, L.; Magistrali, C.F.; Franciosini, M.P. Antimicrobial Susceptibility of Commensal *E. coli* Isolated from Wild Birds in Umbria (Central Italy). *Animals* **2023**, *13*, 1776. [CrossRef]
12. Collignon, P.J.; McEwen, S.A. One Health-Its Importance in Helping to Better Control Antimicrobial Resistance. *Trop. Med. Infect. Dis.* **2019**, *4*, 22. [CrossRef]
13. Ramatla, T.; Mafokwane, T.; Lekota, K.; Monyama, M.; Khasapane, G.; Serage, N.; Nkhebenyane, J.; Bezuidenhout, C.; Thekisoe, O. "One Health" Perspective on Prevalence of Co-Existing Extended-Spectrum β-Lactamase (ESBL)-Producing *Escherichia coli* and Klebsiella Pneumoniae: A Comprehensive Systematic Review and Meta-Analysis. *Ann. Clin. Microbiol. Antimicrob.* **2023**, *22*, 88. [CrossRef]
14. Massé, J.; Lardé, H.; Fairbrother, J.M.; Roy, J.-P.; Francoz, D.; Dufour, S.; Archambault, M. Prevalence of Antimicrobial Resistance and Characteristics of *Escherichia coli* Isolates From Fecal and Manure Pit Samples on Dairy Farms in the Province of Québec, Canada. *Front. Vet. Sci.* **2021**, *8*, 654125. [CrossRef]
15. Cho, S.; Jackson, C.R.; Frye, J.G. Freshwater Environment as a Reservoir of Extended-Spectrum β-Lactamase-Producing *Enterobacteriaceae*. *J. Appl. Microbiol.* **2023**, *134*, lxad034. [CrossRef]
16. Homeier-Bachmann, T.; Kleist, J.F.; Schütz, A.K.; Bachmann, L. Distribution of ESBL/AmpC-*Escherichia coli* on a Dairy Farm. *Antibiotics* **2022**, *11*, 940. [CrossRef]
17. Weber, L.P.; Dreyer, S.; Heppelmann, M.; Schaufler, K.; Homeier-Bachmann, T.; Bachmann, L. Prevalence and Risk Factors for ESBL/AmpC-*E. coli* in Pre-Weaned Dairy Calves on Dairy Farms in Germany. *Microorganisms* **2021**, *9*, 2135. [CrossRef]
18. Elizondo-Salazar, J.; Jones, C.; Heinrichs, A. Evaluation of Calf Milk Pasteurization Systems on 6 Pennsylvania Dairy Farms. *J. Dairy. Sci.* **2010**, *93*, 5509–5513. [CrossRef] [PubMed]

19. EFSA Panel on Biological Hazards (BIOHAZ); Koutsoumanis, K.; Allende, A.; Álvarez-Ordóñez, A.; Bolton, D.; Bover-Cid, S.; Chemaly, M.; de Cesare, A.; Hilbert, F.; Lindqvist, R.; et al. Update of the List of Qualified Presumption of Safety (QPS) Recommended Microorganisms Intentionally Added to Food or Feed as Notified to EFSA. *EFSA J.* **2023**, *21*, e07747. [CrossRef]
20. Munns, K.D.; Selinger, L.B.; Stanford, K.; Guan, L.; Callaway, T.R.; McAllister, T.A. Perspectives on Super-Shedding of *Escherichia coli* O157:H7 by Cattle. *Foodborne Pathog. Dis.* **2015**, *12*, 89–103. [CrossRef] [PubMed]
21. Power, G.M.; Renaud, D.L.; Miltenburg, C.; Spence, K.L.; Hagen, B.N.M.; Winder, C.B. Perceptions of Biosecurity in a Canadian Dairy Context. *J. Dairy. Sci.* **2024**. [CrossRef] [PubMed]
22. Ibrahim, D.R.; Dodd, C.E.R.; Stekel, D.J.; Meshioye, R.T.; Diggle, M.; Lister, M.; Hobman, J.L. Multidrug-Resistant ESBL-Producing *E. coli* in Clinical Samples from the UK. *Antibiotics* **2023**, *12*, 169. [CrossRef] [PubMed]
23. European Medicines Agency. Categorisation of Antibiotics for Use in Animals, 2020. Available online: https://www.ema.europa.eu/en/documents/report/categorisation-antibiotics-european-union-answer-request-european-commission-updating-scientific-advice-impact-public-health-and-animal-health-use-antibiotics-animals_en.pdf (accessed on 25 February 2024).
24. Waade, J.; Seibt, U.; Honscha, W.; Rachidi, F.; Starke, A.; Speck, S.; Truyen, U. Multidrug-Resistant Enterobacteria in Newborn Dairy Calves in Germany. *PLoS ONE* **2021**, *16*, e0248291. [CrossRef] [PubMed]
25. Home | Biocheck.UGent. Available online: https://biocheckgent.com/en (accessed on 29 January 2024).
26. Santman-Berends, I.M.G.A.; Gonggrijp, M.A.; Hage, J.J.; Heuvelink, A.E.; Velthuis, A.; Lam, T.J.G.M.; van Schaik, G. Prevalence and Risk Factors for Extended-Spectrum β-Lactamase or AmpC-Producing *Escherichia coli* in Organic Dairy Herds in the Netherlands. *J. Dairy. Sci.* **2017**, *100*, 562–571. [CrossRef] [PubMed]
27. Achá, S.J.; Kühn, I.; Mbazima, G.; Colque-Navarro, P.; Möllby, R. Changes of Viability and Composition of the *Escherichia coli* Flora in Faecal Samples during Long Time Storage. *J. Microbiol. Methods* **2005**, *63*, 229–238. [CrossRef] [PubMed]
28. Rosa, N.M.; Penati, M.; Fusar-Poli, S.; Addis, M.F.; Tola, S. Species Identification by MALDI-TOF MS and *gap* PCR–RFLP of Non-*aureus Staphylococcus*, *Mammaliicoccus*, and *Streptococcus* spp. Associated with Sheep and Goat Mastitis. *Vet. Res.* **2022**, *53*, 84. [CrossRef]
29. Clinical & Laboratory Standards Institute: CLSI Guidelines. Available online: https://clsi.org/ (accessed on 13 February 2023).
30. Lubbers, B.V.; Clinical and Laboratory Standards Institute (Eds.) *Performance Standards for Antimicrobial Disk and Dilution Susceptibility Tests for Bacteria Isolated from Animals*, 4th ed.; VET08; Clinical and Laboratory Standards Institute, CLSI: Wayne, PA, USA, 2018; ISBN 978-1-68440-010-2.
31. Traczewski, M.M.; Clinical and Laboratory Standards Institute (Eds.) *Methods for Antimicrobial Susceptibility Testing of Infrequently Isolated or Fastidious Bacteria Isolated from Animals*, 1st ed.; VET06; Clinical and Laboratory Standards Institute, CLSI: Wayne, PA, USA, 2017; ISBN 978-1-56238-810-2.
32. Weinstein, M.P.; Ii, J.S.L. *Performance Standards for Antimicrobial Susceptibility Testing*, 34th ed.; M100; Clinical and Laboratory Standards Institute, CLSI: Wayne, PA, USA, 2024; ISBN 978-1-68440-221-2.
33. Eucast: Clinical Breakpoints and Dosing of Antibiotics. Available online: https://www.eucast.org/clinical_breakpoints (accessed on 13 February 2023).
34. Comité de l'antibiogramme de la Société Francaise de Microbiologie (CASFM)2019_V1.0.Pdf. Break Point Suggested by the French Society of Microbiology in Veterinary Medicine. Available online: https://www.sfm-microbiologie.org/wp-content/uploads/2019/01/CASFM_VET2018.pdf (accessed on 25 February 2024).
35. Magiorakos, A.-P.; Srinivasan, A.; Carey, R.B.; Carmeli, Y.; Falagas, M.E.; Giske, C.G.; Harbarth, S.; Hindler, J.F.; Kahlmeter, G.; Olsson-Liljequist, B.; et al. Multidrug-Resistant, Extensively Drug-Resistant and Pandrug-Resistant Bacteria: An International Expert Proposal for Interim Standard Definitions for Acquired Resistance. *Clin. Microbiol. Infect.* **2012**, *18*, 268–281. [CrossRef]
36. Berrazeg, M.; Drissi, M.; Medjahed, L.; Rolain, J.M. Hierarchical Clustering as a Rapid Tool for Surveillance of Emerging Antibiotic-Resistance Phenotypes in *Klebsiella pneumoniae* Strains. *J. Med. Microbiol.* **2013**, *62*, 864–874. [CrossRef]
37. Ward, J.H., Jr. Hierarchical Grouping to Optimize an Objective Function. *J. Am. Stat. Assoc.* **1963**, *58*, 236–244. [CrossRef]

Review

Control of *Escherichia coli* in Poultry Using the In Ovo Injection Technique

Gabriel da Silva Oliveira [1] , Concepta McManus [1] and Vinícius Machado dos Santos [2,*]

[1] Faculty of Agronomy and Veterinary Medicine, University of Brasília, Brasília 70910-900, Brazil; gabriels.unb@gmail.com (G.d.S.O.)
[2] Laboratory of Poultry Science, Federal Institute of Brasília—Campus Planaltina, Brasília 73380-900, Brazil
* Correspondence: vinicius.santos@ifb.edu.br

Abstract: Pathogens, such as *Escherichia coli* (*E. coli*), have been identified as significant causes of poultry mortality. Poultry can serve as potential sources of *E. coli* transmission, even when asymptomatic, posing a substantial threat to food safety and human health. The in ovo administration of antimicrobials is crucial for preventing and/or effectively combating acute and chronic infections caused by poultry pathogens. To achieve this goal, it is critical that antimicrobials are properly injected into embryonic fluids, such as the amnion, to reach target tissues and trigger robust antimicrobial responses. Several protocols based on antimicrobials were evaluated to meet these requirements. This review analyzed the impacts of antimicrobial substances injected in ovo on the control of *E. coli* in poultry. The reduction in infection rates, resulting from the implementation of in ovo antimicrobials, combined with efforts aimed at hygienic-sanitary action plans in poultry sheds, reinforces confidence that *E. coli* can be contained before causing large scale damage. For example, antimicrobial peptides and probiotics have shown potential to provide protection to poultry against infections caused by *E. coli*. Issues related to the toxicity and bacterial resistance of many synthetic chemical compounds represent challenges that need to be overcome before the commercial application of in ovo injection protocols focused on microbiological control.

Keywords: antimicrobials; egg microbiology; in-ovo injection; microbial reduction; poultry microbiology; poultry safety

Citation: Oliveira, G.d.S.; McManus, C.; dos Santos, V.M. Control of *Escherichia coli* in Poultry Using the In Ovo Injection Technique. *Antibiotics* **2024**, *13*, 205. https://doi.org/10.3390/antibiotics13030205

Academic Editor: Nahla O. Eltai

Received: 27 January 2024
Revised: 18 February 2024
Accepted: 20 February 2024
Published: 22 February 2024

1. Introduction

The establishment of microbiota in the eggshell may or may not influence healthy embryonic growth. It has been suggested that embryos may be resistant to bacterial infections originating in the eggshell, thanks to transgenerational immunological benefits [1]. On the other hand, it has been reported that the frequency of dead chick embryos with neck and beak deformities during the late incubation period may be associated with the *Escherichia coli* (*E. coli*) infectious process [2]. This microorganism is commonly found on eggshells [3]. The colonization of the microbiota in the eggshell begins in the hen's oviduct [4], raising questions about the possible negative effect of oviductal bacteria on embryonic development. However, microbiota colonization of freshly laid eggshells has received greater attention given the recognized association with embryonic infections resulting from bacterial penetration [5].

Research has revealed several active agents, mainly with antibacterial effects, for the treatment of hatching eggshells after collection [6–15]. Oliveira et al. [10] reported that one hour after spraying a 0.39% clove essential oil solution on hatching eggs, the count of total aerobic mesophilic bacteria (−1.19 log) and Enterobacteriaceae (−1.19 log) in eggshells significantly reduced. Cantu et al. [9] demonstrated that spraying 3% hydrogen peroxide followed by immediate exposure to UVC light (254 nm) significantly reduced aerobic plate counts (−3.51 log) on the surface of the hatching eggshell. However, before applying sanitizers, it is crucial to consider that a specific microbial load may have penetrated or already have been present in the internal contents during egg formation in the oviduct. Direct treatment of the egg contents may be beneficial, as it is not yet clear what proportion of sanitizer residues applied to eggshells have to penetrate and perform their antimicrobial role internally. Given that the poultry embryo is the most important figure in poultry production, it is essential to guarantee their development away from any microbiological risk that would make their survival unfeasible at any stage. Therefore, it is hypothesized that the injection of antimicrobial substances directly into the internal contents of the egg during embryonic development represents a strategy to ensure more effective protection of embryos against microbial action, thus seeking to protect them from possible infections after hatching.

To address this issue, this review analyzed the impacts of antimicrobial substances injected in ovo on controlling *E. coli* infections in poultry.

2. Consultation of Published Studies

This review was prepared based on bibliographical research, consulting studies indexed on Google Scholar. The terms used were "in ovo injection", "antimicrobial substances in ovo", "eggshell", "eggshell contamination", "microorganisms in eggshells", "eggshell penetration", "eggshell antimicrobial defense", "albumen", "albumen antimicrobial defense", "yolk", "yolk antimicrobial defense", "poultry embryonic infection", " *E. coli* in hatching eggs", "*E. coli* in poultry embryos", "poultry infected by *E. coli*", "in ovo antimicrobials to control *E. coli*", and "humans infected with *E. coli*". The criteria adopted for inclusion included: original articles and reviews written in English or Portuguese; studies that investigated the eggshell; studies focused on administering antimicrobial substances through the in ovo technique; and studies related to microbial contamination of eggs and embryos, specifically with *E. coli*. Any studies that did not meet these inclusion criteria were promptly excluded from the analysis. The literature was consulted until the writing of each topic was finalized.

3. Eggshells and Their Natural Defenses

The eggshell generally has two predominant functions: nourishing and protecting the embryo. This protective function encompasses defense against pathogens, which is effective thanks to the interaction between the physical barrier capacity and the antimicrobial proteins present in the eggshell [16]. In addition to having pores, the eggshell is subdivided into the cuticular, vertical crystal, palisade, and mammillary layers, and the outer and inner membranes (Figure 1) [17]. The cuticle is the upper layer, rich in polysaccharides, hydroxyapatite crystals, lipids, and glycoproteins [18]. The eggshell comprises the lower layer, the vertical crystal, formed by crystals aligned perpendicular to the surface, the palisade layer, composed of calcite crystals embedded in an organic matrix, and the mammillary layer, consisting of calcified columns and cones that penetrate the shell membranes [19]. The inner layers, formed by the outer and inner membranes, represent the basal protective layer of the eggshell, composed of protein fibers [17,19].

Over the years, several studies have explored different possibilities as to how pathogens can overcome eggshell barriers (Table 1). The channels that influence the penetration of microorganisms into the eggshell can be significantly linked to poultry, egg, microorganisms, or environmental conditions.

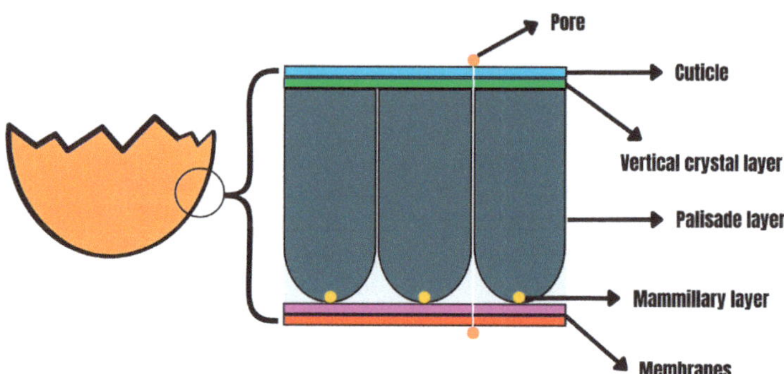

Figure 1. Poultry eggshell structure.

Table 1. Some factors associated with microbial penetration into the eggshell.

Factors	Reference
Absence or partial deposition of the cuticle	[20]
Eggshell pore diameter	[20]
Exposure of the egg to temperature variation regimes	[21]
Translucent eggshell surface	[21]
Genetic origin of Poultry	[22]
Egg dynamic stiffness	[23]
High contamination of the eggshell surface	[23]
Motile and non-clustering properties of some microorganisms	[24]
Poultry housing system	[25]
Poultry feed	[25]
Washing and sanitizing methods	[26]
Egg storage time	[27]
Number of pores in the eggshell	[27]
Eggshell condensation	[28]
Newly laid eggs (immature cuticle)	[29]
Chemical composition of the cuticle	[29]

4. *Escherichia coli* (*E. coli*) as a Threat during and after Embryonic Development

E. coli is a harmful pathogen in avian infections. This Gram-negative bacterium belongs to the Enterobacteriaceae family and can thrive in both aerobic and anaerobic environments, demonstrating adaptability when growing at temperatures ranging from 18–44 °C [30]. It may represent the most predominant bacteria among those isolated from eggs, shell-dead embryos, and newborn chicks [31]. *E. coli* can progress from harmless and asymptomatic colonization of the eggshell to the onset of potentially fatal embryonic diseases [32]. Its pathogenic specificity becomes particularly evident in embryonic infections, where *E. coli* demonstrates a remarkable ability to colonize the eggshell, invade it, and colonize embryonic tissues [33]. The invasion of *E. coli* through the eggshell not only represents a direct threat, but also promotes the invasion of other bacteria, such as *Staphylococcus aureus*, which is associated with high rates of embryonic mortality [34]. Among the main complications resulting from embryonic *E. coli* infection that lead to embryonic death are septicemia, omphalitis, and congenital deformities [2,35,36]. The presence of *E. coli* can result in the death of up to 92% of affected embryos [32]. Wang et al. [33] revealed that chick embryos died 48 h after being infected by *E. coli*. These findings provide an explanation for the decreased hatchability rate of *E. coli*-infected embryos at 18 days of development [37].

Another worrying aspect is the possibility of infection of embryos by *E. coli* through the eggshell, without them showing clinical signs during development. Such symptoms can appear after hatching [38], substantially increasing the risk of cross-contamination outbreaks and widespread mortality in poultry houses. Undesirable effects have been identified in broiler chickens infected by *E. coli*, manifesting through clinical signs and histopathological lesions such as: (1) Ruffled feathers, (2) inappetence, (3) respiratory manifestations, (4) sitting on hocks, (5) yellow and whitish diarrhea, (6) pericarditis, (7) enteritis, (8) airsacculitis, (9) liver and lung congestion, and (10) myocardial degeneration [39].

5. *Escherichia coli* (*E. coli*) as a Threat to Human Health

Although this review does not directly focus on human health, it is imperative to recognize that the seriousness of microbial contamination in hatcheries, poultry farms and slaughterhouses cannot, under any circumstances, be ignored. Human health must always prevail over any poultry production process. Both ingestion and inhalation are crucial routes of direct exposure to microbial contamination in humans, covering both occupational and non-occupational contexts. Hatcheries, poultry farms and slaughterhouses pose potential risks to humans, both in terms of contamination through inhalation and the possibility of ingestion if the final products intended for consumption are contaminated, as these products are considered one of the main reservoirs of *E. coli* [40]. An additional concern arises when products initially supplied to commercial establishments, in accordance with microbiological standards, end up suffering contamination during storage, especially if this occurs under inadequate climatic and sanitary conditions. The consequences resulting from the inhalation or ingestion of *E. coli* can manifest themselves in humans as acute or chronic infections, compromising the integrity of human health. Some such infections include urinary infections that may or may not be associated with cases of bacteremia [41], intestinal problems, including diarrhea [42], and meningitis, associated with significant mortality rates, or with a high risk of developing serious neurological sequelae [42,43] (Figure 2).

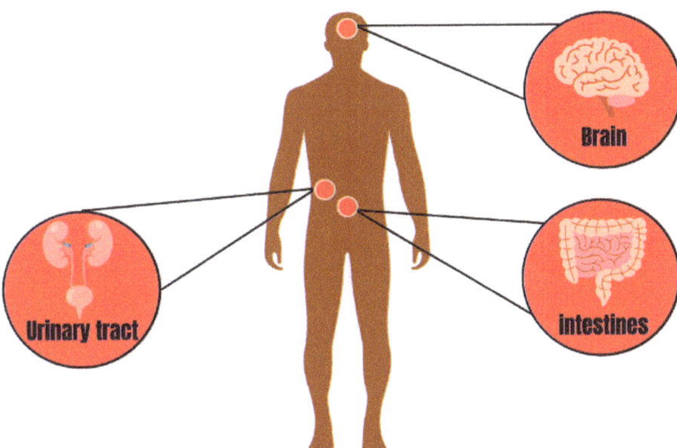

Figure 2. Common target sites of *E. coli* infection in humans.

6. What Is the In Ovo Injection Technique?

Antimicrobials administered into the egg via injections may represent an effective and rapid regimen to ensure microbial suppression during embryonic development and post-hatch. This regime, known as "in ovo injection", aims to deposit a compound of interest in the internal environment of the egg via the intervention of qualified professionals [44]. Approximately forty years ago, researchers tested this regime in the laboratory for vaccinating poultry before hatching [45]. Today, its commercial application around the world continues to prioritize vaccination as its main objective. Based on research already carried out [46–51], in ovo injections offer a range of advantages in poultry farming, such as:

- The in ovo injection technique does not require very complex professional training to be administered.
- It can be considered the best option for early and systemic immunization of poultry, with the absence of pain and stress.

- This technique allows for rapid and effective absorption of the injected medication, leading to faster immunization, or a faster response to treatment.
- It can inhibit bacterial growth and multiplication, thus reducing the cross-spread of bacteria in hatcheries and farms, as well as outbreaks of fatal diseases.
- It can induce long-lasting immunity, ensuring that poultry protection is maintained over time.
- It can favor the achievement of productivity gains related to the effects of the injected compound.
- It precipitates a reduction in operational and treatment costs related to poultry farming.

Firstly, before carrying out the in ovo injection technique, individual safety equipment must be used (Figure 3). The use of syringes with sterile needles and appropriate calibers for eggs is crucial, as it must be a minimally invasive and painless protocol. In general, the amount of substance injected is 0.1 mL [52], although a larger volume may be considered [53]. However, it is essential to highlight that, depending on the nature of the substance, the injection volume cannot exceed 0.4 mL, as this practice may be related to undesirable productive effects [54]. Additionally, 1 mL syringes with 23-G and 1-inch needles have been efficiently used in this practice [55]. After application, sterile paraffin is normally used to seal the pierced egg [56]. Although the recommendations above are not a general rule, the absence of adequate conditions, specifically for each antimicrobial substance, can significantly increase the risk of failures and embryonic mortality in the in ovo injection process [54]. The anatomical region of the egg used to administer antimicrobials is relevant to the safety and effectiveness of treatments developed to prevent or treat avian microbiological complications until post-hatch. Thus, the amnion, an extra-embryonic membrane, has been recommended as a potential site for the direct delivery of antimicrobial substances [48] (Figure 3). This intervention can occur during the prenatal or perinatal phases of embryonic development [51]. After the intervention, an immediate and prolonged microbial reduction is expected.

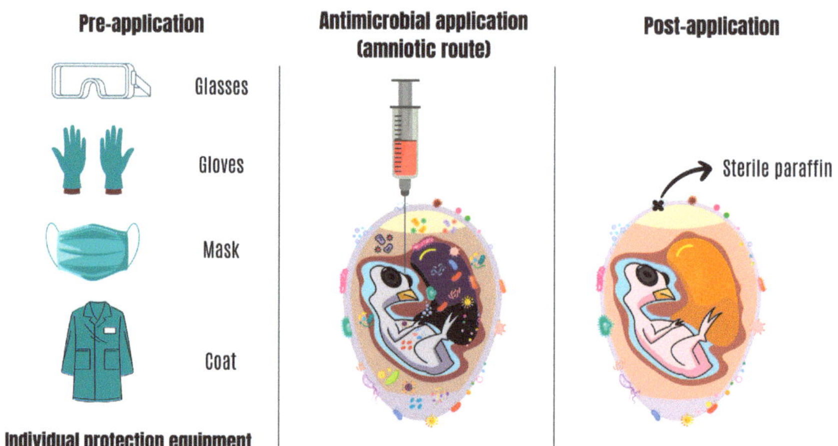

Figure 3. Antimicrobial intervention through in ovo application. Source: Adapted from Oliveira et al. [7].

To understand the in ovo drug administration route, it is necessary to understand the physiology of embryonic development. Among the various in ovo injection routes, the amniotic route, as mentioned earlier, is the most popular approach for in ovo drug administration. The main advantage of the drug administration system via the amniotic route, compared to other in ovo delivery routes (Figure 4), is the rapid distribution of the compound to the embryo. According to Williams [47], after being deposited in the amnion, therapeutic substances are rapidly absorbed orally and through the mucosal surfaces of the embryo's respiratory and digestive tracts. Antibacterial therapies require that pharmacological agents act quickly on the body of the target organism to provide protection and/or treatment. In this context, the amnion stands out as the best option for prevention or treatment against *E. coli* in embryos, since the drugs deposited in it normally have an efficacy rate above 90% [57].

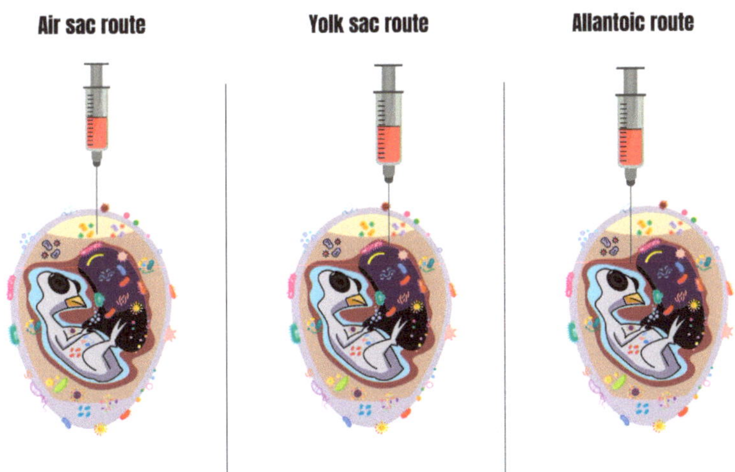

Figure 4. Other routes of drug application in ovo. Source: Adapted from Oliveira et al. [7]

7. In Ovo Injection as a Front Line against *Escherichia coli* (*E. coli*) Infection in Poultry

The management of infectious diseases in poultry requires daily administration of antimicrobials for a period that varies according to the target bacteria and its susceptibility to the antimicrobial, the severity of the infection, the immunological status of the poultry, and administration standards defined by the manufacturer, among others. However, non-adherence to therapy by poultry can lead to recurrence of the disease. Therefore, it is more advantageous to adopt preventive management practices even before signs of avian infection appear. The use of injectable antimicrobial formulations in the egg during embryonic development emerges as an effective preventive practice against microbial infection in poultry, especially by *E. coli* (Table 2). However, it is worth highlighting the importance of being cautious when using antibacterials for this purpose, mainly due to the development of antibacterial resistance. It is hypothesized that this efficiency of the in ovo injection practice is due to the rapid distribution of the antimicrobial throughout the body and its prolonged action. Twenty-four hours after administration into the amnion, the antimicrobial substance may have already spread throughout the embryo's body, including the gastrointestinal tract, respiratory system, and skin [48]. The effectiveness of the antimicrobial administered in ovo can allow the survival of 90% of embryos against *E. coli* infection [58] and ensure the protection of 100% of chicks against yolk sac infection by the same microorganism [59]. Furthermore, the antimicrobial effect of the injectable substance in the egg can be observed in poultry even when they reach 21 days of age [49].

Table 2. Control of *E. coli* in poultry after application of antimicrobials in ovo.

Compound Classification	Concentration	Day of Application in Embryos	Application Location	Effects Found after Application	Study
Immune stimulants					
Cytosine-phosphodiester-guanine oligodeoxynucleotides + polyphosphazene	50 µg/100 µL	E18	Amnion	Increased the immunoprotective effect against *E. coli* infections in poultry	[46]
Cytosine-phosphodiester-guanine oligodeoxynucleotides	50 µg/100 µL	E18	Amnion	It can be used to prevent and control mortality due to yolk sac infection by *E. coli*	[60]
Probiotics					
Intestinal microbial product	3.3×10^5 viable bacteria/egg	E18	Amnion	Reduced the abundance of Enterobacteriaceae (a family that includes *E. coli*) in the intestinal microbiota	[61]
Bacillus spp. probiotic-based	5×10^7 CFU/mL (1×10^7 CFU/200 µL)	E18	Amnion	Reduced the severity of virulent horizontal transmission of *E. coli* and infection of poultry in the incubation cabinet	[62]
Lactic acid microbiota	10^7 CFU/mL	E19	Amnion	Reduced Enterobacteriaceae colonization in poultry after *E. coli* infection	[63]
Bacillus subtilis, *Pediococcus acidilactici*, and *Enterococcus faecium*	10^7 CFU/mL	E18	Amnion	Reduced the intestinal population of *E. coli* in poultry	[64]
Antimicrobial peptides					
Avian antimicrobial peptides	30 µg peptide/100 µL PBS/embryo	E18	Amnion	Effective protection against yolk sac infection caused by *E. coli*	[65]
Chicken cathelicidin analog DCATH-2	4.4 mg/mL/100 µL PBS/embryo	E18	Amnion	Protected poultry against *E. coli* infection	[48]
Prebiotics					
Raffinose and stachyose	5 and 10%	E17	Amnion	The concentration of *E. coli* in the intestinal content of poultry did not show significant variations	[66]
Nanomaterials					
Green Silver Nanoparticles	0.17 mg/mL	E17.5	Amnion	Reduced *E. coli* counts in the cecal content of poultry	[67]
Bacteriophages					
Phage cocktail	100 µL of the phage cocktail (5.2×10^8 PFU/mL) or DPBS	E16	Allantois	Prevented the development of avian colibacillosis	[50]
Synbiotics					
Lactobacillus plantarum + Astragalus polysaccharide	200 µL of the *Lactobacillus plantarum* + 2 mg/egg Astragalus polysaccharide	E18.5	Amnion	Increased colonization of *Lactobacillus* spp. and *Bifidobacterium* spp. and decreased the population of *E. coli* in the avian cecum.	[49]
Natural extract and vitamins					
Grape seed extract and vitamin C	Grape seed extract (3, 4.5 or 6 mg/egg), and vitamin C (3 mg/egg)	E18	Air sac	Decrease in the population of *E. coli* in the ileum	[68]
Amino acids					
L-arginine	1–0.5%	E14	Amnion	Reduced *E. coli* in the cecum of poultry	[69]

As noted previously, successful antimicrobials demonstrate high efficacy in long-lasting prevention and rapid treatment of poultry infection-causing pathogens such as *E. coli*, providing systematic protection that effectively limits or prevents the spread of infection in farming systems. Many antimicrobials demonstrate success in combating *E. coli* due to their action mechanisms that result in the death of this bacterium. It has been elucidated that natural antimicrobials may have the ability to cause damage to the cell membrane of *E. coli*, resulting in the leakage of proteins and nucleic acids (Figure 5). This phenomenon triggers the destabilization of metabolic activity, ultimately culminating in bacterial cell death [70]. In the same way, synthetic chemical antimicrobials can also induce disturbances in the cell walls and membranes of *E. coli*, reducing its protection and resulting in the loss of intracellular content [71]. This is the most elucidated antibacterial mechanism.

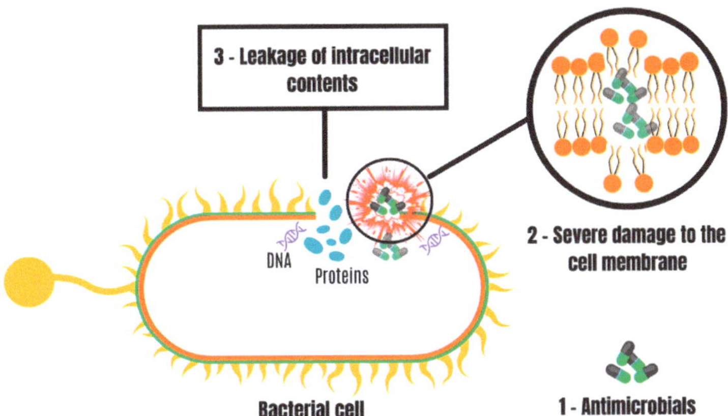

Figure 5. One of the mechanisms of action of antibiotics on bacteria.

Advances in preventing or treating *E. coli* infections through in ovo delivery of substances have primarily focused on the use of antimicrobial peptides and probiotics. A peptide is a chain of amino acids that generally does not exceed 50 amino acids, linked together by peptide bonds [72]. Identified sources of peptides include microorganisms, plants, animals, and humans [73]. Peptides present a cocktail of attractive characteristics, such as compatibility with poultry safety [65]. Furthermore, they have pharmacological aspects, including activity against gram-positive and gram-negative bacteria [65,74]. The implementation of antibacterial peptides in poultry farming can significantly contribute to solving several problems related to poultry productivity and health [75]. Two important families of antimicrobial peptides with potential application in poultry farming are β-defensins and cathelicidins [76].

Probiotics are beneficial live microorganisms that, in certain concentrations, exert a broad spectrum of biological activities. This includes antibacterial properties, which have played a significant role in increasing interest in opening new therapeutic horizons in poultry farming [77,78]. A review carried out by Cox and Dalloul [79] on the role of probiotics in poultry concluded that probiotics are beneficial for improving performance, maintaining healthy balance of the intestinal microbiota, and neutralizing adverse effects of infectious diseases. Several microorganisms have physiological and technological characteristics that classify them as probiotics. Among them are *Lactobacillus acidophilus*, *Lactobacillus casei*, *Lactobacillus fermentum*, *Lactobacillus gasseri*, *Lactobacillus johnsonii*, *Lactobacillus lactis*, *Lactobacillus paracasei*, *Lactobacillus plantarum*, *Lactobacillus reuteri*, *Lactobacillus rhamnosus*, *Lactobacillus salivarius*, *Bifidobacterium longum*, *Bifidobacterium bifidum*, *Bifidobacterium breve*, *Bifidobacterium animalis*, and *Streptococcus thermophilus* [80].

In short, it has been observed that the antibacterial compounds injected into the egg act mainly to reduce the bacterial load of the poultry, protecting them against infections before or after hatching. Furthermore, they beneficially modulate the intestinal microbiota and strengthen the poultry immune response, minimizing cases of mortality.

8. Is In Ovo Injection Harmful to Hatchability?

Hatchability is the gold parameter for evaluating antimicrobial techniques involving embryos and hatchery performance. It represents the proportion of chicks born alive for a specific sample of eggs [51]. A recent bibliographic mapping addressing the relationship between hatchability and the in ovo injection technique [51] showed that, in general, the practice of in ovo injection tends to improve hatchability. However, the review highlighted that this technique has a more significant impact on poultry health parameters than on hatchability itself, and that association of the technique with possible loss of hatchability was observed in specific cases [51]. Therefore, it is more interesting for the poultry sector to adopt sanitary procedures, with the potential not only to ensure poultry is free from bacterial infections but also to at least preserve hatchability, given that poultry yields depend significantly on this index and high-quality standards of the poultry. Choosing a multifunctional antimicrobial can also minimize costs that could otherwise make the adoption of in ovo infection unfeasible. A wide repertoire of antimicrobial solutions, such as carbohydrate/electrolyte + potassium chloride + theophylline, tripotassium citrate + potassium chloride + theophylline, creatine + potassium chloride + theophylline [81], the nano form of zinc, copper, or selenium [82], vitamin A, vitamin E, vitamin D3, folic acid, [83] L-Arginine, and L-Threonine [84], were not associated with harm in hatchability.

9. Antimicrobials and Hygiene Practices in the Poultry Sector

Eggshell contamination by *E. coli* often originates in breeding sheds. Subsequently, this contamination can be transmitted horizontally to the embryo, persisting until after hatching. Furthermore, poultry can be directly contaminated by *E. coli* present in the shed environment. Therefore, poultry houses with unsanitary and microbiologically compromised conditions can negatively affect the quality of poultry and act as sources of inoculum for pathogenic microorganisms, such as *E. coli*, which can cause significant damage to poultry production and the safety of poultry food products. These unsanitary conditions also have the potential to obstruct trade in poultry products in both national and international markets. To prevent infectious outbreaks caused by *E. coli* and poultry health emergencies at regional, national, or international levels, it is essential to implement preventive microbial control programs. This includes effective safety management before, during and after production. The use of antimicrobials selected based on antibiograms, under the guidance of qualified professionals and in partnership with poultry companies, is a key component of these programs, ensuring effective disease prevention.

Ahmed et al. [85] showed that the application of 250 mL of chlorine dioxide (ClO_2) for fumigation in a broiler shed at the end of 5 weeks of rearing resulted in a significant reduction in the concentration of *E. coli* in just 10 min. This reduction remained significantly effective up to 12 h after application, without causing any adverse effects to the health of the poultry. Likewise, Jiang et al. [86] presented results indicating that spraying a poultry house with a sanitizer containing aldehydes, quaternary ammonium salt, and alcohol (ratio 1:1500) resulted in a significant reduction in the relative abundance of pathogens of the genus *Escherichia-Shigella*. Based on these studies, the importance of a detailed management plan that incorporates antimicrobial actions in poultry sheds is reinforced. However, the efficiency of the plan depends on the daily execution of these actions, as well as the training of the professionals responsible for their execution [13].

10. Conclusions and Future Perspectives

The in ovo injection technique allows the development of personalized protocols to overcome specific challenges in the effective administration of antimicrobials and combating *E. coli* infections. The integration of this technique with practices already established in the poultry sector, such as rigorous hygienic-sanitary maintenance in sheds, can enhance the efficiency and precision of treatment, ensuring a more targeted and effective approach to poultry care. Although this technology has great potential, it is crucial to address some issues before its full implementation in industrial poultry environments. For example, additional research is essential to evaluate the effectiveness of combining eggshell sanitation and in ovo antimicrobial administration, specifically to improve the practice of in ovo injection in poultry production, especially in the treatment of *E. coli*. Furthermore, before proceeding with its large-scale adoption in poultry farming, it is vital that the combined technology undergoes rigorous testing to ensure its safety for poultry, humans, and the environment. It is hoped that this review will provide poultry researchers and professionals with a clear perspective on how careful selection of antimicrobials, combined with a refined in ovo application protocol, can constitute an effective strategy to significantly optimize yields in the poultry sector.

In the practice of in ovo injection in the poultry industry, it is expected that, in addition to vaccination, there will be a routine dedication to the in ovo delivery of antimicrobials, with the main objective of controlling bacterial proliferation. However, this requires a careful and comprehensive approach to several issues, such as:

- Over the past few decades, several protocols have been developed for the delivery of substances in ovo in the poultry field. Some of these protocols have been specifically designed to protect poultry against bacterial infections. Within these protocols, the use of antimicrobial peptides and probiotics has been the subject of intense investigation and reporting. The implementation of these protocols, centered on such compounds, takes priority in commercial production, given the concentrated database available that supports their characteristics of simplicity, cost-benefit, ease of in ovo application, and compatibility with poultry safety. In addition, the toxicity and bacterial resistance of many synthetic chemical compounds have been considered.

- The chosen in ovo delivery route may influence the effectiveness of antibacterials for poultry. Therefore, studies have proposed the amniotic route as the most effective to guarantee avian protection. These results will contribute to the development of commercial protocols utilizing a more advantageous in ovo delivery route.

- Some tested compounds may exhibit antibacterial specificity for a specific group of bacteria, meaning that the compound does not have a broad antibacterial spectrum. Although this review focuses on the control of *E. coli*, the search for compounds with broad-spectrum antibacterial properties represents a promising avenue for in ovo injection protocols. This requires further investigation.

- When developing in ovo application protocols, it is crucial to consider the associated economic cost and environmental damage. High costs can create barriers to commercial application, while the use of toxic synthetic chemicals can pose a threat to the environment.

- Many of the compounds tested in ovo were only evaluated under laboratory conditions. Therefore, testing under commercial conditions is essential, since the results obtained in the laboratory may encounter several limitations, even if minimal, due to the different realities faced in practice.

Author Contributions: Conceptualization, G.d.S.O. and V.M.d.S.; writing—original draft preparation, G.d.S.O.; writing—review and editing, G.d.S.O., C.M. and V.M.d.S.; visualization, G.d.S.O., C.M. and V.M.d.S.; supervision, C.M. and V.M.d.S. All authors have read and agreed to the published version of the manuscript.

Funding: This research was funded by Coordenação de Aperfeiçoamento de Pessoal de Nível Superior (CAPES) grant number 001. It received support from the Fundação de Amparo à Pesquisa do Distrito Federal (FAPDF) for scientific publication.

Institutional Review Board Statement: Not applicable.

Informed Consent Statement: Not applicable.

Data Availability Statement: Not applicable.

Acknowledgments: The authors thank Coordenação de Aperfeiçoamento de Pessoal de Nível Superior (CAPES) for the granted scholarship to the doctoral student G.d.S.O.

Conflicts of Interest: The authors declare no conflicts of interest.

References

1. Soler, J.J.; Peralta-Sánchez, J.M.; Flensted-Jensen, E.; Martín-Platero, A.M.; Møller, A.P. Innate humoural immunity is related to eggshell bacterial load of European birds: A comparative analysis. *Naturwissenschaften* **2011**, *98*, 807–813. [CrossRef]
2. Hananeh, W.M.; Al-Natour, M.Q.; Alaboudi, A.R.; Abo-Shehada, M.N.; Ismail, Z.A.B. Congenital abnormalities in dead-in-shell chicks associated with mixed bacterial infections. *Heliyon* **2021**, *7*, e06272. [CrossRef]
3. Oliveira, G.d.S.; dos Santos, V.M.; McManus, C. Propolis: Effects on the Sanitisation of Hatching Eggs. *World's Poult. Sci. J.* **2022**, *78*, 261–272. [CrossRef]
4. Lee, S.; La, T.-M.; Lee, H.-J.; Choi, I.-S.; Song, C.-S.; Park, S.-Y.; Lee, J.-B.; Lee, S.-W. Characterization of microbial communities in the chicken oviduct and the origin of chicken embryo gut microbiota. *Sci. Rep.* **2019**, *9*, 6838. [CrossRef] [PubMed]
5. Ahmed, B.; de Boeck, C.; Dumont, A.; Cox, E.; de Reu, K.; Vanrompay, D. First Experimental Evidence for the Transmission of Chlamydia psittaci in Poultry through Eggshell Penetration. *Transb. Emerg. Dis.* **2017**, *64*, 167–170. [CrossRef] [PubMed]
6. Uçan, U.S.; Gök, A. Efficacy of a Water-Based Disinfectant on Reduction of Eggshell Bacterial Contamination. *Eur. J. Vet. Sci.* **2012**, *28*, 57–59.
7. Oliveira, G.d.S.; McManus, C.; de Araújo, M.V.; de Sousa, D.E.R.; de Macêdo, I.L.; de Castro, M.B.; dos Santos, V.M. Sanitizing Hatching Eggs with Essential Oils: Avian and Microbiological Safety. *Microorganisms* **2023**, *11*, 1890. [CrossRef]
8. Mousa-Balabel, T.M.; Al-Midani, S.A.; Al-Refaay, M.A.; Kerady, S.M. Coliform Bacteria and Hatching Egg Disinfectants. *Int. J. Sci. Basic Appl. Res.* **2017**, *33*, 151–163.
9. Cantu, K.; Archer, G.S.; Tucker, Z.S.; Coufal, C.D. Effectiveness of Duck Hatching Egg Sanitization with the Combination of Hydrogen Peroxide and Ultraviolet Light. *J. Appl. Poult. Res.* **2019**, *28*, 301–306. [CrossRef]
10. Oliveira, G.S.; Nascimento, S.T.; dos Santos, V.M.; Silva, M.G. Clove Essential Oil in the Sanitation of Fertile Eggs. *Poult. Sci.* **2020**, *99*, 5509–5516. [CrossRef]
11. Oliveira, G.d.S.; dos Santos, V.M.; Nascimento, S.T. Essential Oils as Sanitisers for Hatching Eggs. *World's Poult. Sci. J.* **2021**, *77*, 605–617. [CrossRef]
12. Oliveira, G.S.; Nascimento, S.T.; dos Santos, V.M.; Dallago, B.S.L. Spraying Hatching Eggs with Clove Essential Oil Does Not Compromise the Quality of Embryos and One-Day-Old Chicks or Broiler Performance. *Animals* **2021**, *11*, 2045. [CrossRef]
13. Oliveira, G.d.S.; McManus, C.; dos Santos, V.M. Garlic as Active Principle of Sanitiser for Hatching Eggs. *World's Poult. Sci. J.* **2022**, *78*, 1037–1052. [CrossRef]
14. Oliveira, G.d.S.; McManus, C.; Salgado, C.B.; dos Santos, V.M. Effects of Sanitizers on Microbiological Control of Hatching Eggshells and Poultry Health during Embryogenesis and Early Stages after Hatching in the Last Decade. *Animals* **2022**, *12*, 2826. [CrossRef] [PubMed]
15. Oliveira, G.S.; dos Santos, V.M.; Nascimento, S.T.; Rodrigues, J.C. Alternative Sanitizers to Paraformaldehyde for Incubation of Fertile Eggs. *Poult. Sci.* **2020**, *99*, 2001–2006. [CrossRef]
16. Hincke, M.T.; Da Silva, M.; Guyot, N.; Gautron, J.; McKee, M.D.; Guabiraba-Brito, R.; Rehault-Godbert, S. Dynamics of Structural Barriers and Innate Immune Components during Incubation of the Avian Egg: Critical Interplay between Autonomous Embryonic Development and Maternal Anticipation. *J. Innate Immun.* **2019**, *11*, 111–124. [CrossRef] [PubMed]
17. Gautron, J.; Stapane, L.; Le Roy, N.; Nys, Y.; Rodriguez-Navarro, A.B.; Hincke, M.T. Avian eggshell biomineralization: An update on its structure, mineralogy and protein tool kit. *BMC Mol. Cell Biol.* **2021**, *22*, 11. [CrossRef] [PubMed]
18. Wellman-Labadie, O.; Picman, J.; Hincke, M.T. Antimicrobial Activity of the Anseriform Outer Eggshell and Cuticle. *Comp. Biochem. Physiol. Part B Biochem. Mol. Biol.* **2008**, *149*, 640–649. [CrossRef] [PubMed]
19. Gautron, J.; Nys, Y. Function of eggshell matrix proteins. In *Bioactive Egg Compounds*; Huopalahti, R., Lopez-Fandino, R., Anton, M., Schade, R., Eds.; Springer: Berlin/Heidelberg, Germany, 2007; pp. 109–115, ISBN 978-3-540-37883-9.
20. Chen, X.; Li, X.; He, Z.; Hou, Z.; Xu, G.; Yang, N.; Zheng, J. Comparative Study of Eggshell Antibacterial Effectivity in Precocial and Altricial Birds Using Escherichia Coli. *PLoS ONE* **2019**, *14*, e0220054. [CrossRef]
21. Chousalkar, K.K.; Flynn, P.; Sutherland, M.; Roberts, J.R.; Cheetham, B.F. Recovery of Salmonella and Escherichia coli from commercial egg shells and effect of translucency on bacterial penetration in eggs. *Int. J. Food Microbiol.* **2010**, *142*, 207–213. [CrossRef]
22. Rathgeber, B.M.; McCarron, P.; Budgell, K.L. Salmonella penetration through eggshells of chickens of different genetic backgrounds. *Poult. Sci.* **2013**, *92*, 2457–2462. [CrossRef]
23. Messens, W.; Grijspeerdt, K.; De Reu, K.; De Ketelaere, B.; Mertens, K.; Bamelis, F.; Kemps, B.; De Baerdemaeker, J.; Decuypere, E.; Herman, L. Eggshell penetration of various types of hens' eggs by Salmonella enterica serovar Enteritidis. *J. Food Prot.* **2007**, *70*, 623–628. [CrossRef] [PubMed]
24. De Reu, K.; Grijspeerdt, K.; Messens, W.; Heyndrickx, M.; Uyttendaele, M.; Debevere, J.; Herman, L. Eggshell factors influencing eggshell penetration and whole egg contamination by different bacteria, including *Salmonella* Eenteritidis. *Int. J. Food Microbiol.* **2006**, *112*, 253–260. [CrossRef] [PubMed]

25. Messens, W.; Grijspeerdt, K.; Herman, L. Intrinsic and extrinsic factors influencing eggshell penetration by Salmonella Enteritidis. In Proceedings of the XVII European Symposium on the Quality of Poultry Meat and XI European Symposium on the Quality of Eggs and Egg Products, Golden Tulip Parkhotel Doorwerth, Doorwerth, The Netherlands, 23–26 May 2005; pp. 23–26.
26. Sauter, E.A.; Petersen, C.F. The effect of egg shell quality on penetration by *Pseudomonas fluorescens*. *Poult. Sci.* **1969**, *48*, 1525–1528. [CrossRef]
27. Vlčková, J.; Tůmová, E.; Ketta, M.; Englmaierová, M.; Chodová, D. Effect of housing system and age of laying hens on eggshell quality, microbial contamination, and penetration of microorganisms into eggs. *Czech J. Anim. Sci.* **2018**, *63*, 51–60. [CrossRef]
28. De Reu, K.; Grijspeerdt, K.; Heyndrickx, M.; Messens, W.; Uyttendaele, M.; Debevere, J.; Herman, L. Influence of eggshell condensation on eggshell penetration and whole egg contamination with Salmonella enterica serovar enteritidis. *J. Food Prot.* **2006**, *69*, 1539–1545. [CrossRef]
29. Munoz, A.; Dominguez-Gasca, N.; Jimenez-Lopez, C.; Rodriguez-Navarro, A.B. Importance of eggshell cuticle composition and maturity for avoiding trans-shell *Salmonella* contamination in chicken eggs. *Food Control* **2015**, *55*, 31–38. [CrossRef]
30. Swelum, A.A.; Elbestawy, A.R.; El-Saadony, M.T.; Hussein, E.O.S.; Alhotan, R.; Suliman, G.M.; Taha, A.E.; Ba-Awadh, H.; El-Tarabily, K.A.; El-Hack, M.E.A. Ways to minimize bacterial infections, with special reference to *Escherichia coli*, to cope with the first-week mortality in chicks: An updated overview. *Poult. Sci.* **2021**, *100*, 101039. [CrossRef]
31. Abd El-Daye, G.A.; Moustafa, S.S. Effect of using lactobacillus acidophilus on *E. coli* causing embryonic death and low hatchability in balady hatcheries at dakahlia governorate. *Assiut Vet. Med. J.* **2013**, *59*, 170–176.
32. Ozaki, H.; Murase, K.Y. Virulence of Escherichia coli Isolates Obtained from Layer Chickens with Colibacillosis Associated with Pericarditis, Perihepatitis, and Salpingitis in Experimentally Infected Chicks and Embryonated Eggs. *Avian Dis.* **2018**, *62*, 233–236. [CrossRef]
33. Wang, C.; Pors, S.E.; Olsen, R.H.; Bojesen, A.M. Transmission and pathogenicity of *Gallibacterium anatis* and *Escherichia coli* in embryonated eggs. *Vet. Microbiol.* **2018**, *217*, 76–81. [CrossRef]
34. Kosecka-Strojek, M.; Trzeciak, J.; Homa, J.; Trzeciak, K.; Władyka, B.; Trela, M.; Międzobrodzki, J.; Lis, M.W. Effect of *Staphylococcus aureus* infection on the heat stress protein 70 (HSP70) level in chicken embryo tissues. *Poult. Sci.* **2021**, *100*, 101119. [CrossRef]
35. Kabir, S.M.K. Avian colibacillosis and salmonellosis: A closer look at epidemiology, pathogenesis, diagnosis, control and public health concerns. *Int. J. Environ. Res. Public Health* **2010**, *7*, 89–114. [CrossRef]
36. Kalita, N.; Pathak, N.; Saikia, G.; Ahmed, M. Prevalence and pathology of dead-in-shell embryos of Vanaraja egg. *Indian J.* **2013**, *37*, 104–105.
37. Fatemi, S.A.; Lindsey, L.L.; Evans, J.D.; Elliott, K.E.C.; Leigh, S.A.; Robinson, K.J.; Mousstaaid, A.; Gerard, P.; Peebles, E.D. Effects of the in ovo injection of an *Escherichia coli* vaccine on the hatchability and quality characteristics of commercial layer hatchlings. *Poult. Sci.* **2023**, *102*, 103057. [CrossRef] [PubMed]
38. Nolan, L.K.; John Barnes, H.; Vaillancourt, J.-P.; Abdul-Aziz, T.; Logue, C.M. Colibacillosis. In *Diseases of Poultry*; Swayne, D.E., Ed.; John Wiley & Sons, Inc.: Hoboken, NJ, USA, 2013.
39. Elnagar, R.; Elkenany, R.; Younis, G. Interleukin gene expression in broiler chickens infected by different *Escherichia coli* serotypes. *Vet. World* **2021**, *14*, 2727. [CrossRef] [PubMed]
40. Bergeron, C.R.; Prussing, C.; Boerlin, P.; Daignault, D.; Dutil, L.; Reid-Smith, R.J.; Zhanel, G.G.; Manges, A.R. Chicken as reservoir for extraintestinal pathogenic *Escherichia coli* in humans, Canada. *Emerg. Infect. Dis.* **2012**, *18*, 415–421. [CrossRef] [PubMed]
41. Bonten, M.; Johnson, J.R.; Van Den Biggelaar, A.H.J.; Georgalis, L.; Geurtsen, J.; De Palacios, P.I.; Gravenstein, S.; Verstraeten, T.; Hermans, P.; Poolman, J.T. Epidemiology of *Escherichia Coli* Bacteremia: A Systematic Literature Review. *Clin. Infect. Dis.* **2021**, *72*, 1211–1219. [CrossRef] [PubMed]
42. Croxen, M.A.; Finlay, B.B. Molecular mechanisms of *Escherichia coli* pathogenicity. *Nat. Rev. Microbiol.* **2010**, *8*, 26–38. [CrossRef] [PubMed]
43. Kaper, J.B.; Nataro, J.P.; Mobley, H.L. Pathogenic *Escherichia coli*. *Nat. Rev. Microbiol.* **2004**, *2*, 123–140. [CrossRef]
44. Uni, Z.; Ferket, P.R.; Tako, E.; Kedar, O. In Ovo Feeding Improves Energy Status of Late-Term Chicken Embryos. *Poult. Sci.* **2005**, *84*, 764–770. [CrossRef]
45. Williams, C.J. In ovo vaccination for disease prevention. *Int. J. Poult. Sci.* **2007**, *15*, 7–9.
46. Taghavi, A.; Allan, B.; Mutwiri, G.; Foldvari, M.; Kessel, A.; Willson, P.; Babiuk, L.; Potter, A.; Gomis, S. Enhancement of immunoprotective effect of CpG-ODN by formulation with polyphosphazenes against *E. coli* septicemia in neonatal chickens. *Curr. Drug Deliv.* **2009**, *6*, 76–82. [CrossRef] [PubMed]
47. Williams, C.J. In ovo vaccination and chick quality. *Int. Hatch. Prac.* **2011**, *19*, 7–13.
48. Cuperus, T.; van Dijk, A.; Matthijs, M.G.; Veldhuizen, E.J.; Haagsman, H.P. Protective effect of in ovo treatment with the chicken cathelicidin analog D-CATH-2 against avian pathogenic *E. coli*. *Sci. Rep.* **2016**, *6*, 26622. [CrossRef] [PubMed]
49. Duan, A.Y.; Ju, A.Q.; Zhang, Y.N.; Qin, Y.J.; Xue, L.G.; Ma, X.; Luan, W.M.; Yang, S.B. The Effects of In Ovo Injection of Synbiotics on the Early Growth Performance and Intestinal Health of Chicks. *Front. Vet. Sci.* **2021**, *8*, 658301. [CrossRef] [PubMed]
50. Nicolas, M.; Faurie, A.; Girault, M.; Lavillatte, S.; Menanteau, P.; Chaumeil, T.; Riou, M.; Velge, P.; Schouler, C. In ovo administration of a phage cocktail partially prevents colibacillosis in chicks. *Poult. Sci.* **2023**, *102*, 102967. [CrossRef] [PubMed]
51. Oliveira, G.D.S.; McManus, C.; Salgado, C.B.; Dos Santos, V.M. Bibliographical Mapping of Research into the Relationship between In Ovo Injection Practice and Hatchability in Poultry. *Vet. Sci.* **2023**, *10*, 296. [CrossRef] [PubMed]

52. Zhao, M.; Li, J.; Shi, Q.; Shan, H.; Liu, L.; Geng, T.; Yu, L.; Gong, D. The Effects of In Ovo Feeding of Selenized Glucose on Selenium Concentration and Antioxidant Capacity of Breast Muscle in Neonatal Broilers. *Biol. Trace Elem. Res.* **2023**, *201*, 5764–5773. [CrossRef]

53. Gupta, V.; Ncho, C.M.; Goel, A.; Jeong, C.-M.; Choi, Y.-H. Effects of In Ovo Injection of α-Ketoglutaric Acid on Hatchability, Growth, Plasma Metabolites, and Antioxidant Status of Broilers. *Antioxidants* **2022**, *11*, 2102. [CrossRef]

54. Zhai, W.; Rowe, D.E.; Peebles, E.D. Effects of commercial in ovo injection of carbohydrates on broiler embryogenesis. *Poult. Sci.* **2011**, *90*, 1295–1301. [CrossRef]

55. Ncho, C.M.; Goel, A.; Jeong, C.M.; Youssouf, M.; Choi, Y.H. In ovo injection of gaba can help body weight gain at hatch, increase chick weight to egg weight ratio, and improve broiler heat resistance. *Animals* **2021**, *11*, 1364. [CrossRef]

56. Yang, S.B.; Qin, Y.J.; Ma, X.; Luan, W.M.; Sun, P.; Ju, A.Q.; Duan, A.Y.; Zhang, Y.N.; Zhao, D.H. Effects of in ovo injection of Astragalus polysaccharide on the intestinal development and mucosal immunity in broiler chickens. *Front. Vet. Sci.* **2021**, *8*, 738816. [CrossRef]

57. Wakenell, P.S.; Bryan, T.; Schaeffer, J.; Avakian, A.; Williams, C.; Whitfill, C. Effect of in ovo vaccine delivery route on HVT/SB1 efficacy and viremia. *Avian Dis.* **2002**, *46*, 274–280. [CrossRef]

58. Nicolas, M.; Trotereau, A.; Culot, A.; Moodley, A.; Atterbury, R.; Wagemans, J.; Lavigne, R.; Velge, P.; Schouler, C. Isolation and Characterization of a Novel Phage Collection against Avian-Pathogenic *Escherichia coli*. *Microbiol. Spectr.* **2023**, *11*, e04296-22. [CrossRef] [PubMed]

59. Sarfraz, M.; Nguyen, T.T.T.; Wheler, C.; Köster, W.; Gerdts, V.; Dar, A. Characterization of dosage levels for in ovo administration of innate immune stimulants for prevention of yolk sac infection in chicks. *Vet. Sci.* **2022**, *9*, 203. [CrossRef] [PubMed]

60. Allan, B.; Wheler, C.; Koster, W.; Sarfraz, M.; Potter, A.; Gerdts, V.; Dar, A. In Ovo Administration of Innate Immune Stimulants and Protection from Early Chick Mortalities due to Yolk Sac Infection. *Avian Dis.* **2018**, *62*, 316–321. [CrossRef] [PubMed]

61. Pedroso, A.A.; Batal, A.B.; Lee, M.D. Effect of in ovo administration of an adult-derived microbiota on establishment of the intestinal microbiome in chickens. *Am. J. Vet. Res.* **2016**, *77*, 514–526. [CrossRef]

62. Arreguin-Nava, M.A.; Graham, B.D.; Adhikari, B.; Agnello, M.; Selby, C.M.; Hernandez-Velasco, X.; Vuong, C.N.; Solis-Cruz, B.; Hernandez-Patlan, D.; Latorre, J.D.; et al. Evaluation of in ovo *Bacillus* spp. based probiotic administration on horizontal transmission of virulent *Escherichia coli* in neonatal broiler chickens. *Poult. Sci.* **2019**, *98*, 6483–6491. [CrossRef] [PubMed]

63. Arreguin-Nava, M.A.; Graham, B.D.; Adhikari, B.; Agnello, M.; Selby, C.M.; Hernandez-Velasco, X.; Vuong, C.N.; Solis-Cruz, B.; Hernandez-Patlan, D.; Latorre, J.D.; et al. In ovo administration of defined lactic acid bacteria previously isolated from adult hens induced variations in the cecae microbiota structure and enterobacteriaceae colonization on a virulent escherichia coli horizontal infection model in broiler chicken. *Front. Vet. Sci.* **2020**, *7*, 489. [CrossRef]

64. Majidi-Mosleh, A.; Sadeghi, A.; Mousavi, S.; Chamani, M.; Zarei, A. Ileal MUC2 gene expression and microbial population, but not growth performance and immune response, are influenced by in ovo injection of probiotics in broiler chickens. *Br. Poult. Sci.* **2017**, *58*, 40–45. [CrossRef]

65. Nguyen, T.T.T.; Allan, B.; Wheler, C.; Köster, W.; Gerdts, V.; Dar, A. Avian antimicrobial peptides: In vitro and in ovo characterization and protection from early chick mortality caused by yolk sac infection. *Sci. Rep.* **2021**, *11*, 2132. [CrossRef] [PubMed]

66. Pacifici, S.; Song, J.; Zhang, C.; Wang, Q.; Glahn, R.; Kolba, N.; Tako, E. Intra amniotic administration of raffinose and stachyose affects the intestinal brush border functionality and alters gut microflora populations. *Nutrients* **2017**, *9*, 304. [CrossRef] [PubMed]

67. Ahmed, M.M.; Ismail, Z.S.; Elwardany, I.; Lohakare, J.; Abdel-Wareth, A.A. In Ovo Feeding Techniques of Green Nanoparticles of Silver and Probiotics: Evaluation of Performance, Physiological, and Microbiological Responses of Hatched One-Day-Old Broiler Chicks. *Animals* **2023**, *13*, 3725. [CrossRef] [PubMed]

68. Hajati, H.; Hassanabadi, A.; Golian, A.; Nassiri-Moghaddam, H.; Nassiri, M.R. The effect of in ovo injection of grape seed extract and vitamin C on hatchability, antioxidant activity, yolk sac absorption, performance and ileal micro flora of broiler chickens. *Res. Opin. Anim. Vet. Sci.* **2014**, *4*, 633–638.

69. Omidi, S.; Ebrahimi, M.; Janmohammadi, H.; Moghaddam, G.; Rajabi, Z.; Hosseintabar-Ghasemabad, B. The impact of in ovo injection of l-arginine on hatchability, immune system and caecum microflora of broiler chickens. *J. Anim. Physiol. Anim. Nutr.* **2020**, *104*, 178–185. [CrossRef]

70. Wang, X.; Shen, Y.; Thakur, K.; Han, J.; Zhang, J.G.; Hu, F.; Wei, Z.J. Antibacterial activity and mechanism of ginger essential oil against *Escherichia coli* and *Staphylococcus aureus*. *Molecules* **2020**, *25*, 3955. [CrossRef]

71. Ding, N.; Li, Z.W.; Jiang, L.; Liu, H.; Zhang, Y.P.; Sun, Y.X. Kinetics and mechanisms of bacteria disinfection by performic acid in wastewater: In comparison with peracetic acid and sodium hypochlorite. *Sci. Total Environ.* **2023**, *878*, 162606. [CrossRef]

72. Chen, C.H.; Lu, T.K. Development and challenges of antimicrobial peptides for therapeutic applications. *Antibiotics* **2020**, *9*, 24. [CrossRef]

73. Savitskaya, A.; Masso-Silva, J.; Haddaoui, I.; Shymaa, E. Exploring the arsenal of antimicrobial peptides: Mechanisms, diversity, and applications. *Biochimie* **2023**, *214*, 216–227. [CrossRef]

74. Vila-Farres, X.; Giralt, E.; Vila, J. Update of peptides with antibacterial activity. *Curr. Med. Chem.* **2012**, *19*, 6188–6198. [CrossRef]

75. Xie, Z.; Zhao, Q.; Wang, H.; Wen, L.; Li, W.; Zhang, X.; Lin, W.; Li, H.; Xie, Q.; Wang, Y. Effects of antibacterial peptide combinations on growth performance, intestinal health, and immune function of broiler chickens. *Poult. Sci.* **2020**, *99*, 6481–6492. [CrossRef] [PubMed]

76. Nazeer, N.; Uribe-Diaz, S.; Rodriguez-Lecompte, J.C.; Ahmed, M. Antimicrobial peptides as an alternative to relieve antimicrobial growth promoters in poultry. *Br. Poult. Sci.* **2021**, *62*, 672–685. [CrossRef] [PubMed]
77. Neveling, D.P.; van Emmenes, L.; Ahire, J.J.; Pieterse, E.; Smith, C.; Dicks, L.M.T. Effect of a Multi-Species Probiotic on the Colonisation of Salmonella in Broilers. *Probiotics Antimicrob. Proteins* **2019**, *12*, 896–905. [CrossRef] [PubMed]
78. El-Hack, M.E.A.; El-Saadony, M.T.; Shafi, M.E.; Qattan, S.Y.A.; Batiha, G.E.; Khafaga, A.F.; Abdel-Moneim, A.E.; Alagawany, M. Probiotics in poultry feed: A comprehensive review. *J. Anim. Physiol. Anim. Nutr.* **2020**, *104*, 1835–1850. [CrossRef]
79. Cox, C.M.; Dalloul, R.A. Immunomodulatory role of probiotics in poultry and potential in ovo application. *Benef. Microbes* **2015**, *6*, 45–52. [CrossRef]
80. Kopp-Hoolihan, L. Prophylactic and therapeutic uses of probiotics: A review. *J. Am. Diet. Assoc.* **2001**, *101*, 229–238. [CrossRef]
81. McGruder, B.M.; Zhai, W.; Keralapurath, M.M.; Gerard, P.D.; Peebles, E.D. Effects of in Ovo Injection of Theophylline and Electrolyte Solutions on Hatchability and Growth of Broilers from Day 0 to Day 10 Post-Hatch. *Int. J. Poult. Sci.* **2011**, *10*, 927–932. [CrossRef]
82. Joshua, P.P.; Valli, C.; Balakrishnan, V. Effect of in ovo supplementation of nano forms of zinc, copper, and selenium on post-hatch performance of broiler chicken. *Vet. World* **2016**, *9*, 287–294. [CrossRef]
83. Gouda, A.; Tolba, S.A.; El-Moniary, M.M. Impact of in Ovo Injection of Certain Vitamins to Improve the Physiological Conditions of Hatching Chicks. *Pak. J. Biol. Sci.* **2021**, *24*, 268–273. [CrossRef]
84. Nabi, F.; Arain, M.A.; Bhutto, Z.A.; Shah, Q.A.; Bangulzai, N.; Ujjan, N.A.; Fazlani, S.A. Effect of Early Feeding of L-Arginine and L-Threonine on Hatchability and Post-Hatch Performance of Broiler Chicken. *Trop. Anim. Health Prod.* **2022**, *54*, 380. [CrossRef] [PubMed]
85. Ahmed, S.T.; Bostami, A.R.; Mun, H.S.; Yang, C.J. Efficacy of chlorine dioxide gas in reducing *Escherichia coli* and *Salmonella* from broiler house environments. *J. Appl. Poult. Res.* **2017**, *26*, 84–88. [CrossRef]
86. Jiang, L.; Li, M.; Tang, J.; Zhao, X.; Zhang, J.; Zhu, H.; Yu, X.; Li, Y.; Feng, T.; Zhang, X. Effect of Different Disinfectants on Bacterial Aerosol Diversity in Poultry Houses. *Front. Microbiol.* **2018**, *9*, 2113. [CrossRef] [PubMed]

Article

Whole-Genome Investigation of Zoonotic Transmission of Livestock-Associated Methicillin-Resistant *Staphylococcus aureus* Clonal Complex 398 Isolated from Pigs and Humans in Thailand

Pawarut Narongpun [1,†], Pattrarat Chanchaithong [2,†], Junya Yamagishi [3], Jeewan Thapa [1], Chie Nakajima [1,4,*] and Yasuhiko Suzuki [1,4,5,*]

1   Division of Bioresources, Hokkaido University International Institute for Zoonosis Control, Sapporo 001-0020, Japan; npawarut@czc.hokudai.ac.jp (P.N.); jeewan@czc.hokudai.ac.jp (J.T.)
2   Department of Veterinary Microbiology, Faculty of Veterinary Science, Chulalongkorn University, Bangkok 10330, Thailand; pattrarat.c@chula.ac.th
3   Division of Collaboration and Education, Hokkaido University International Institute for Zoonosis Control, Sapporo 001-0020, Japan; junya@czc.hokudai.ac.jp
4   International Collaboration Unit, Hokkaido University International Institute for Zoonosis Control, Sapporo 001-0020, Japan
5   Institute for Vaccine Research and Development, Hokkaido University, Sapporo 001-0020, Japan
*   Correspondence: cnakajim@czc.hokudai.ac.jp (C.N.); suzuki@czc.hokudai.ac.jp (Y.S.); Tel.: +81-011-706-7315 (Y.S.); Fax: +81-0-706-7310 (Y.S.)
†   These authors contributed equally to this work.

Citation: Narongpun, P.; Chanchaithong, P.; Yamagishi, J.; Thapa, J.; Nakajima, C.; Suzuki, Y. Whole-Genome Investigation of Zoonotic Transmission of Livestock-Associated Methicillin-Resistant *Staphylococcus aureus* Clonal Complex 398 Isolated from Pigs and Humans in Thailand. *Antibiotics* **2023**, *12*, 1745. https://doi.org/10.3390/antibiotics12121745

Academic Editor: Jonathan Frye

Received: 3 November 2023
Revised: 5 December 2023
Accepted: 14 December 2023
Published: 16 December 2023
Corrected: 23 July 2024

Abstract: Livestock-associated methicillin-resistant *Staphylococcus aureus* (LA-MRSA) has been widespread globally in pigs and humans for decades. Nasal colonization of LA-MRSA is regarded as an occupational hazard to people who are regularly involved in livestock production. Our previous study suggested pig-to-human transmission caused by LA-MRSA clonal complex (CC) 398, using traditional molecular typing methods. Instead, this study aimed to investigate the zoonotic transmission of LA-MRSA CC398 using whole genome sequencing (WGS) technologies. A total of 63 LA-MRSA isolates were identified and characterized in Thailand. Further, the 16 representatives of LA-MRSA CC9 and CC398, including porcine and worker isolates, were subjected to WGS on the Illumina Miseq platform. Core-genome single nucleotide polymorphism (SNP)-based analyses verify the zoonotic transmission caused by LA-MRSA CC398 in two farms. WGS-based characterization suggests the emergence of a novel staphylococcal cassette chromosome (SCC) *mec* type, consisting of multiple cassette chromosome recombinase (*ccr*) gene complexes via genetic recombination. Additionally, the WGS analyses revealed putative multi-resistant plasmids and several cross-resistance genes, conferring resistance against drugs of last resort used in humans such as quinupristin/dalfopristin and linezolid. Significantly, LA-MRSA isolates, in this study, harbored multiple virulence genes that may become a serious threat to an immunosuppressive population, particularly for persons who are in close contact with LA-MRSA carriers.

Keywords: livestock-associated methicillin-resistant *Staphylococcus aureus*; LA-MRSA; antimicrobial resistance; whole-genome sequencing; WGS; pigs; workers; occupational diseases; Thailand

1. Introduction

Over the past two decades, pigs have been regarded as crucial reservoirs of livestock-associated *Staphylococcus aureus* (LA-MRSA). Pig-associated MRSA can be occupationally transmitted to individuals who are frequently in close contact with colonized animals, for example, husbandry workers, veterinarians, and slaughterhouse workers [1–3]. Eventually, these groups of people may become asymptomatic carriers of LA-MRSA. Nasal LA-MRSA carriage in humans is associated with a greater risk of developing infections; moreover, carriers can potentially transfer LA-MRSA from farms to their household members, although it has been studied that human-to-human transmission caused by LA-MRSA transmission rarely occurs [4–7]. To date, it is believed that LA-MRSA tends to be less virulent than the other two strains in healthcare settings and communities, hospital-associated MRSA (HA-MRSA) and community-associated MRSA (CA-MRSA) [8,9]. As previously reported, for infections in healthy hosts, LA-MRSA generally develops mild or local infections; however, severe infections, occasionally leading to death, can be presented in fragile population such as children, the elderly, and patients with immunosuppression or chronic diseases [10–12].

At present, LA-MRSA has spread widely in swine farming and pork industrial production, and it has become a growing threat to public health worldwide [13]. In North America and Europe, the most prevalent LA-MRSA lineage belongs to clonal complex (CC) 398. On the other hand, most of the pig population in Asian countries has been successfully occupied by CC9 for years [13]; however, recently, the incidence rate of LA-MRSA CC398 among pig herds has increased dramatically to 87.5% and 96.3% in Thailand and China, respectively [2,14]. According to our previous study in Thailand, conventional molecular typing methods exhibited that LA-MRSA isolates from swine workers and their pigs possessed identical molecular characteristics, and they were also phenotypically resistant to the same antimicrobial agents [2]. These observations suggested inter-species transmission caused by CC398; however, a study exploring higher genetic relatedness among LA-MRSA from both species has never been conducted in Thailand. Currently, several molecular typing methods are used to approach epidemiological studies and outbreak investigations of LA-MRSA, such as staphylococcal cassette chromosome (SCC) *mec* typing, *spa* typing, pulsed-field gel electrophoresis (PFGE), and multi-locus sequence typing (MLST). These typing methods, however, present some limitations. Whole genome sequencing (WGS) is another alternative technique that allows us to access greater resolution and DNA sequence dissimilarities at the nucleotide level. Due to its several benefits, this study aimed to use WGS approaches to illustrate the zoonotic transmission of LA-MRSA CC398 between pigs and farm workers in the central region of Thailand and to characterize their genomic features using public international databases.

2. Results

2.1. WGS-Based Characteristics

The most prevalent genotype (sequence type-SCC*mec* type-*spa* type) was ST398-SCC*mec*V-t034 (9/16), followed by ST398-SCC*mec* composite island (CI)-t034 (4/16), ST9-SCC*mec*IX-t337 (2/16), and ST4576-SCC*mec*IX-t337 (1/16) (Table 1). The type V SCC*mec* element of the ST398-t034 strain contained a class C2 *mec* gene complex with a type 5 cassette chromosome recombinase (*ccr*) gene complex (*ccrC1*). The SCC*mec* type IX carried by the members of CC9, ST9-t337, and ST4576-t337 clones, consisted of the class C2 *mec* gene complex and a type 1 *ccr* gene complex (*ccrA1B1*). Interestingly, the composite SCC*mec* elements identified in the 4 ST398-t034 isolates were organized by a combination of multiple *ccr* gene complexes, *ccrA1B1* and *ccrC1*, and the *mec* class C2 complex.

Table 1. WGS-based characteristics, including *ccr* gene complex(es) and *mec* complex class identified in each representative of LA-MRSA isolates.

Strain Name	Farm	Location	ST	CC	SCC*mec* Type	*spa* Type	*mec* Complex Class	*ccr* Gene Complex			
								Type	*ccrA1*	*ccrB1*	*ccrC1*
Q10.1	6	PB	9	9	IX	t337	C2	1	+	+	-
Y1.3	8	PB	4576	9	IX	t337	C2	1	+	+	-
BA3.1	2	RB	9	9	IX	t337	C2	1	+	+	-
J101.2	11	PB	398	398	CI	t034	C2	NT	+	+	+
L3.1	1	NP	398	398	CI	t034	C2	NT	+	+	+
L43.2	1	NP	398	398	CI	t034	C2	NT	+	+	+
Z19.1	1	NP	398	398	CI	t034	C2	NT	+	+	+
AA3.1	4	SB	398	398	V	t034	C2	5	-	-	+
M3.1	3	SB	398	398	V	t034	C2	5	-	-	+
M31.1	3	SB	398	398	V	t034	C2	5	-	-	+
Y1.2	8	PB	398	398	V	t034	C2	5	-	-	+
S2.1	7	PB	398	398	V	t034	C2	5	-	-	+
H49.1	10	NR	398	398	V	t034	C2	5	-	-	+
G2.1	9	NR	398	398	V	t034	C2	5	-	-	+
D16.1	5	PB	398	398	V	t034	C2	5	-	-	+
X1.1	7	PB	398	398	V	t034	C2	5	-	-	+

PB: Prachin Buri, RB: Ratchaburi, NP: Nakhon Pathom, SB: Suphanburi, NR: Nakhon Ratchasima, NT: non-typable, +: detected, -: not detected.

2.2. Core-Genome Single Nucleotide Polymorphism (SNP)-Based Analyses and Transmission of LA-MRSA CC398

Figure 1 presents the phylogenetic tree based on single nucleotide polymorphisms (SNPs) in the core genome of the 13 LA-MRSA CC398 isolates. The isolates were divided into two distinct phylogenetic clades, namely, Clade I and Clade II. Almost all internal nodes had bootstrap support of 90% or greater. Clade I was occupied by the 4 LA-MRSA isolates carrying the composite SCC*mec* elements from Farm 1 and Farm 11. Two porcine isolates and one human isolate from Farm 1 apparently were clustered in the same sub-clade, elucidating a high degree of genome relatedness among them; however, one human isolate from Farm 11 was distantly related to those from Farm 1. Clade II was completely clustered by all LA-MRSA isolates carrying the SCC*mec* type V. The LA-MRSA isolates from Farm 3 and Farm 4, located in Suphanburi, were placed into the same sub-clade showing a geographical specificity (Supplementary Materials Figure S1). Like Farm 1, one porcine isolate from Farm 3 was clustered tightly with a human isolate obtained from the same location. The other six porcine isolates in Clade II, obtained from five different farms in Prachin Buri and Nakhon Ratchasima, resided together in a discrete sub-clade.

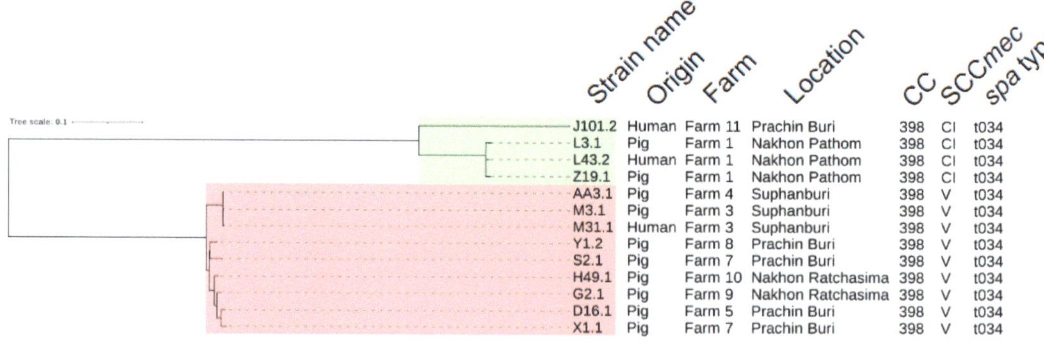

Figure 1. Phylogenetic analysis of the 13 LA-MRSA CC398 isolates based on core genome SNPs. The evolutionary tree was reconstrued using maximum likelihood inference and a bootstrap value of 1000 repetitions. The LA-MRSA isolates of CC398, derived from live pigs (n = 10) and healthy swine workers (n = 3), diverged into two monophyletic groups, namely, Clade I (green) and Clade II (red). Importantly, the phylogenetic tree depicted two possible events of zoonotic transmission in Farm 1 and Farm 3.

To verify all transmission events detected in this study, a pairwise SNP distance matrix made from the core genome alignment was investigated (Supplementary Materials Table S1). In Clade I, the pairwise SNP distances between the two porcine strains L3.1 and Z19.1 from Farm 1 were 13-core genome SNPs. These porcine strains were even closer to the human strain L43.2 from the same farm with 11 and 10 SNPs, respectively. On top of that, in Clade II, the porcine strain M3.1 from Farm 3 was only three SNPs distant from the human strain M31.1 from the same farm. This indicates that both isolates were genomically related and originated from the same root. For Farm 4, the porcine strain AA3.1 was separated from both isolates from Farm 3 (the strains M3.1 and M31.1) by 2 and 3 SNPs, respectively. The remaining porcine strains in this clade, collected from Prachin Buri and Nakhon Ratchasima, shared a common ancestor with a different number of SNP distances ranging from 14 to 38 SNPs.

2.3. Antimicrobial Resistance-Associated Genes and Stress Genes Profile

In total, we identified 32 antimicrobial resistance-associated genes, mediating resistance against 14 antimicrobial groups: aminoglycosides ($n = 6$), beta-lactams ($n = 3$), fluoroquinolones ($n = 2$), fosfomycin ($n = 3$), glycopeptide antibiotics ($n = 1$), lincosamides ($n = 1$), lincosamides-pleuromutilins-streptogramin A compounds (PLS$_A$) ($n = 3$), macrolide-lincosamide-streptogramin B compounds (MLS$_B$) ($n = 3$), phenicols ($n = 2$), phenicols-oxazolidinones-lincosamides-pleuromutilins-streptogramin A compounds (PhLOPS$_A$) ($n = 1$), rifampicin ($n = 1$), tetracyclines ($n = 4$), trimethoprim ($n = 1$), and multidrug efflux MATE (multidrug and toxic compound extrusion) transporter ($n = 1$) (Table 2). It is noteworthy that PLS$_A$, MLS$_B$, and PhLOPS$_A$ phenotypes are related to cross-resistance, which refers to an ability of resistance to several classes of antimicrobials by a single mechanism [15,16].

It is revealed that all LA-MRSA isolates could be defined as multidrug-resistant (MDR) MRSA due to resistance against at least three classes of antimicrobials. In addition, all of them tested positive for *mecA*, S84L mutation in *gyrA*, *tet*(38), *tet*(M), *dfrG*, and *mepA*. In contrast, the vancomycin resistance gene *vanA* and the mutation of *rpoB* gene were not presented in any isolates.

The *in-silico* detection of antimicrobial resistance gene patterns among LA-MRSA isolates within each genotype were highly similar. Within the CC398 subpopulation, the ST398-SCC*mec*CI-t034 genotype did not exhibit a significant difference with the ST398-SCC*mec*V-t034 genotype, except for *aac(6′)-Ie/aph(2″)-Ia*, mutations in *parC* and *erm* genes; however, it should be mentioned that, when comparing the patterns of gene carriage and mutations between the CC9-t337 and CC398-t034 subpopulations, we observed the distribution of some resistance genes or mutations that was evidently associated with a particular subpopulation. As shown in Table 2, the streptomycin resistance gene *str* was localized only in the CC9-t337 clone, but the other three aminoglycosides-related genes, *ant(6)-Ia*, *ant(9)-Ia*, and *spw*, were specifically found in the CC398-t034 clone. While the CC9-t337 strains mediated resistance to fosfomycin by encoding protein FosB, all LA-MRSA with the CC398-t034 genotype conferred resistance through mutations of *glpT* A100V/F31I and *murA* D278E/E291D [17,18]. For the PLS$_A$ resistance phenotype, the CC398-t034 clone expressed resistance through the encoding *lsa*(E) gene; however, the CC9-t337 clone harbored *vga*(A) genes conferring resistance against PLS$_A$ antibiotics. It is important to highlight that some resistance phenotypes were limited to either the CC9-t337 or CC398-t034 subpopulation. The *lnu*(B) gene, involving resistance to lincosamides, was present in all isolates with the CC398-t034 genotype. On the one hand, the strain Q10.1 from the CC9-t337 clone possessed the *catA* gene, showing resistance to phenicols. Also, the strain BA3.1 carried the phenicols resistance gene *fexA* as well as the *cfr* gene exhibiting resistance to PhLOPS$_A$.

We also examined the absence/presence of antimicrobial resistance genes distributed among LA-MRSA isolates involved in zoonotic transmission. Undoubtedly, LA-MRSA isolates from swine workers showed an almost 100% identity of antimicrobial resistance gene carriage with those from pigs isolated from the same farm origin.

Table 2. Distribution of 32 antimicrobial resistance-associated genes within the 16 LA-MRSA CC9 and CC398. Each gene was classified into 14 antimicrobial groups in accordance with their phenotypes. The grey color indicates the presence of genes or mutations in each isolate predicted by the AMRFinderPlus database.

	Strain Name	Q10.1	Y1.3	BA3.1	J101.2	L3.1	L43.2	Z19.1	AA3.1	M3.1	M31.1	Y1.2	S2.1	H49.1	G2.1	D16.1	X1.1
	Origin	Pig	Pig	Pig	Human	Pig	Human	Pig	Pig	Pig	Human	Pig	Pig	Pig	Pig	Pig	Pig
	Farm	6	8	2	11	1	1	1	4	3	3	8	7	10	9	5	7
	Location	PB	PB	RB	PB	NP	NP	NP	SB	SB	SB	PB	PB	NR	NR	PB	PB
	ST	9	4576	9	398	398	398	398	398	398	398	398	398	398	398	398	398
	CC	9	9	9	398	398	398	398	398	398	398	398	398	398	398	398	398
	SCC*mec* type	IX	IX	IX	CI	CI	CI	CI	V	V	V	V	V	V	V	V	V
	spa type	t337	t337	t337	t034	t034	t034	t034	t034	t034	t034	t034	t034	t034	t034	t034	t034
AMGs	*aac(6′)-Ie/aph(2″)-Ia*																
	aadD1																
	ant(6)-Ia																
	ant(9)-Ia																
	spw																
	str																
BLs	*blaPC1*																
	blaZ																
	mecA																
FQs	Mutation of *gyrA* S84L																
	Mutation of *parC* E84G																
	Mutation of *parC* S80F																
	Mutation of *parC* S80Y																
FOS	*fosB*																
	Mutation of *glpT* A100V																
	Mutation of *glpT* F3I																
	Mutation of *murA* D278E																
	Mutation of *murA* E291D																
GPAs	*vanA*																
LINs	*lnu*(B)																
PLS_A	*lsa*(E)																
	vga(A)																
	vga(A)-LC																
MLS_B	*erm*(A)																
	erm(B)																
	erm(C)																

Table 2. *Cont.*

	Strain Name	Q10.1	Y1.3	BA3.1	J101.2	L3.1	L43.2	Z19.1	AA3.1	M3.1	M31.1	Y1.2	S2.1	H49.1	G2.1	D16.1	X1.1
PHEs	*catA*																
	fexA																
PhLOPS$_A$	*cfr*																
RIF	Mutation of *rpoB*																
TCs	*tet*(38)																
	tet(K)																
	tet(L)																
	tet(M)																
TMP	*dfrG*																
MATE transporter	*mepA*																

AMGs: aminoglycosides, BLs: beta-lactams, FQs: fluoroquinolones, FOS: fosfomycin, GPAs: glycopeptide antibiotics, LINs: lincosamides, PLS$_A$: lincosamides-pleuromutilins-streptogramin A compounds, MLS$_B$: macrolide-lincosamide-streptogramin B compounds, PHEs: phenicols, PhLOPS$_A$: phenicols-oxazolidinones-lincosamides-pleuromutilins-streptogramin A compounds, RIF: rifampicin, TCs: tetracyclines, TMP: trimethoprim, MATE: multidrug and toxic compound extrusion.

Besides screening for antimicrobial resistance genes, stress genes, including biocide and metal resistance genes, were also determined. A total of five stress genes, including *arsB, arsC, lmrS, mco,* and *qacG,* were detected (Supplementary Materials Table S2). The most prevalent genes were *lmrS* and *mco,* carried by all LA-MRSA isolates. The arsenic resistance-related genes, *arsB* and *arsC,* were significantly associated with all LA-MRSA belonging to ST398-SCC*mec*CI-t034 and CC9-SCC*mec*IX-t337 genotypes; however, *qacG* resulted positive for the three strains Q10.1, Y1.3, and S2.1 from both CC9-t337 and CC398-t034 subpopulations.

2.4. Virulence Gene Repertoire

To assess whether pig-associated LA-MRSA would become a serious issue in human medicine, a total of 76 virulence genes were analyzed and further classified into 5 categories, according to their functions: adherence ($n = 13$), exoenzymes ($n = 10$), host immune evasion ($n = 16$), iron uptake and metabolism ($n = 8$), and toxins and type IV secretion systems ($n = 29$) (Table 3) [19]. Most of the virulence genes were distributed homogeneously in all subpopulations, except for some genes that were restrictedly occupied by either the CC9-t337 or CC398-t034 subpopulation. For example, the collagen adhesion gene (*cna*) and coagulase gene (*coa*) were exclusively found in the CC398-t034 strain. Conversely, *map, sdrD, aur, cap8F, essC, esxB, esxC,* staphylococcal enterotoxin (SE) genes *sei, sem, sen, seo, seu* and *sey,* and staphylococcal enterotoxins-like toxin (SEL) genes *sel27, sel28,* and *selX* were specifically found in the CC9-t337 strains; however, *sel26* was the only SEL gene widespread in all isolates.

Notably, nine adherence genes (*clfA, clfB, ebp, icaA, icaB, icaC, icaD, icaR,* and *sdrE*) and seven genes encoding the iron-regulated surface determinant protein (Isd) A-G (*isdA, isdB, isdC, isdD, isdE, isdF,* and *isdG*) resulted positive for all LA-MRSA isolates. The following virulence genes were, however, totally absent in all isolates: *sak, scn, chp, lukF-PV, lukS-PV, sea, sep,* and *tsst-1* genes.

Table 3. Distribution of 76 virulence determinants among the 16 LA-MRSA CC9 and CC398. The grey color indicates the presence of genes harbored by each strain.

Strain Name		Q10.1	Y1.3	BA3.1	J101.2	L3.1	L43.2	Z19.1	AA3.1	M3.1	M31.1	Y1.2	S2.1	H49.1	G2.1	D16.1	X1.1
Origin		Pig	Pig	Pig	Human	Pig	Human	Pig	Pig	Pig	Human	Pig	Pig	Pig	Pig	Pig	Pig
Farm		6	8	2	11	1	1	1	4	3	3	8	7	10	9	5	7
Location		PB	PB	RB	PB	NP	NP	NP	SB	SB	SB	PB	PB	NR	NR	PB	PB
ST		9	4576	9	398	398	398	398	398	398	398	398	398	398	398	398	398
CC		9	9	9	398	398	398	398	398	398	398	398	398	398	398	398	398
SCC*mec* type		IX	IX	IX	CI	CI	CI	CI	V	V	V	V	V	V	V	V	V
spa type		t337	t337	t337	t034	t034	t034	t034	t034	t034	t034	t034	t034	t034	t034	t034	t034
	clfA																
	clfB																
	cna																
	ebp																
	icaA																
	icaB																
Virulence genes related to adherence	*icaC*																
	icaD																
	icaR																
	map																
	sdrC																
	sdrD																
	sdrE																

Table 3. *Cont.*

Strain Name		Q10.1	Y1.3	BA3.1	J101.2	L3.1	L43.2	Z19.1	AA3.1	M3.1	M31.1	Y1.2	S2.1	H49.1	G2.1	D16.1	X1.1
Virulence genes related to exoenzymes	*adsA*																
	aur																
	Coa																
	geh																
	hysA																
	Lip																
	sak *																
	sspA																
	sspB																
	sspC																
Virulence genes related to host immune evasion	*cap8A*																
	cap8B																
	cap8C																
	cap8D																
	cap8E																
	cap8F																
	cap8G																
	cap8L																
	cap8M																
	cap8N																
	cap8O																
	cap8P																
	chp																
	scn *																
	sbi																
	spa																

Table 3. *Cont.*

Strain Name		Q10.1	Y1.3	BA3.1	J101.2	L3.1	L43.2	Z19.1	AA3.1	M3.1	M31.1	Y1.2	S2.1	H49.1	G2.1	D16.1	X1.1
Virulence genes related to iron uptake and metabolism	isdA																
	isdB																
	isdC																
	isdD																
	isdE																
	isdF																
	isdG																
	srtB																
Virulence genes related to toxins and type IV secretion	esaA																
	esaB																
	essA																
	essB																
	essC																
	esxA																
	esxB																
	esxC																
	hla																
	hlb																
	hld																
	hlgA																
	hlgB																
	hlgC																
	lukF-PV																
	lukS-PV																
	sea *																
	sep *																
	sei *																
	sel26*																

Table 3. *Cont.*

Strain Name		Q10.1	Y1.3	BA3.1	J101.2	L3.1	L43.2	Z19.1	AA3.1	M3.1	M31.1	Y1.2	S2.1	H49.1	G2.1	D16.1	X1.1
	sel27 *																
	sel28 *																
	selx *																
Virulence genes related to toxins and type IV secretion	*sem* *																
	sen *																
	seo *																
	seu *																
	sey *																
	tsst-1																

* Virulence genes additionally predicted by the AMRFinderPlus.

2.5. Mobile Genetic Elements

MobileElementFinder discovered 10 plasmid replicon sequences with diverse combinations of carriage. The most widespread plasmid replicon belongs to types rep7a and repUS43 (16/16), followed by rep19b (9/16), and rep10 and rep21 (8/16). Furthermore, a total of four transposon elements, including Tn*551* (4/16), Tn*554* (4/16), Tn*558* (1/16), and Tn*6009* (9/16), were identified in CC398-t034 isolates, as demonstrated in Table 4.

The plasmid replicon repUS43 carried by different strains was obviously linked to various genetic determinants. It was shown to be associated with *mecA* and *tet*(M) genes in the CC9-t337 subpopulation, whereas the ST398-SCC*mec*CI-t034 and ST398-SCC*mec*V-t034 strains carried this plasmid coupled with the *tet*(M) gene and transposon Tn*6009*, respectively. Similarly, the plasmid replicon rep7a carrying *tet*(K) was predominately present in all CC398-t034 isolates, except for the strain M3.1. On the other hand, the rep7a carrying *str* was detected in all CC9-t337 isolates, one of which (the strain Q10.1) also harbored the chloramphenicol resistance gene *cat*(pC221) on this plasmid. Among the CC398-t034 isolates, both Tn*551* and Tn*554* were carried by all isolates with the composite SCC*mec* element. The *erm*(A) gene, together with *ant(9)-la*, was located on the transposon Tn*554*, but the *erm*(B) gene resided on the transposon Tn*551*. In contrast, the rep7a plasmid associated with the *erm*(C) gene was mainly restricted to the 8 CC398-t034 with the SCC*mec* type V.

Importantly, it is worth nothing that we found three interesting carriages of antimicrobial resistance genes with mobile genetic elements in two isolates. The resistance genes *aadD*, *erm*(B), and *tet*(L) were co-localized on the plasmid sequence of the strain BA3.1. On top of that, this strain also tested positive for the transposon Tn*558* carrying *fexA*. Another multi-resistance gene cluster resided on the rep22 belonging to the strain J101.2. It contained two aminoglycosides resistance genes, *aadD* and *ant(6)-la*, the *blaZ* gene, the lincosamides resistance gene *lnu*(B), and the *lsa*(E) encoding PLS$_A$ phenotype.

Apart from the plasmid replicons and transposons, other mobile genetic elements were also determined. Within the CC398-t034 subpopulation, the insertion sequence IS*256* was present in only one isolate, the strain J101.2. The IS*Sau8* were predictably harbored by the following three strains: L43.2, D16.1, and X1.1; however, we found no association between these insertion sequences and antimicrobial resistance genes.

Table 4. Distribution of mobile genetic elements with their associated antimicrobial resistance genes carried by the 16 LA-MRSA CC9 and CC398.

Strain Name	Origin	Farm	Location	ST	CC	SCC*mec* Type	*spa* Type	ARG-Associated Plasmid Replicon and Transposon	Insertion Sequence
Q10.1	Pig	6	PB	9	9	IX	t337	rep5d-*vga*(A)-LC	-
								rep7a-*str-cat*(pC221)	
								rep13-*qacG*	
								rep21	
								repUS43-*mecA-tet*(M)	
Y1.3	Pig	8	PB	4576	9	IX	t337	rep5d-*vga*(A)-LC	-
								rep7a-*str*	
								rep13-*qacG*	
								rep21	
								repUS43-*mecA-tet*(M)	
BA3.1	Pig	2	RB	9	9	IX	t337	rep7a-*str*	-
								rep10b-*vga*(A)-LC	
								repUS18-*aadD-erm*(B)-*tet*(L)	
								repUS43-*mecA-tet*(M)	
								Tn*558*	
J101.2	Human	11	PB	398	398	CI	t034	rep7a-*tet*(K)	*IS256*
								rep22-*aadD-ant*(6)-*la-blaZ* *-*lnu*(B)-*lsa*(E)	
								repUS43-*tet*(M)	
								Tn*551-erm*(B)	
								Tn*554-ant*(9)-*la-erm*(A)	
L3.1	Pig	1	NP	398	398	CI	t034	rep7a-*tet*(K)	-
								rep22-*aadD*	
								repUS43-*tet*(M)	
								Tn*551-erm*(B)	
								Tn*554-ant*(9)-*la-erm*(A)	

Table 4. *Cont.*

Strain Name	Origin	Farm	Location	ST	CC	SCC*mec* Type	*spa* Type	ARG-Associated Plasmid Replicon and Transposon	Insertion Sequence
L43.2	Human	1	NP	398	398	CI	t034	rep7a-*tet*(K) rep22-*aadD* repUS43-*tet*(M) Tn*551*-*erm*(B) Tn*554*-*ant*(9)-*la*-*erm*(A)	*IS*Sau8
Z19.1	Pig	1	NP	398	398	CI	t034	rep7a-*tet*(K) rep22-*aadD* repUS43-*tet*(M) Tn*551*-*erm*(B) Tn*554*-*ant*(9)-*la*-*erm*(A)	-
AA3.1	Pig	4	SB	398	398	V	t034	rep7a-*tet*(K) rep10-*erm*(C) rep19b-*blaZ* rep21 repUS43-*tet*(M)-Tn*6009*	-
M3.1	Pig	3	SB	398	398	V	t034	rep7a rep10-*erm*(C) rep19b rep21 repUS43-*tet*(M)-Tn*6009*	-
M31.1	Human	3	SB	398	398	V	t034	rep7a-*tet*(K) rep10-*erm*(C) rep19b-*blaZ* rep21 repUS43-*tet*(M)-Tn*6009*	-

Table 4. *Cont.*

Strain Name	Origin	Farm	Location	ST	CC	SCC*mec* Type	*spa* Type	ARG-Associated Plasmid Replicon and Transposon	Insertion Sequence
Y1.2	Pig	8	PB	398	398	V	t034	rep7a-*tet*(K) rep10 rep19b repUS43-*tet*(M)-Tn*6009*	-
S2.1	Pig	7	PB	398	398	V	t034	rep7a-*tet*(K) rep10-*erm*(C) rep13-*qacG* rep19b-*blaZ* rep21 repUS43-*tet*(M)-Tn*6009*	-
H49.1	Pig	10	NR	398	398	V	t034	rep7a-*tet*(K) rep10-*erm*(C) rep19b-*blaZ* repUS43-*tet*(M)-Tn*6009*	-
G2.1	Pig	9	NR	398	398	V	t034	rep7a-*tet*(K) rep10-*erm*(C) rep19b-*blaZ* rep21 repUS43-*tet*(M)-Tn*6009*	-

Table 4. *Cont.*

Strain Name	Origin	Farm	Location	ST	CC	SCC*mec* Type	*spa* Type	ARG-Associated Plasmid Replicon and Transposon	Insertion Sequence
D16.1	Pig	5	PB	398	398	V	t034	rep7a-*tet*(K) rep10-*erm*(C) rep13 rep19b-*blaZ* rep21 repUS43-*tet*(M)-Tn*6009*	*IS*Sau8
X1.1	Pig	7	PB	398	398	V	t034	rep7a-*tet*(K) rep10-*erm*(C) rep19b-*blaZ* repUS43-*tet*(M)-Tn*6009*	*IS*Sau8

* This gene was not detected in the human strain J101.2 using the AMRFinderPlus tool. ARG: antimicrobial resistance gene, -: not detected.

3. Discussion

This study, to the best of our knowledge, is the first documentation of a whole-genome investigation of zoonotic transmission caused by LA-MRSA CC398 in Thailand. The nasal carriage of LA-MRSA among pigs and farm workers, suggesting the potential of human colonization and zoonotic transmission, has been previously defined by several reports in different parts of Thailand [1,2,20,21]. These studies used various typing methods based on molecular characterization, for example, classical MLST and SCC*mec* typing. Instead, in this study, WGS approaches were used for the genotypic characterization of LA-MRSA. Overall, the results of WGS-based characterization are nearly in accordance with those of molecular characterization in our previous study [2]. At that time, one porcine strain Z19.1 was typed as SCC*mec* non-typeable (NT) using multiplex PCR assays. Additionally, the PCR-based SCC*mec* typing could not specify an allotype of *ccrC* gene complex; however, these uncertainties were clarified by the WGS analyses in our present study. Together, these findings indicate a higher discriminatory power of WGS beyond other general typing approaches.

The occurrence of the SCC*mec*Cl in this study implies an empirical impact on animal movement through international trading. As seen in Table 1, the composite SCC*mec* elements of the 4 CC398-t034 isolates contained both *ccrA1B1* and *ccrC1* gene complexes, which were detected in CC9-SCC*mec*IX-t337 and CC398-SCC*mec*V-t034 subpopulations, respectively; moreover, it should be noted that LA-MRSA CC398 with the type V SCC*mec* element detected in several Asian countries seems to be strongly associated with the international pig trade. To exemplify this, several LA-MRSA CC398 harboring SCC*mec* type V were detected in imported pigs from two different countries during the quarantine period in Japan [22]. On top of that, the CC398-SCC*mec*V-t034 genotype in pigs was documented by a national survey in South Korea. That country has reportedly imported breeding pigs from Canada, Denmark, and the U.S. [23]. More importantly, one previous study demonstrated that six LA-MRSA isolates with the CC398-SCC*mec*V-t034 genotype, from retail pork and a slaughtered pig in the central region of Thailand, were closely related to Danish LA-MRSA, showing the same characteristics [24]. That study also mentioned that the import of live pigs from Denmark to Thailand would constitute the introduction of the LA-MRSA CC398 strain into the Thai pig population. All things considered, it can be hypothesized that LA-MRSA CC398-t034 with SCC*mec*V had been introduced into Thailand due to the cross-border movements of pigs. Subsequently, it would acquire the additional gene complex (*ccrA1B1*) from the domestic strain (CC9-SCC*mec*IX-t337) via genetic recombination, leading to the emergence of a unique composite of the SCC*mec* element [25].

In recent years, the carriage of multiple *ccr* gene complexes has been presented by a Chinese LA-MRSA ST9 from clinical and porcine isolates [26]. In addition, another study from China has reported the presence of coagulase-negative staphylococci (CoNS) carrying multiple SCC*mec* elements [27]. It demonstrated that the isolates with multiple SCC*mec* had a more stable capability to continue *mecA* gene transcription involved in cell wall synthesis. As a result, these isolates did not lose Gram positivity under antibiotic exposure when compared to the ones with a single SCC*mec* element [27]; however, we did not perform any phenotypic analyses regarding multiple SCC*mec* elements.

We elucidate and verify zoonotic transmission caused by LA-MRSA CC398 in two farms using core-genome SNP-based analyses. The phylogenetic tree reconstruction illustrated a high genome similarity between porcine and human isolates. Genomic characteristics support the phylogenetic tree as well as imply that the origin of human LA-MRSA CC398 would be pigs. Ultimately, the pairwise SNP analysis does not only confirm pig-to-human transmission; it also indicates evolutionary changes in the core genome of LA-MRSA CC398 after zoonotic spillover and colonization on human nares. One study in Denmark also used WGS to investigate the zoonotic transmission of LA-MRSA between livestock and farmers. A phylogenetic tree based on core genome SNPs revealed that animal isolates differed from Danish worker isolates by only 3–5 SNPs [28]. The small number of pairwise SNPs is very similar to those in the present study.

Additionally, phylogenetic analysis clearly revealed two distinct clades of LA-MRSA CC398 with their own specific SCC*mec* types. LA-MRSA isolates from the same province or their neighboring province were phylogenetically grouped into the same clade or sub-clade. It is exemplified by the six porcine isolates from Prachin Buri and Nakhon Ratchasima. Nakhon Ratchasima shares its provincial borders with Prachin Buri (Supplementary Materials Figure S1). Unsurprisingly, all porcine isolates from these two provinces in Clade II resided together in the same sub-clade; therefore, it can be stated that these clues speculate as to local transmission events of pig-associated LA-MRSA within the same province or between provinces.

Like the phenotypic susceptibility testing described earlier [2], the WGS analyses exhibited that all LA-MRSA isolates in the present study were identified as MDR as well as harboring several cross-resistance genes. The overall distribution of antimicrobial resistance genes suggests the genetic homogeneity of LA-MRSA in each subpopulation. In other words, the isolates sharing the same genotypic characteristics have a highly similar pattern of antimicrobial resistance gene carriage, although they were collected from different locations or hosts. In addition, LA-MRSA CC9 and CC398 possess different resistance mechanisms against a particular antimicrobial group such as a resistance mechanism against fosfomycin or PLS$_A$. It is also important to emphasize that several cross-resistance genes carried by our LA-MRSA isolates do not only confer resistance against antimicrobial agents widely used in livestock, but also antibiotics of last resort in human medicine such as quinupristin/dalfopristin and linezolid; however, analysis of antimicrobial resistance gene carriage using WGS in LA-MRSA CC398 has been explored by a study collecting porcine specimens from a province in central Thailand [24,29], and it is consistent with the results of our study.

Although the *lmrS* gene encoding a multidrug efflux pump is categorized as a stress gene according to the AMRFinderPlus database, it has been proved that this gene is able to implicate in resistance to linezolid, aminoglycosides, macrolides, and phenicols, which is relatively similar to the PhLOPS$_A$ phenotype expressed by the *cfr* gene [30]; hence, it can be assumed that all LA-MRSA isolates in this study are highly likely to potentiate resistance against oxazolidinones via the lincomycin resistance protein of *Staphylococcus aureus* (LmrS). Three LA-MRSA strains could develop quaternary ammonium compounds (QACs) resistance due to harboring the *qacG* gene. QAC-based disinfectants are commonly used in livestock farms to chemically kill microorganisms on non-living surfaces; therefore, this gene would probably protect LA-MRSA from commercially available disinfectants promoting bacterial persistence in farm environment.

In this study, WGS approaches also disclosed an abundance and diversity of virulence determinants imposed by LA-MRSA isolates. As expected, toxic shock syndrome, the toxin-1 (TSST-1)-encoding gene (*tsst-1*), Panton–Valentine leucocidin (PVL)-encoding genes (*lukF-PV* and *lukS-PV*), as well as the immune evasion gene cluster (IEC), including *sak*, *scn*, *chp*, *sea*, and *sep*, were absent in all LA-MRSA isolates because they were rarely detected in livestock-derived isolates [31–34]. In addition, they are well known as crucial virulence determinants promoting the pathogenesis or severity of MRSA infections in humans. Furthermore, only a few SE and SEL genes were found in our LA-MRSA isolates.

There are, however, multiple genes which were carried by our LA-MRSA isolates, including *clfA-B*, *cna*, *icaA-D*, *icaR*, *isdA-G*, and *sdrC-E*. Specifically, the *clfB* genes encoding clumping factors B (ClfB) has been determined to play an important role in nasal colonization in humans and experimental animals [35–37]. It could bind to the upper layer of epithelial cells in anterior nares by a ligand-receptor interaction with loricrin and was able to interact with cytokeratin 10 (CK10) expressed on skin epithelium. Further, it has been proved to have an effect on skin and soft tissue infection (SSTI) during the early stage of pathogenesis [38,39]. Seven strains had the *cna* gene, which is responsible for bacterial adherence to collagen in host tissue. With this function, the collagen-binding protein could influence staphylococcal infection in collagen-rich tissues such as heart, joints, and bones, and the cornea [40–43]. The *ica* operon, consisting of *icaA-D* genes and its negative regulator gene *icaR*, is involved in biofilm production. The slime formation does not only protect bacterial communities against host immunity, but also mediates the release of dispersed cells to new sites of infection and avoids penetration of antimicrobial agents through the staphylococcal biofilm [44,45]. Besides being involved in heme-iron acquisition systems, the IsdA is also able to bind to several host receptors such as loricrin and CK10, supporting the adherence of *S. aureus* to desquamated nasal epithelial cells in humans and colonization in animal experiments [46]. Only a few LA-MRSA harbored *sdrC-D* genes even though *sdrE* gene was identified in all strains. The *sdrC-D* genes are related to adhesion to human epithelial cells; however, the *sdrE* gene is recognized as an inhibitor of both classical and alternative complement pathways [46,47].

Altogether, it can be summarized that all LA-MRSA isolates in our study pose virulence genes that can potentiate intercellular adhesion and colonization in human nostrils, facilitate biofilm formation, and promote a wide spectrum of staphylococcal infections. These findings support the conclusion that LA-MRSA can serve as an invasive pathogen threatening human health, especially in patients with immunocompromised or underlying medical conditions [5].

In-silico prediction of antimicrobial resistance gene-associated mobile genetic elements revealed two unique multi-resistant plasmids on draft genomes of LA-MRSA strains BA3.1 and J101.2; nevertheless, previous studies have reported the emergence of MRSA possessing multi-resistance gene clusters either on chromosome or plasmid. To illustrate this, the co-existence of numerous antimicrobial resistance genes on the LA-MRSA genome J101.2, including *aadD, blaZ, lnu*(B), and *lsa*(E) genes, had been detected earlier on a chromosome of MRSA ST9 isolated from frozen food in China [48]. Additionally, a multi-resistance gene cluster on a plasmid was identified in MRSA ST9 from a Chinese pig [49]. Apart from the aminoglycosides, beta-lactams, lincosamides, and PLS$_A$ resistance genes, the tetracycline resistance gene *tet*(L) and MLS$_B$ resistance *erm*(B) were co-localized on this plasmid, which were also found on our LA-MRSA strain BA3.1. The co-carriage of diverse antimicrobial resistance genes on the plasmids demonstrated in our study raises an awareness of significant risk related to antimicrobial resistance gene transfer [49].

We have, however, encountered some challenges in using short-read sequencing technologies. First, we could not access whole sequences of SCC*mec*CI elements carried by our LA-MRSA CC398; therefore; we cannot clarify whether the additional *ccr* gene complex was a result of two discrete integrated SCC*mec* elements [50]. Another limitation is that we were not able to visualize the genomic organization of multi-resistance gene clusters located on putative plasmid contigs. This would highlight that several antimicrobial resistance genes and other resistance genes can be horizontally transferred in a single event. To tackle these problems, the so-called hybrid genome assembly produced from long-read and short-read sequencing technologies will be implemented in order to obtain the reconstruction of a complete bacterial chromosome, SCC*mec* cassette, or plasmid in our future strategies.

4. Materials and Methods

4.1. LA-MRSA Isolates

A total of 16 representative isolates of LA-MRSA, which were previously characterized using molecular typing methods in 2019, were selected for WGS analyses in this study [2]. Nasal swabs were obtained from healthy swine farmers ($n = 3$) and live pigs ($n = 13$) in 11 swine farms located in the central region of Thailand, designated as Farm 1–Farm 11, 2015–2017 [2]. Bacterial stock was kept in tryptic soy broth (BD Difco, Franklin Lakes, NJ, USA) with 30% glycerol at -80 °C before recovery on 5% sheep blood agar for further processes.

4.2. DNA Extraction

The heat-killed bacterial solution of all representative isolates was transported to the Hokkaido University International Institute for Zoonosis Control, Japan, for DNA extraction. The genomic DNA (gDNA) was extracted by the bead-beating technique described previously [24]. Further, the concentration and quality of DNA were assessed using Qubit 3.0 Fluorometer and NanoDrop One Spectrophotometer (Thermo Fisher Scientific, Waltham, MA, USA).

4.3. Library Preparation, WGS and Genome Assembly

Paired-end libraries were prepared using the Nextera XT DNA Library Preparation Kit (Illumina Inc., San Diego, CA, USA). The concentration of each extracted gDNA sample was diluted to 0.2 ng/μL and subsequently sequenced on an Illumina MiSeq with 2×300 bp reads. The generated raw reads were quantified using FastQC v0.11.9 [51]. Then, the low-quality reads with a Phred quality score <20 were filtered out by Trim Reads Tool from CLC Genomics Workbench v22.0.1 (Qiagen, Hilden, Germany). The trimmed reads were assembled into contigs using SPAdes v3.15.4 on the Galaxy platform [52,53]. The assembly statistics were evaluated using QUAST v5.2.0 [54]. An N50 value of at least 50,000 was required for downstream analyses [55].

4.4. Genomic Characterization and In-Silico Identification of Antimicrobial Resistance, Stress, and Virulence Genes

In-silico MLST was performed to assign a sequence type to each LA-MRSA isolate using MLST v2.0 [56]. SCCmecFinder v1.2 with a minimum gene coverage of 80% and minimum identity cut-off of 90% was used to determine SCC*mec* types; spaTyper v1.0 was applied to identify *spa* types [57,58]. Acquired antimicrobial resistance genes, mutations in genes associated with antimicrobial resistance, and stress genes were *in-silico* screened using NCBI AMRFinderPlus v3.10.42 [59]. VFDB was primarily used for *in-silico* detection of virulence genes on the BV-BRC platform v3.28.5 [60,61]; however, the virulence genes that were not included in VFDB were additionally predicted by AMRFinderPlus. The minimum length and percentage identity of these two bioinformatic tools were adjusted to 80% and 90%, respectively, to determine the absence/presence of a particular gene.

4.5. Core Genome Alignment and Phylogenetic Tree Reconstruction

The trimmed paired-end reads of each LA-MRSA were aligned against the LA-MRSA ST398 reference genome (strain S0385; GenBank accession no. AM990992) using Snippy pipeline v4.6.0. Snippy-core v4.6.0 was used for SNP calling from core-genome alignment [62]. Gubbins was run to eliminate the polymorphic sites of recombination in the alignment [63]. Further, the number of pair-wise SNP distances was computed using Snp-Dists v0.8.2 [64]. The phylogenetic tree of LA-MRSA CC398 based on SNPs in the core genome was reconstructed on MEGA v11 [65]. The maximum-likelihood inference, together with Kimura's two-parameter substitution model (K2P), was utilized to infer the evolutionary tree [66]. The bootstrap support values of 1000 replicates were calculated to assess the robustness of each node of the resulting tree. The final phylogenetic tree was visualized and integrated with a set of metadata using iTOL v6.8 [67].

5. Conclusions

Our study demonstrates how to apply WGS technologies for the epidemiological investigation of zoonotic transmission occupationally caused by LA-MRSA CC398 in Thailand. The WGS analyses with a high-resolution genomic approach also reveal genetic recombination through the evolutionary process, influenced by the introduction of an exotic strain of LA-MRSA CC398. Diverse antimicrobial resistance-related genes are widespread in our LA-MRSA isolates. Cross-resistance genes emphasize the judicious usage of antimicrobials in livestock production. The co-existence of several antimicrobial resistance genes on plasmids and the virulence gene repertoire reflects the robustness of biosecurity-associated strategies to confine horizontal gene transfer among bacterial communities outside agricultural areas as well as to reduce the risk of transmission at pre-harvest. These also accentuate the primary role of the One Health approach, collaboratively addressing antimicrobial resistance issues.

Supplementary Materials: The following supporting information can be downloaded at: https://www.mdpi.com/article/10.3390/antibiotics12121745/s1. Table S1: pairwise SNP distance matrix of the 13 LA-MRSA CC398 isolates based on core-genome alignment; Table S2: distribution of five stress genes identified by the AMRFinderPlus database among the 16 LA-MRSA CC9 and CC398; Figure S1: map focusing on the central region of Thailand. Refs. [68–71] are cited in the Supplementary Materials.

Author Contributions: P.N., P.C., C.N. and Y.S. contributed to conceptualization; all authors contributed to the methodology; P.C., C.N. and Y.S. contributed to validation; all authors contributed to investigation; P.C. and P.N. contributed to resources; P.N. and J.Y. contributed to data curation and formal analysis; P.N., P.C., C.N. and Y.S. contributed to writing—original draft; all authors contributed to reviewing and editing the manuscript; J.T., C.N. and Y.S. contributed to supervision and project administration; Y.S. contributed to funding acquisition. All authors have read and agreed to the published version of the manuscript.

Funding: This work was supported in part by the Japan Agency for Medical Research and Development (AMED) under Grant Numbers JP23wm0125008 and JP233fa627005 to YS.

Institutional Review Board Statement: Not applicable.

Informed Consent Statement: Not applicable.

Data Availability Statement: Data are contained within the article and Supplementary Materials.

Acknowledgments: We would like to express our sincere gratitude to the Ministry of Education, Culture, Sports, Science and Technology, Japan (MEXT), as well as the Department of Veterinary Microbiology, Faculty of Veterinary Science, Chulalongkorn, for bacterial recovery and sample preparation.

Conflicts of Interest: The authors declare no conflict of interest.

References

1. Patchanee, P.; Tadee, P.; Arjkumpa, O.; Love, D.; Chanachai, K.; Alter, T.; Hinjoy, S.; Tharavichitkul, P. Occurrence and Characterization of Livestock-Associated Methicillin-Resistant *Staphylococcus aureus* in Pig Industries of Northern Thailand. *J. Vet. Sci.* **2014**, *15*, 529–536. [CrossRef] [PubMed]

2. Chanchaithong, P.; Perreten, V.; Am-in, N.; Lugsomya, K.; Tummaruk, P.; Prapasarakul, N. Molecular Characterization and Antimicrobial Resistance of Livestock-Associated Methicillin-Resistant *Staphylococcus aureus* Isolates from Pigs and Swine Workers in Central Thailand. *Microbial. Drug Resist.* **2019**, *25*, 1382–1389. [CrossRef] [PubMed]

3. Chen, C.; Wu, F. Livestock-Associated Methicillin-Resistant *Staphylococcus aureus* (LA-MRSA) Colonisation and Infection among Livestock Workers and Veterinarians: A Systematic Review and Meta-Analysis. *Occup. Environ. Med.* **2021**, *78*, 530. [CrossRef] [PubMed]

4. Voss, A.; Loeffen, F.; Bakker, J.; Klaassen, C.; Wulf, M. Methicillin-Resistant *Staphylococcus aureus* in Pig Farming. *Emerg. Infect. Dis. J.* **2005**, *11*, 1965. [CrossRef] [PubMed]

5. The Danish Health Authority. *Guidance on Preventing the Spread of MRSA*, 3rd ed.; The Danish Health Authority: Copenhagen, Denmark, 2016.

6. Lee, A.S.; de Lencastre, H.; Garau, J.; Kluytmans, J.; Malhotra-Kumar, S.; Peschel, A.; Harbarth, S. Methicillin-Resistant *Staphylococcus aureus*. *Nat. Rev. Dis. Primers* **2018**, *4*, 18033. [CrossRef] [PubMed]

7. Bosch, T.; Verkade, E.; van Luit, M.; Landman, F.; Kluytmans, J.; Schouls, L.M. Transmission and Persistence of Livestock-Associated Methicillin-Resistant *Staphylococcus aureus* among Veterinarians and Their Household Members. *Appl. Environ. Microbiol.* **2015**, *81*, 124–129. [CrossRef] [PubMed]

8. Mutters, N.T.; Bieber, C.P.; Hauck, C.; Reiner, G.; Malek, V.; Frank, U. Comparison of Livestock-Associated and Health Care–Associated MRSA—Genes, Virulence, and Resistance. *Diagn. Microbiol. Infect. Dis.* **2016**, *86*, 417–421. [CrossRef]

9. Crespo-Piazuelo, D.; Lawlor, P.G. Livestock-Associated Methicillin-Resistant *Staphylococcus aureus* (LA-MRSA) Prevalence in Humans in Close Contact with Animals and Measures to Reduce on-Farm Colonisation. *Ir. Vet. J.* **2021**, *74*, 21. [CrossRef]

10. van Alen, S.; Ballhausen, B.; Peters, G.; Friedrich, A.W.; Mellmann, A.; Köck, R.; Becker, K. In the Centre of an Epidemic: Fifteen Years of LA-MRSA CC398 at the University Hospital Münster. *Vet. Microbiol.* **2017**, *200*, 19–24. [CrossRef]

11. Larsen, J.; Petersen, A.; Larsen, A.R.; Sieber, R.N.; Stegger, M.; Koch, A.; Aarestrup, F.M.; Price, L.B.; Skov, R.L.; ohansen, H.K.; et al. Emergence of Livestock-Associated Methicillin-Resistant *Staphylococcus aureus* Bloodstream Infections in Denmark. *Clin. Infect. Dis.* **2017**, *65*, 1072–1076. [CrossRef]

12. European Food Safety Authority (EFSA). Assessment of the Public Health Significance of Meticillin Resistant *Staphylococcus aureus* (MRSA) in Animals and Foods. *EFSA J.* **2009**, *7*, 993. [CrossRef]

13. Sahreena, L.; Kunyan, Z. Methicillin-Resistant *Staphylococcus aureus*: Molecular Characterization, Evolution, and Epidemiology. *Clin. Microbiol. Rev.* **2018**, *31*, e00020-18. [CrossRef]

14. Cui, M.; Ali, T.; Li, J.; Song, L.; Shen, S.; Li, T.; Zhang, C.; Cheng, M.; Zhao, Q.; Wang, H. New Clues about the Global MRSA ST398: Emergence of MRSA ST398 from Pigs in Qinghai, China. *Int. J. Food Microbiol.* **2022**, *378*, 109820. [CrossRef] [PubMed]

15. Petinaki, E.; Papagiannitsis, C. Resistance of Staphylococci to Macrolides-Lincosamides-Streptogramins B (MLSB): Epidemiology and Mechanisms of Re-sistance. In *Staphylococcus Aureus*; Hemeg, H., Ozbak, H., Afrin, F., Eds.; IntechOpen: Rijeka, Croatia, 2018. [CrossRef]

16. Feßler, A.T.; Wang, Y.; Wu, C.; Schwarz, S. Mobile Lincosamide Resistance Genes in Staphylococci. *Plasmid* **2018**, *99*, 22–31. [CrossRef] [PubMed]

17. Thompson, M.K.; Keithly, M.E.; Sulikowski, G.A.; Armstrong, R.N. Diversity in Fosfomycin Resistance Proteins. *Perspect. Sci.* **2015**, *4*, 17–23. [CrossRef]

18. Fu, Z.; Ma, Y.; Chen, C.; Guo, Y.; Hu, F.; Liu, Y.; Xu, X.; Wang, M. Prevalence of Fosfomycin Resistance and Mutations in MurA, GlpT, and UhpT in Methicillin-Resistant *Staphylococcus aureus* Strains Isolated from Blood and Cerebrospinal Fluid Samples. *Front. Microbiol.* **2016**, *6*, 1544. [CrossRef] [PubMed]

19. Naushad, S.; Naqvi, S.A.; Nobrega, D.; Luby, C.; Kastelic, J.P.; Barkema, H.W.; De Buck, J. Comprehensive Virulence Gene Profiling of Bovine Non-Aureus Staphylococci Based on Whole-Genome Sequencing Data. *mSystems* **2019**, *4*, 00098-18. [CrossRef]

20. Anukool, U.; O'Neill, C.E.; Butr-Indr, B.; Hawkey, P.M.; Gaze, W.H.; Wellington, E.M.H. Meticillin-Resistant *Staphylococcus aureus* in Pigs from Thailand. *Int. J. Antimicrob. Agents* **2011**, *38*, 86–87. [CrossRef]

21. Sinlapasorn, S.; Lulitanond, A.; Angkititrakul, S.; Chanawong, A.; Wilailuckana, C.; Tavichakorntrakool, R.; Chindawong, K.; Seelaget, C.; Krasaesom, M.; Chartchai, S.; et al. SCCmec IX in Meticillin-Resistant *Staphylococcus aureus* and Meticillin-Resistant Coagulase-Negative Staphylococci from Pigs and Workers at Pig Farms in Khon Kaen, Thailand. *J. Med. Microbiol.* **2015**, *64*, 1087–1093. [CrossRef]

22. Furuno, M.; Uchiyama, M.; Nakahara, Y.; Uenoyama, K.; Fukuhara, H.; Morino, S.; Kijima, M. A Japanese Trial to Monitor Methicillin-Resistant *Staphylococcus aureus* (MRSA) in Imported Swine during the Quarantine Period. *J. Glob. Antimicrob. Resist.* **2018**, *14*, 182–184. [CrossRef]

23. Lim, S.-K.; Nam, H.-M.; Jang, G.-C.; Lee, H.-S.; Jung, S.-C.; Kwak, H.-S. The First Detection of Methicillin-Resistant *Staphylococcus aureus* ST398 in Pigs in Korea. *Vet. Microbiol.* **2012**, *155*, 88–92. [CrossRef] [PubMed]

24. Tanomsridachchai, W. Genetic Characterization of Methicillin-Resistant *Staphylococcus aureus* Isolated from Pigs and Pork Meat in Thailand. Ph.D. Thesis, Hokkaido University, Sapporo, Japan, 2023. Volume 14717. [CrossRef]

25. Zong, Z.; Peng, C.; Lü, X. Diversity of SCCmec Elements in Methicillin-Resistant Coagulase-Negative Staphylococci Clinical Isolates. *PLoS ONE* **2011**, *6*, e20191. [CrossRef] [PubMed]

26. Chen, Y.; Sun, L.; Hong, Y.; Chen, M.; Zhang, H.; Peng, Y.; Liao, K.; Wang, H.; Zhu, F.; Zhuang, H.; et al. Exploring the Third-Generation Tetracycline Resistance of Multidrug-Resistant Livestock-Associated Methicillin-Resistant *Staphylococcus aureus* ST9 across Healthcare Settings in China. *J. Antimicrob. Chemother.* **2023**, *78*, 1871–1881. [CrossRef] [PubMed]

27. Chen, X.-P.; Li, W.-G.; Zheng, H.; Du, H.-Y.; Zhang, L.; Zhang, L.; Che, J.; Wu, Y.; Liu, S.-M.; Lu, J.-X. Extreme Diversity and Multiple SCCmec Elements in Coagulase-Negative Staphylococcus Found in the Clinic and Community in Beijing, China. *Ann. Clin. Microbiol. Antimicrob.* **2017**, *16*, 57. [CrossRef] [PubMed]

28. Harrison, E.M.; Paterson, G.K.; Holden, M.T.G.; Larsen, J.; Stegger, M.; Larsen, A.R.; Petersen, A.; Skov, R.L.; Christensen, J.M.; Bak Zeuthen, A.; et al. Whole Genome Sequencing Identifies Zoonotic Transmission of MRSA Isolates with the Novel MecA Homologue MecC. *EMBO Mol. Med.* **2013**, *5*, 509–515. [CrossRef] [PubMed]

29. Tanomsridachchai, W.; Changkaew, K.; Changkwanyeun, R.; Prapasawat, W.; Intarapuk, A.; Fukushima, Y.; Yamasamit, N.; Flav Kapalamula, T.; Nakajima, C.; Suthienkul, O.; et al. Antimicrobial Resistance and Molecular Characterization of Methicillin-Resistant *Staphylococcus aureus* Isolated from Slaughtered Pigs and Pork in the Central Region of Thailand. *Antibiotics* **2021**, *10*, 206. [CrossRef] [PubMed]

30. Floyd, J.L.; Smith, K.P.; Kumar, S.H.; Floyd, J.T.; Varela, M.F. LmrS Is a Multidrug Efflux Pump of the Major Facilitator Superfamily from *Staphylococcus aureus*. *Antimicrob. Agents Chemother.* **2010**, *54*, 5406–5412. [CrossRef]

31. Cuny, C.; Abdelbary, M.; Layer, F.; Werner, G.; Witte, W. Prevalence of the Immune Evasion Gene Cluster in *Staphylococcus aureus* CC398. *Vet. Microbiol.* **2015**, *177*, 219–223. [CrossRef]

32. Durand, G.; Bes, M.; Meugnier, H.; Enright, M.C.; Forey, F.; Liassine, N.; Wenger, A.; Kikuchi, K.; Lina, G.; Vandenesch, F.; et al. Detection of New Methicillin-Resistant *Staphylococcus aureus* Clones Containing the Toxic Shock Syndrome Toxin 1 Gene Responsible for Hospital- and Community-Acquired Infections in France. *J. Clin. Microbiol.* **2006**, *44*, 847–853. [CrossRef]

33. van Wamel, W.J.B.; Rooijakkers, S.H.M.; Ruyken, M.; van Kessel, K.P.M.; van Strijp, J.A.G. The Innate Immune Modulators Staphylococcal Complement Inhibitor and Chemotaxis Inhibitory Protein of *Staphylococcus aureus* Are Located on β-Hemolysin-Converting Bacteriophages. *J. Bacteriol.* **2006**, *188*, 1310–1315. [CrossRef]

34. Price, L.B.; Stegger, M.; Hasman, H.; Aziz, M.; Larsen, J.; Andersen, P.S.; Pearson, T.; Waters, A.E.; Foster, J.T.; Schupp, J.; et al. *Staphylococcus aureus* CC398: Host Adaptation and Emergence of Methicillin Resistance in Livestock. *mBio* **2012**, *3*, 00305-11. [CrossRef] [PubMed]

35. Schaffer, A.C.; Solinga, R.M.; Cocchiaro, J.; Portoles, M.; Kiser, K.B.; Risley, A.; Randall, S.M.; Valtulina, V.; Speziale, P.; Walsh, E.; et al. Immunization with *Staphylococcus aureus* Clumping Factor B, a Major Determinant in Nasal Carriage, Reduces Nasal Colonization in a Murine Model. *Infect. Immun.* **2006**, *74*, 2145–2153. [CrossRef] [PubMed]

36. Mulcahy, M.E.; Geoghegan, J.A.; Monk, I.R.; O'Keeffe, K.M.; Walsh, E.J.; Foster, T.J.; McLoughlin, R.M. Nasal Colonisation by *Staphylococcus aureus* Depends upon Clumping Factor B Binding to the Squamous Epithelial Cell Envelope Protein Loricrin. *PLoS Pathog.* **2012**, *8*, e1003092. [CrossRef] [PubMed]

37. Wertheim, H.F.L.; Walsh, E.; Choudhurry, R.; Melles, D.C.; Boelens, H.A.M.; Miajlovic, H.; Verbrugh, H.A.; Foster, T.; van Belkum, A. Key Role for Clumping Factor B in *Staphylococcus aureus* Nasal Colonization of Humans. *PLoS Med.* **2008**, *5*, e17. [CrossRef]

38. Sakr, A.; Brégeon, F.; Mège, J.-L.; Rolain, J.-M.; Blin, O. *Staphylococcus aureus* Nasal Colonization: An Update on Mechanisms, Epidemiology, Risk Factors, and Subsequent Infections. *Front. Microbiol.* **2018**, *9*, 2419. [CrossRef]

39. Lacey, K.A.; Mulcahy, M.E.; Towell, A.M.; Geoghegan, J.A.; McLoughlin, R.M. Clumping Factor B Is an Important Virulence Factor during *Staphylococcus aureus* Skin Infection and a Promising Vaccine Target. *PLoS Pathog.* **2019**, *15*, e1007713. [CrossRef]

40. Elasri, M.O.; Thomas, J.R.; Skinner, R.A.; Blevins, J.S.; Beenken, K.E.; Nelson, C.L.; Smelter, M.S. *Staphylococcus aureus* Collagen Adhesin Contributes to the Pathogenesis of Osteomyelitis. *Bone* **2002**, *30*, 275–280. [CrossRef]

41. Patti, J.M.; Bremell, T.; Krajewska-Pietrasik, D.; Abdelnour, A.; Tarkowski, A.; Rydén, C.; Höök, M. The *Staphylococcus aureus* Collagen Adhesin Is a Virulence Determinant in Experimental Septic Arthritis. *Infect. Immun.* **1994**, *62*, 152–161. [CrossRef]

42. Rhem, M.N.; Lech, E.M.; Patti, J.M.; McDevitt, D.; Höök, M.; Jones, D.B.; Wilhelmus, K.R. The Collagen-Binding Adhesin Is a Virulence Factor in *Staphylococcus aureus* Keratitis. *Infect. Immun.* **2000**, *68*, 3776–3779. [CrossRef]

43. Hienz, S.A.; Schennings, T.; Heimdahl, A.; Flock, J.-I. Collagen Binding of *Staphylococcus aureus* Is a Virulence Factor in Experimental Endocarditis. *J. Infect. Dis.* **1996**, *174*, 83–88. [CrossRef]

44. Lister, J.L.; Horswill, A.R. *Staphylococcus aureus* Biofilms: Recent Developments in Biofilm Dispersal. *Front. Cell Infect. Microbiol.* **2014**, *4*, 178. [CrossRef] [PubMed]

45. Nguyen, H.T.T.; Nguyen, T.H.; Otto, M. The Staphylococcal Exopolysaccharide PIA—Biosynthesis and Role in Biofilm Formation, Colonization, and Infection. *Comput. Struct. Biotechnol. J.* **2020**, *18*, 3324–3334. [CrossRef] [PubMed]

46. Mulcahy, M.E.; McLoughlin, R.M. Host–Bacterial Crosstalk Determines *Staphylococcus aureus* Nasal Colonization. *Trends Microbiol.* **2016**, *24*, 872–886. [CrossRef] [PubMed]

47. Paharik, A.E.; Horswill, A.R. The Staphylococcal Biofilm: Adhesins, Regulation, and Host Response. *Microbiol. Spectr.* **2016**, *4*, 529–566. [CrossRef] [PubMed]

48. Zhang, F.; Wu, S.; Lei, T.; Wu, Q.; Zhang, J.; Huang, J.; Dai, J.; Chen, M.; Ding, Y.; Wang, J.; et al. Presence and Characterization of Methicillin-Resistant *Staphylococcus aureus* Co-Carrying the Multidrug Resistance Genes Cfr and Lsa(E) in Retail Food in China. *Int. J. Food Microbiol.* **2022**, *363*, 109512. [CrossRef] [PubMed]

49. Wendlandt, S.; Li, B.; Ma, Z.; Schwarz, S. Complete Sequence of the Multi-Resistance Plasmid PV7037 from a Porcine Methicillin-Resistant *Staphylococcus aureus*. *Vet. Microbiol.* **2013**, *166*, 650–654. [CrossRef] [PubMed]

50. International Working Group on the Classification of Staphylococcal Cassette Chromosome Elements (IWG-SCC). Classification of Staphylococcal Cassette Chromosome *Mec* (SCC*Mec*): Guidelines for Reporting Novel SCC*Mec* Elements. *Antimicrob. Agents Chemother.* **2009**, *53*, 4961–4967. [CrossRef]

51. Andrews, S. FASTQC. A Quality Control Tool for High Throughput Sequence Data. 2010. Available online: https://www.bioinformatics.babraham.ac.uk/projects/fastqc/ (accessed on 31 March 2022).

52. Bankevich, A.; Nurk, S.; Antipov, D.; Gurevich, A.A.; Dvorkin, M.; Kulikov, A.S.; Lesin, V.M.; Nikolenko, S.I.; Pham, S.; Prjibelski, A.D.; et al. SPAdes: A New Genome Assembly Algorithm and Its Applications to Single-Cell Sequencing. *J. Comput. Biol.* **2012**, *19*, 455–477. [CrossRef]

53. Blankenberg, D.; Coraor, N.; Von Kuster, G.; Taylor, J.; Nekrutenko, A.; Galaxy Team. Integrating Diverse Databases into an Unified Analysis Framework: A Galaxy Approach. *Database* **2011**, *2011*, bar011. [CrossRef]

54. Gurevich, A.; Saveliev, V.; Vyahhi, N.; Tesler, G. QUAST: Quality Assessment Tool for Genome Assemblies. *Bioinformatics* **2013**, *29*, 1072–1075. [CrossRef]

55. Wei, W.; Michelle, B.; Yue, H.; Jin, X.; Dajin, Y.; Alexandre, M.-G.; Ning, X.; Hui, L.; Shaofei, Y.; Menghan, L.; et al. Whole-Genome Sequencing and Machine Learning Analysis of *Staphylococcus aureus* from Multiple Heterogeneous Sources in China Reveals Common Genetic Traits of Antimicrobial Resistance. *mSystems* **2021**, *6*. [CrossRef]

56. Larsen, M.V.; Cosentino, S.; Rasmussen, S.; Friis, C.; Hasman, H.; Marvig, R.L.; Jelsbak, L.; Sicheritz-Pontén, T.; Ussery, D.W.; Aarestrup, F.M.; et al. Multilocus Sequence Typing of Total-Genome-Sequenced Bacteria. *J. Clin. Microbiol.* **2012**, *50*, 1355–1361. [CrossRef] [PubMed]

57. Kaya, H.; Hasman, H.; Larsen, J.; Stegger, M.; Johannesen, T.B.; Allesøe, R.L.; Lemvigh, C.K.; Aarestrup, F.M.; Lund, O.; Larsen, A.R. SCC*Mec*Finder, a Web-Based Tool for Typing of Staphylococcal Cassette Chromosome *Mec* in *Staphylococcus aureus* Using Whole-Genome Sequence Data. *mSphere* **2018**, *3*. [CrossRef] [PubMed]

58. Bartels, M.D.; Petersen, A.; Worning, P.; Nielsen, J.B.; Larner-Svensson, H.; Johansen, H.K.; Andersen, L.P.; Jarløv, J.O.; Boye, K.; Larsen, A.R.; et al. Comparing Whole-Genome Sequencing with Sanger Sequencing for *Spa* Typing of Methicillin-Resistant *Staphylococcus aureus*. *J. Clin. Microbiol.* **2014**, *52*, 4305–4308. [CrossRef] [PubMed]
59. Feldgarden, M.; Brover, V.; Haft, D.H.; Prasad, A.B.; Slotta, D.J.; Tolstoy, I.; Tyson, G.H.; Zhao, S.; Hsu, C.-H.; McDermott, P.F.; et al. Validating the AMRFinder Tool and Resistance Gene Database by Using Antimicrobial Resistance Genotype-Phenotype Correlations in a Collection of Isolates. *Antimicrob. Agents Chemother.* **2019**, *63*. [CrossRef] [PubMed]
60. Chen, L.; Yang, J.; Yu, J.; Yao, Z.; Sun, L.; Shen, Y.; Jin, Q. VFDB: A Reference Database for Bacterial Virulence Factors. *Nucleic Acids Res.* **2005**, *33* (Suppl. S1), D325–D328. [CrossRef] [PubMed]
61. Olson, R.D.; Assaf, R.; Brettin, T.; Conrad, N.; Cucinell, C.; Davis, J.J.; Dempsey, D.M.; Dickerman, A.; Dietrich, E.M.; Kenyon, R.W.; et al. Introducing the Bacterial and Viral Bioinformatics Resource Center (BV-BRC): A Resource Combining PATRIC, IRD and ViPR. *Nucleic Acids Res.* **2023**, *51*, D678–D689. [CrossRef]
62. Seemann, T. Snippy: Fast Bacterial Variant Calling from NGS Reads. 2015. Available online: https://github.com/tseemann/snippy (accessed on 16 November 2022).
63. Croucher, N.J.; Page, A.J.; Connor, T.R.; Delaney, A.J.; Keane, J.A.; Bentley, S.D.; Parkhill, J.; Harris, S.R. Rapid Phylogenetic Analysis of Large Samples of Recombinant Bacterial Whole Genome Sequences Using Gubbins. *Nucleic Acids Res.* **2015**, *43*, e15. [CrossRef]
64. Seemann, T.; Klotzl, F.; Page, A.J. Snp-Dists. Pairwise SNP Distance Matrix from a FASTA Sequence Alignment. 2018. Available online: https://github.com/tseemann/snp-dists (accessed on 16 November 2022).
65. Tamura, K.; Stecher, G.; Kumar, S. MEGA11: Molecular Evolutionary Genetics Analysis Version 11. *Mol. Biol. Evol.* **2021**, *38*, 3022–3027. [CrossRef]
66. Kimura, M. A Simple Method for Estimating Evolutionary Rates of Base Substitutions through Comparative Studies of Nucleotide Sequences. *J. Mol. Evol.* **1980**, *16*, 111–120. [CrossRef]
67. Letunic, I.; Bork, P. Interactive Tree Of Life (ITOL) v4: Recent Updates and New Developments. *Nucleic Acids Res.* **2019**, *47*, W256–W259. [CrossRef] [PubMed]
68. Wickham, H. *Ggplot2: Elegant Graphics for Data Analysis*; Springer-Verlag: New York, NY, USA, 2016; Available online: https://ggplot2.tidyverse.org (accessed on 30 July 2023).
69. Hijmans, R. Raster: Geographic Data Analysis and Modeling. 2023. Available online: https://rspatial.org/raster (accessed on 30 July 2023).
70. Pebesma, E.J. Simple Features for R: Standardized Support for Spatial Vector Data. *R J.* **2018**, *10*, 439–446. [CrossRef]
71. Wickham, H. Tidyverse: Easily Install and Load the "Tidyverse.". 2017. Available online: https://CRAN.R-project.org/package=tidyverse (accessed on 30 July 2023).

Article

Assessing the Effect of Oxytetracycline on the Selection of Resistant *Escherichia coli* in Treated and Untreated Broiler Chickens

Ekaterina Pokrant [1,2,3,†] , María Belén Vargas [1,2,†] , María José Navarrete [1], Karina Yévenes [1,2], Lina Trincado [2,3], Paula Cortés [3], Aldo Maddaleno [3], Lisette Lapierre [4,*] and Javiera Cornejo [1,3,*]

[1] Laboratory of Food Safety, Department of Preventive Animal Medicine, Faculty of Veterinary and Animal Sciences, University of Chile, Santiago 8820808, Chile; katiavalerievna@uchile.cl (E.P.); maria.vargas.s@ug.uchile.cl (M.B.V.); maria.navarrete.g@ug.uchile.cl (M.J.N.); kyevenes@ug.uchile.cl (K.Y.)

[2] Doctorate Program of Forestry, Agriculture, and Veterinary Sciences (DCSAV), University of Chile, Av. Santa Rosa 11315, La Pintana, Santiago 8820808, Chile; lina.trincado@ug.uchile.cl

[3] Laboratory of Veterinary Pharmacology (FARMAVET), Faculty of Veterinary and Animal Sciences, University of Chile, Santiago 8820808, Chile; paula.cortes@ug.uchile.cl (P.C.); amaddaleno@veterinaria.uchile.cl (A.M.)

[4] Laboratory of Bacterial Pathogens Diagnostic and Antimicrobial Resistance, Department of Preventive Animal Medicine, Faculty of Veterinary and Animal Sciences, University of Chile, Santiago 8820808, Chile

* Correspondence: llapierre@uchile.cl (L.L.); jacornej@uchile.cl (J.C.); Tel.: +56-2-29785615 (L.L.); +56-2-29785609 (J.C.)

† These authors contributed equally to this work.

Abstract: Oxytetracycline (OTC) is administered in the poultry industry for the treatment of digestive and respiratory diseases. The use of OTC may contribute to the selection of resistant bacteria in the gastrointestinal tract of birds or in the environment. To determine the effect of OTC on the selection of resistant *Escherichia coli* strains post-treatment, bacteria were isolated from droppings and litter sampled from untreated and treated birds. Bacterial susceptibility to tetracyclines was determined by the Kirby–Bauer test. A total of 187 resistant isolates were analyzed for the presence of *tet*(A), (B), (C), (D), (E), and (M) genes by PCR. Fifty-four strains were analyzed by PFGE for subtyping. The proportion of tetracycline-resistant *E. coli* strains isolated was 42.88%. The susceptibility of the strains was treatment-dependent. A high clonal diversity was observed, with the *tet*(A) gene being the most prevalent, followed by *tet*(C). Even at therapeutic doses, there is selection pressure on resistant *E. coli* strains. The most prevalent resistance genes were *tet*(A) and *tet*(C), which could suggest that one of the main mechanisms of resistance of *E. coli* to tetracyclines is through active efflux pumps.

Keywords: *Escherichia coli*; antibiotic resistance; droppings; broiler litter; oxytetracycline; resistance genes

Citation: Pokrant, E.; Vargas, M.B.; Navarrete, M.J.; Yévenes, K.; Trincado, L.; Cortés, P.; Maddaleno, A.; Lapierre, L.; Cornejo, J. Assessing the Effect of Oxytetracycline on the Selection of Resistant *Escherichia coli* in Treated and Untreated Broiler Chickens. *Antibiotics* **2023**, *12*, 1652. https://doi.org/10.3390/antibiotics12121652

Academic Editor: Carlos M. Franco

Received: 17 October 2023
Revised: 10 November 2023
Accepted: 18 November 2023
Published: 23 November 2023

1. Introduction

One of the most important side effects of using antimicrobials in productive animals is the selection of resistant bacteria. Efforts have been made to control the improper or excessive use of antimicrobials, highlighting the global action plan on antimicrobial resistance adopted at the World Health Assembly in May 2015, where different objectives related to this issue were established [1]. The use of tetracyclines has led to the emergence of resistant bacterial variants, in particular those containing *tet* genes, which are generally associated with mobile genetic elements or conjugative transposons [2–4]. These elements code for different resistance mechanisms, such as efflux pumps, ribosomal protection, enzymatic inactivation, and mutations, such as the one described in the 30S ribosomal subunit. In Gram-negative bacteria, the efflux pump systems are encoded by *tet* genes, mostly by *tet*(A), *tet*(B), *tet*(C), and *tet*(E), while the *tet*(M) gene, which is one of the most studied determinants, encodes for ribosomal protection [5].

Antibiotic resistance genes (ARGs) are currently considered to be emerging contaminants because they have been detected throughout the environment, including soils, river sediments, watercourses, and wastewater [6,7]. The presence of ARGs in different environments is a risk because it confers selection pressure on pathogenic and commensal strains at the genetic level, leading to the emergence of antimicrobial resistance. The genes can be acquired by other pathogenic bacteria that affect humans and animals, which has a socio-economic impact due to increased treatment costs and production losses due to the spread of ARGs [8,9].

Escherichia coli is a Gram-negative bacillus that is part of the normal intestinal microbiota of animals and humans. However, there are some pathogenic strains that can cause fatal diseases in the host [10]. These bacteria are used as indicator bacteria for antimicrobial resistance levels in different productive species, as they have been described as a reservoir of resistance genes, which could be transmitted to pathogenic and zoonotic bacteria [10–13]. Antimicrobial use increases the risk of antimicrobial resistance in *E. coli* in pigs. However, there is little information concerning the impact of dose or concentration and or the effects of antimicrobial use over time [14]. Chantziaras et al. [15] evaluated the correlation between antibiotic use and the prevalence of *E. coli* strains isolated from pigs, poultry, and cattle. Their results indicated that there is a correlation between the use of specific antimicrobials and the level of resistance of these microorganisms. They concluded that more detailed data collection and harmonization are needed due to data restrictions in their study. Although it has been previously observed that oral administration of tetracycline does not induce significant changes in the cecal bacterial community of chickens, a relationship between its use as a growth promotor and an increase in the population of tetracycline-resistant *E. coli* harboring *tet*(A) or *tet*(B) has been shown. However, the effect of therapeutic doses of oxytetracycline (OTC) on the selection of tetracycline-resistant *E. coli* under controlled conditions has not been demonstrated [15].

Da Costa et al. [16] investigated variations in the prevalence of antibiotic resistance in *E. coli* isolated from the fecal matter of broilers raised with three different commercial antimicrobial treatments administered through their water supply. After the administration of different antibiotics, resistance grew quickly. The sharp rise in antimicrobial resistance rates following drug administration was a direct result of the formation of new antimicrobial resistance patterns rather than the amplification of previously resistant organisms [16].

Herrero-Fresno et al. [17] determined that the intestinal microbiota of apramycin-treated pigs showed resistance selection from treatment, resulting in noticeably greater counts of resistant strains than untreated pigs [17]. More recently, Das et al. [18] analyzed the incidence and distribution of oxytetracycline- and ciprofloxacin-resistant *E. coli* isolated from live broilers and the farm environment. They determined that 100% of commensal *E. coli* strains isolated from chickens and the environment were resistant to tetracyclines, with the majority harboring the *tet*(A) gene [18].

In an in vivo study, Shah et al. [19] determined that the use of antimicrobials such as oxytetracycline as growth promoters in the diet has no detrimental effect on beneficial bacteria, but an alteration in the growth of harmful bacteria was observed [19]. However, there are no known controlled studies that determine the effect of OTC administration, at therapeutic doses, in broilers on the selection of post-treatment resistant strains of *E. coli*. The objective of this study was to determine the presence of *E. coli* isolates not susceptible to OTC and resistance genes in broiler droppings and litter from treated and untreated birds and to determine if there is a relationship between treatment and the selection of resistant bacteria. The purpose of this study focused on generating further scientific knowledge regarding whether the use of OTC at therapeutic doses predisposed to the selection of resistant *E. coli* in a controlled environment by determining phenotypic and genotypic resistance in droppings and poultry litter, as these by-products could be a potential source of reservoir and future dissemination of ARGs into the environment.

2. Results

2.1. Isolation and Confirmation of E. coli

E. coli were isolated from samples collected the day before treatment to determine a base resistance profile. *E. coli* was only isolated from manure samples. *E. coli* isolates were confirmed by biochemical tests. No microorganisms were observed in the litter samples prior to treatment.

Post-treatment, *E. coli* colonies were isolated and confirmed by biochemical tests and PCR. Six manure samples and six litter samples were analyzed for each experimental group on days 1, 7, 14, and 21 post-treatment, and six litter samples for each experimental group on days 7 and 14 post-slaughter. Five typical colonies on MacConkey agar were obtained from each sample and confirmed by the IMViC test: positive for ornithine, motility, and methyl red and negative for Voges–Proskauer and citrate (Figure 1).

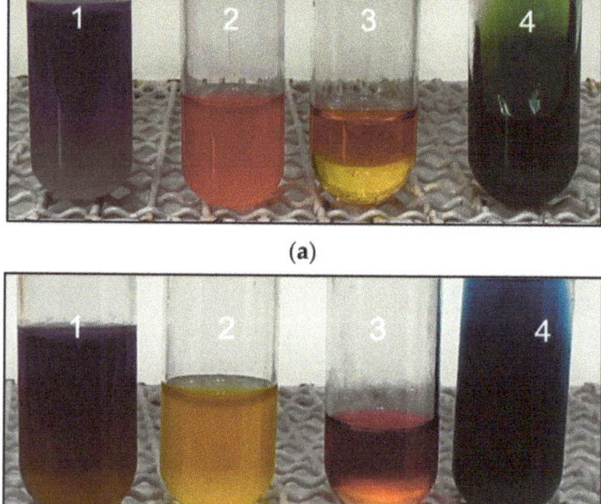

(a)

(b)

Figure 1. Representative images of IMViC for the confirmation of *E. coli* strains: (**a**) positive reaction to *E. coli*. [1: MIO agar (+); 2: methyl red (+); 3: Voges–Proskauer broth (−); 4: citrate agar (−)] and (**b**) negative reaction to *E. coli*. [1: MIO agar (−); 2: methyl red (−); 3: Voges–Proskauer broth (+); 4: citrate agar (+)].

2.2. Determination of Antimicrobial Susceptibility

Susceptible and resistant strains were detected at all sample points in both cloacal and litter samples. Strains with intermediate sensitivity were categorized as resistant. Non-susceptible strains were found in both the treated and untreated groups, with the highest percentage in the treatment group on day one post-treatment in litter and droppings (Tables 1 and 2).

The OTC treatment in poultry had a significant effect on tetracycline *E. coli* resistance isolates of droppings, where the chi-squared test showed an association between the treated group and *E. coli* resistance (Table 1). On another hand, McNemar's test indicated an increase in *E. coli* resistance from the same individual's droppings pre-treatment relative to post-treatment ($p < 0.05$).

In relation to the isolates detected in poultry litters, there were significant differences between the treated and untreated groups, showing that the treatment affected the resistance of *E. coli* to tetracyclines isolated from litter (Table 2).

The emergence of tetracycline resistance among commensal organisms such as *E. coli* may result in a non-response to antibiotic therapy, as these microorganisms can be reservoirs of ARGs, which can be transferred to avian pathogens, increasing mortality and thus economic losses at the production level.

Table 1. Percentage of non-susceptible isolates detected in bird droppings, according to experimental groups after oxytetracycline treatment.

Experimental Groups	Strains	Post-Treatment Days					*p*-Value [1]
		0	1	7	14	21	
Treated	Total number of isolates	16	20	19	22	23	
	Non-susceptible strains (%)	37.5	100	84.21	63.63	34.8	
	Susceptible strains (%)	62.5	0.0	15.78	36.36	65.2	3.5×10^{-7}
Untreated	Total number of isolates	30	13	17	12	14	
	Non-susceptible strains (%)	0	76.92	41.17	8.33	28.57	
	Susceptible strains (%)	100	23.07	58.82	91.66	71.42	

[1] Significance difference $p < 0.05$ (chi-squared test).

Table 2. Percentage of non-susceptible isolates detected in poultry litter, according to experimental groups after oxytetracycline treatment.

Experimental Groups	Strains	Post-Treatment Days						*p*-Value [1]
		1	7	14	21	29	36	
Treated	Total number of isolates	29.0	12.0	18.0	17.0	22.0	25.0	
	Non-susceptible strains (%)	82.8	75.0	72.2	17.6	68.2	92.0	
	Susceptible strains (%)	17.2	25.0	27.8	82.4	31.8	8.0	2.2×10^{-16}
Untreated	Total number of isolates	52	15	15	10	10	25	
	Non-susceptible strains (%)	3.8	0.0	0.0	30.0	30.0	8.0	
	Susceptible strains (%)	96.2	100.0	100.0	70.0	70.0	92.0	

[1] Significance difference $p < 0.05$ (chi-squared test).

2.3. Detection of Resistance Genes

The presence of *tet*(A), *tet*(B), *tet*(C), *tet*(E), and *tet*(M) genes in the resistant *E. coli* strains was determined by conventional PCR. Figure 2 shows the amplification products for the detected genes.

The most prevalent resistance gene was *tet*(A), followed by *tet*(C), *tet*(B), *tet*(M), and *tet*(E).

Molecularly, *tet* genes are resistance determinants, which encode inactivation mechanisms comprising the efflux pump system and ribosomal protection. The *tet*(A) gene was one of the most frequently detected genes in this study, which is consistent with the literature. Studies describe that this gene is one of the most frequently detected genes in both Gram-positive and Gram-negative bacteria and encodes for efflux pump systems that contribute to tetracycline resistance by constantly reducing the antibiotic concentration inside the cell [2]. Thus, this gene can be a potential risk of dissemination within the production chain and contribute to antimicrobial resistance.

2.4. Isolate Subtyping

Fifty-four *E. coli* strains isolated pre-treatment and on day 1 post-treatment were analyzed by PFGE genotyping. The criteria for isolate selection for PFGE are based on the fact that OTC at therapeutic doses immediately eliminates sensitive bacteria and selects for resistant strains; when there is a predominance of one clone, it would continue to spread during treatment. The PFGE technique was also used to observe whether there was a predominance of a single clone or multiple clones.

A phylogenetic analysis was performed using the PFGE pulsotypes and was compared to the antibiotic resistance profile (Figure 3). The PFGE pulsotypes were classified into 42 types. The Simpson diversity index was 0.66, which means that the isolated *E. coli* was highly diverse. The phylogenetic dendrogram classified the *E. coli* strains into nine clusters. The samples from droppings were grouped mainly in clusters I, II, VII, and IX, while those from the litter were grouped in clusters III, IV, V, VI, and VIII. *tet*(A) genes were found in almost all isolates, both before and after treatment. *tet*(D) and *tet*(E) genes were not found. Isolates from clusters III, IV, V, and VI contained *tet*(B), *tet*(C), and *tet*(M) genes and were isolated exclusively from litter samples. Only eight isolates, also all litter samples, were negative for the presence of the *tet* genes.

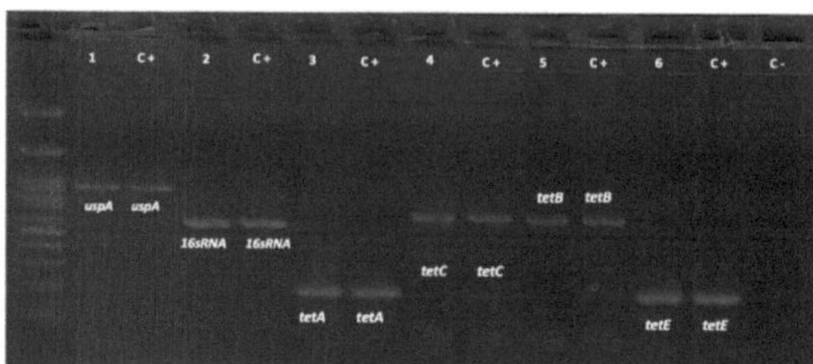

Figure 2. Representative image of the PCR products observed on 2% agarose gel. *tet*(A), 210 bp; *tet*(B), 659 bp; *tet*(C), 418 bp; *tet*(E), 278 bp; *uspA*, 884 bp; and *16s*RNA, 585 bp. Lanes 1 to 6 correspond to target gene amplicons, and lane C+ corresponds to the positive control for each gene.

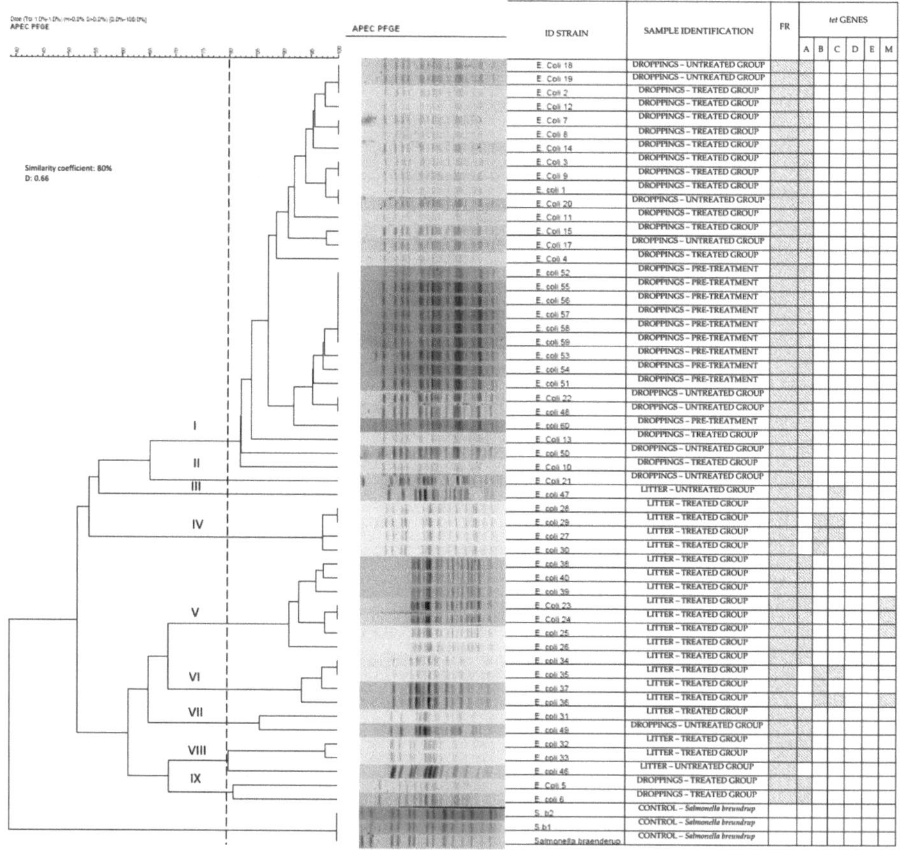

Figure 3. Dendrogram of the cluster analysis of *E. coli* strains generated by Gel Compar II software, version 5.10, using the unweighted pair group arithmetic mean method (UPGMA), with 1% tolerance. The resistance genes detected in each strain are marked with a color. The analyzed strains were isolated from the samples prior to the start of treatment and the first sampling point (day 1 post-treatment). From I to IX, corresponds to the different clusters. Shaded areas indicate positivity. D: Simpson diversity index.

3. Discussion

In this study, the highest prevalence of resistant isolates came from droppings and litter samples from OTC-treated poultry. Although it was not possible to statistically correlate the concentrations with the selection of resistant bacteria, a higher prevalence of non-susceptible *E. coli* was determined in the first sampling point, where the concentrations in both matrices were significantly higher than in the subsequent sampling. Therefore, it is inferred that the isolated *E. coli* population in the first point underwent a higher selection pressure by the excreted antimicrobial, where the concentrations of OTC and its epimer detected were 22,742 µg kg^{-1} in litter and 2,087 µg kg^{-1} in droppings of treated animals [20].

High concentrations of antimicrobials have been shown to produce selection pressure on resistant microbiota [21]. Berge et al. [22], who studied bacterial resistance patterns in *E. coli* from cattle fecal samples after a single dose of florfenicol, observed an increase in the number of resistant *E. coli* isolates [22]. Similarly, Fairchild et al. [23] investigated the effects of tetracycline administration on commensal bacteria from commercial poultry and found that *Enterococcus* spp. and *E. coli* were resistant to tetracyclines with 32.2% harboring *tet*(A) and 30.5% containing *tet*(B) resistance genes [23].

Low residue concentrations, even below the detection limit (LOD), may also be associated with the presence of resistant microorganisms [24–26]. Resistance resulting from even low doses of antibiotics is of worldwide concern since, currently, OTC is not only used for the treatment of productive animals but also in the poultry industry at subtherapeutic doses through feed to promote growth. Use to promote growth is still allowed in some countries, such as Brazil and China [27].

The presence of pre-treatment *E. coli* in manure samples can be attributed to the fact that this bacterium is a commensal microorganism, so birds are colonized during their first days of life; however, the detection of resistant strains could be due to selection pressure contributing to the emergence of resistance genes and their potential dissemination through these production systems. On the other hand, prior to treatment, the wood shavings used for litter were sampled, and according to the analyses, no *E. coli* was isolated from this product. Therefore, and considering that the wood shavings have a very low water activity, we can assure that there was no external contamination from this litter component.

The phenotype for tetracycline resistance matched the genotypic resistance, determined by *tet* gene positivity, in almost all isolates in 75 non-susceptible *E. coli* isolates from droppings at days 1, 7, 14, and 21 post-treatment and 97 *E. coli* isolates from litter samples at days 1, 14, 21, 29, and 36 post-treatment. Most of the genes present in the isolates from droppings and litter were *tet*(A), followed by *tet*(C), *tet*(B), *tet*(M), and *tet*(E). None of the isolates harbored *tet*(D). Only 17 resistant isolates had none of the resistance genes analyzed; therefore, the resistance of these strains could be mediated by a different gene. This discrepancy between phenotypic and genotypic resistance could be due to mutations in resistance genes. In a previous study, where this difference between genotypic versus phenotypic evidence of rifampicin resistance in *Mycobacterium tuberculosis* was observed, mutations in the resistance regions were determined [28].

Moreover, many tetracycline resistance genes have been described. The *tet* genes encode for different resistance mechanisms, such as efflux pumps, ribosomal protection, enzymatic modification, and other unknown mechanisms [2]. Therefore, it is possible that a *tet* gene, other than *tet*(A), (B), (C), (D), (E), and (M), may be mediating the resistance of isolates showing phenotypic resistance by disc diffusion test. It is also possible to attribute this resistance to a new, undescribed gene. Davis et al. [29] observed a new aminoglycoside resistance gene, *rmtE*, that belongs to the 16S ribosomal RNA methylase gene family [29].

In this study, a high clonal diversity was observed that could be the result of the adaptability of *E. coli* strains. Resistant *E. coli* isolates are highly represented by these resistance genes. This finding is consistent with the results of previous studies, where the most prevalent genes reported have been *tet*(A) and *tet*(B) [30,31]. In a current study, Sreejith et al. [32] determined through susceptibility analysis that 77% of the *E. coli* isolates were tetracycline resistant, where 85.18% of the isolates had *tet*(A) genes and 22.22% had *tet*(B) genes [32]. The high prevalence of *tet*(B) could be explained by the gene's ability to reside on highly mobile genetic elements that efficiently transfer the gene between bacteria, like plasmids. The ability of *tet*(A) to spread freely and rapidly in farm animals and in the environment has also been reported [33].

In this study, the fourth most prevalent gene was *tet*(M), which was found in 14% of the isolates. Other studies have determined a prevalence of *tet*(M) between 5% and 13% in *E. coli* strains [31,34,35].

In this study, PFGE grouped the isolates containing the *tet*(B) gene in clusters IV and VI, which came exclusively from the litter. This could be explained by the selective pressure that occurs in the litter, where *tet*(B) was detected in isolates shortly after treatment. Sreejith et al. [32] found that the presence of antibiotics in feed and in the farm environment can help *tet*(A) and *tet*(B) persist in the microbiome for a long time. The presence of antibiotics, even at low concentrations, ensures the persistence of these resistant genes, which can be expressed dominantly in the microbial community [32]. This aspect becomes relevant as these bacteria could be a reservoir of resistance genes, which can be transmitted to other microorganisms [36].

Previous studies have shown the effects of the use of antibiotics at subtherapeutic doses on the selection of resistant bacteria [24,25]. We found an association between the therapeutic treatment given to the birds and the presence of resistant bacteria. Others have shown the presence of resistant bacteria and resistance genes is due to the strong selective pressure provided by the presence of antibiotic treatment [37]. Our results show the use of antimicrobials at therapeutic doses in poultry production can lead to the selection and persistence of resistant *E. coli* strains, which can be a risk to both human and animal health. It is important to highlight that resistant bacteria with transferable resistance genes were detected up to 36 days post-treatment in the chicken litter. We consider that one of the limitations of our study is that only one concentration of the antimicrobial was studied, so we cannot conclude similarities with respect to the lower doses under the same conditions; however, this study lays the foundation for further research in this area and provides scientific information that supports the need to monitor and control the antimicrobials used in the animal production industry.

The results provide scientific information that supports the growing concern about the use of antimicrobials in animal production and the constant worldwide effort to reduce the use of these veterinary drugs to preserve their efficacy since different antimicrobials are critical in veterinary and human medicine. For this reason, the constant monitoring and responsible use of these veterinary drugs is essential.

4. Materials and Methods

4.1. Experimental Animals

Commercial male broilers from Ross 308 genetic line (Ross®, Aviagen Inc., Huntsville, AL, USA), which is characterized by high yields, strong disease resistance, and weight gain, were raised from birth in an experimental unit specially designed to carry out this study. In this experiment, the birds were kept in pens of 1.5 m² surface area with clean shavings that later became part of the birds' litter. Temperature (25 ± 5 °C), humidity (50–60%), and ventilation were controlled in the unit. The animals were kept with ad libitum access to water and non-medicated feed. This study was approved by the Institutional Animal Care and Use Committee (CICUA by its Spanish acronym) through the certificate No. 18187-VET-UCH-E1. Handling and euthanasia were based on Directive 2010/63/EU and the AVMA Guidelines for the Euthanasia of Animals: 2020 Edition [38,39].

The birds were separated into a treated group and an untreated group. Group A (treatment) was 6 birds treated orally with a pharmaceutical formulation containing OTC at 10% at a therapeutical dose of 80 mg kg^{-1} for 10 consecutive days. The antibiotic was administered by orogastric tube to ensure the complete intake of the dose for each bird. Group B (non-treatment) had 6 untreated birds under the same conditions. To avoid cross-contamination, we followed the biosafety measures established by the CONICYT Biosafety Standards Manual [40], along with the biosafety standards instituted by the FAVET Biosafety Committee.

4.2. Sampling Collection

Cloacal and litter samples were collected at 5 sampling points: before treatment and on days 1, 7, 14, and 21 post-treatment. In addition, litter samples were analyzed one and two weeks after the birds were slaughtered (corresponding to days 29 and 36 post-treatment). Cloacal samples were collected from each bird with sterile cotton swabs and were stored in sterile polypropylene tubes. Ten grams of litter were collected from each group and were stored in sterile plastic bags. All samples were processed immediately.

4.3. E. coli Identification and Isolation

One gram of litter was homogenized with 9 mL peptone water for *E. coli* isolation. The cloacal samples were homogenized with 4.5 mL buffered peptone water (Huankai Microbial®, Guangzhou, China). Three loops of the enriched sample were streaked on MacConkey agar (OXOID®, Hants, UK) plates and incubated at 37 °C for 24 h. Triplicates of each sample were performed. After incubation, five typical colonies per plate were collected, and biochemical identification was performed to confirm *E. coli* colonies using IMViC test [41,42]. In addition, PCR analysis was carried out for *uspA* gene detection for confirmation of *E. coli* identity [43]. Confirmed isolates were stored at −20 °C in 20% glycerol.

4.4. Antimicrobial Susceptibility Testing

All confirmed *E. coli* isolates were analyzed by the Kirby–Bauer disk diffusion method, which was performed using antimicrobial susceptibility test discs in Mueller–Hinton agar (Sigma Aldrich®, Saint Louis, MI, USA) according to the recommendation of the Clinical and Laboratory Standards Institute (CLSI). First, a suspension of fresh, pure culture was prepared, and the turbidity of the bacterial suspension was adjusted with 0.85% saline solution until an OD_{600nm} between 0.08 and 0.1 was reached, which is equivalent to 0.5 McFarland turbidity. The adjusted suspension was inoculated into Mueller–Hinton agar plates, and tetracycline disks (30 µg) (OXOID®, Hants, UK) were positioned over the inoculated plate. The plates were incubated inverted at 35 °C for 16–18 h, and the inhibition halos were measured. The diameter of the inhibition zone of each disc was compared with the interpretation criteria of the Clinical and Laboratory Standards Institute guidelines. *E. coli* ATCC 25922 was used as a quality control [44].

4.5. PCR Detection of Tet Genes

The presence of resistance *tet* genes was determined by conventional polymerase chain reaction or PCR. After *E. coli* identification and isolation, template DNA was extracted from MacConkey agar plates using heat treatment [45] and was quantified by spectrophotometry (NANO-400 microspectrophotometer, Hangzhou Allsheng instruments Co., Hangzhou, China). Samples that exhibited an absorbance ratio of 260/280 nm close to the optimal range (1.8–2.0) were analyzed by PCR. The genes analyzed were *tet*(A), *tet*(B), *tet*(C), *tet*(D), *tet*(E), and *tet*(M) [46]; the 16S rRNA gene was included for confirmation of DNA presence [47] (Table 3).

For the identification of the genes *tet*(A), *tet*(B), *tet*(C), *tet*(D), and *tet*(E), PCR multiplex reactions were performed using GoTaq® Green Master Mix following manufacturer's instructions (Promega, Madison, WI, USA). For a 25 µL reaction, a mixture of 12.5 µL of GoTaq® Green Master Mix, 1 µL of nuclease-free water, 1 µL of primer *tet*(A)-F, 1 µL of primer *tet*(A)-R, 1 µL of primer *tet*(B)-F, 1 µL of primer *tet*(B)-R, 1 µL of primer *tet*(C)-F, 1 µL of primer *tet*(C)-R, 1 µL of primer *tet*(D)-F, 1 µL of primer *tet*(D)-R, 1 µL of primer *tet*(E)-F, 1 µL of primer *tet*(E)-R, and 1 µL of DNA sample was prepared.

Table 3. Primers and conditions for each tetracycline resistance gene.

Gene	Sequence (5′ > 3′)	Annealing Temperature (°C)	Size	Reference
tet(A)	F: GCTACATCCTGCTTGCCTTC R: CATAGATCGCCGTGAAGAGG	54	210	[46]
tet(B)	F: TTGGTTAGGGGCAAGTTTTG R: GTAATGGGCCAATAACACCG	54	659	[46]
tet(C)	F: CTTGAGAGCCTTCAACCCAG R: ATGGTCGTCATCTACCTGCC	54	418	[46]
tet(D)	F: AAACCATTACGGCATTCTGC R: GACCGGATACACCATCCATC	54	787	[46]
tet(E)	F: AAACCACATCCTCCATACGC R: AAATAGGCCACAACCGTCAG	54	278	[46]
tet(M)	F: GTGGACAAAGGTACAACGAG R: CGGTAAAGTTCGTCACACAC	54	406	[46]
16sRNA	F: GACCTCGGTTTAGTTCACAGA R: CACACGCTGACGCTGACCA	54	585	[47]
E. coli uspA	F: CCGATACGCTGCCAATCAGT R: ACGCAGACCGTAGGCCAGAT	54	884	[43]

For the identification of the genes *tet*(M), PCR single reactions were performed. For a 25 µL reaction, a mixture of 12.5 µL of GoTaq® Green Master Mix, 9 µL of nuclease-free water, 1.25 µL of primer *tet*(M)-F, 1.25 µL of primer *tet*(M)-R, and 1 µL of DNA sample was prepared.

Nuclease-free water was used as a negative control. On the other hand, previously sequenced DNA from strains positive for the genes studied was used as a positive control.

The PCR protocol included an initial denaturation step at 94 °C for 5 min, followed by 35 cycles of denaturation (94 °C for 1 min), annealing (*tet*: 54 °C for 1 min; *uspA* and 16S rRNA: 58 °C for 1 min), and elongation (72 °C for 1 min), with a final extension step at 72 °C for 10 min. Five µL of the PCR product was visualized on electrophoresis gels (2% (*w/v*) agarose in 1 × TAE buffer), previously stained with SafeView Plus (Fermelo Biotec, Santiago, Chile). The bands were visualized by ultraviolet transillumination, and the sizes of the PCR products were determined using the 100 base pair (bp) size scale (Maestrogen Hsinchu, Taiwan). Isolated strains previously sequenced for the genes studied were used as positive controls.

4.6. Pulsed-Field Gel Electrophoresis

Genetic relatedness among the tetracycline-resistant *E. coli* isolates was established from their *XbaI*-digested chromosomal DNA fragments. The clonality of 54 *E. coli* strains isolated from litter and droppings samples was determined by pulsed-field gel electrophoresis (PFGE) subtyping. Resistant strains from sampling point 0 (pre-treatment day) and post-treatment day 1 from the two experimental groups were analyzed. The first sampling point was considered the most representative as it was carried out immediately after the end of treatment.

The PFGE technique was performed according to the Centers for Disease Control and Prevention (CDC) using the Standard Operating Procedure for PulseNet PFGE of *E. coli* non-O157 (STEC) [48] with minor modifications. Bacterial isolates were suspended in cell suspension buffer (CSB; 100 mM Tris:100 mM EDTA, pH 8.0) at a wavelength of 420 nm (OD = 0.4). The bacterial suspension was mixed with proteinase K (20 mg/mL) and 1% melted SeaKem Gold agarose for cell lysis and carefully transferred into plug molds, which were then cooled to 4 °C. The solid plugs were washed and digested overnight with restriction enzyme *XbaI* (Thermo-Fisher Scientific, Waltham, MA, USA). After restriction digestion, electrophoresis was performed with the CHEF-DR III (Bio-Rad) system using

1% pulsed-field certified agarose (Bio-Rad, Hercules, CA, USA) in 0.5 X TBE (45 mM Tris-HCl, 45 mM boric acid, 1 mM EDTA). The electrophoresis conditions were as follows: initial switch time 2.16 s, final switch time 54.7 s, run time 20 h, included angle 120°, gradient 6 V/cm, and temperature 14 °C. The gel was stained for 30 min with Gelred®, and the fingerprinting profile was observed by an illuminated UV wave to the gel. DNA bands on agarose gels were pictured and saved in TIFF format. DNA band profiles were analyzed by GelCompar II Software v 5.1 (Applied Maths, Sint-Martens-Latem, Belgium).

The PFGE profiles were compared using a Dice similarity coefficient and UPGMA analysis to create the dendrogram; a band position tolerance of 1% was used. A cutoff point of 80% was used to analyze genetic relatedness and establish genetic patterns. Genetic diversity was quantified using the Simpson diversity index.

4.7. Data Analysis

To determine the effect of OTC treatment on tetracycline *E. coli* resistance from the same individual's droppings, the McNemar's test was performed, where the classification criteria were antimicrobial susceptibility (susceptible and non-susceptible) and condition (pre-treatment and post-treatment). Moreover, to determine the effect of OTC treatment on tetracycline *E. coli* resistance from litter and droppings between groups, the chi-squared test was performed. Frequencies corresponded to the percentages of non-susceptible and susceptible isolates. Rstudio® V0.99.903 was used for analysis. A statistically significant difference was considered when the p-value < 0.05. For the genotypic resistance analysis, a genetic profile was performed using the results of the conventional PCR for the design of the different genetic patterns obtained.

5. Conclusions

The highest proportion of non-susceptible *E. coli* isolates (resistant and of intermediate sensitivity) to tetracyclines was detected in dropping and litter samples from the group treated with OTC. Therefore, even at therapeutic doses, there is selection pressure on *E. coli* strains resistant to tetracyclines isolated from litter. The *tet*(A) and *tet*(C) genes were most frequently identified; therefore, we conclude that the main mechanism of resistance in the *E. coli* isolates from our study was mediated by active efflux pumps. The PFGE analysis showed high clonal diversity; however, some clonal strains were isolated from the excreta of both the treated and untreated group, suggesting a local spread of these microorganisms. These results lay the foundation for future controlled studies considering different antimicrobial doses and distance between experimental groups to determine the dissemination and persistence of resistant bacteria and resistance determinants in both animals and the environment.

Author Contributions: Conceptualization, E.P., L.L. and J.C.; methodology, E.P., M.J.N., M.B.V., K.Y., L.T., P.C. and A.M.; software, M.B.V.; validation, E.P. and M.J.N.; formal analysis, E.P., M.J.N. and M.B.V.; investigation, E.P. and M.J.N.; resources, E.P. and J.C.; data curation, E.P., M.J.N. and M.B.V.; writing—original draft preparation, E.P. and M.B.V.; writing—review and editing, K.Y., P.C., A.M. and L.L.; visualization, L.L. and J.C.; supervision, L.L. and J.C.; project administration, A.M. and J.C.; funding acquisition, E.P. and J.C. All authors have read and agreed to the published version of the manuscript.

Funding: This research was funded by the International Atomic Energy Agency (IAEA) under the coordinated research activities number: D52041, Grant Number: Research Contract 22180, and the Agencia Nacional de Investigación y Desarrollo (Conicyt-ANID), Doctorado Nacional, Grant Number 21180769.

Institutional Review Board Statement: The animal study protocol was approved by the Institutional Animal Care and Use Committee (CICUA by its Spanish acronym) of the University of Chile (Protocol code: 18187-VET-UCH-E1, approval date: 5 July 2019).

Informed Consent Statement: Not applicable.

Data Availability Statement: The data presented in this study are available in the article.

Antibiotics 2023, 12, 1652

Acknowledgments: We acknowledge the International Atomic Energy Agency (IAEA) for their support during the research and James Sasanya for technical support.

Conflicts of Interest: The authors declare no conflict of interest.

References

1. Sumpradit, N.; Wongkongkathep, S.; Malathum, K.; Janejai, N.; Paveenkittiporn, W.; Yingyong, T.; Chuxnum, T.; Vijitleela, A.; Boonyarit, P.; Akaleephan, C.; et al. Thailand's national strategic plan on antimicrobial resistance: Progress and challenges. *Bull. World Health Organ.* **2021**, *99*, 661. [CrossRef]
2. Chopra, I.; Roberts, M. Tetracycline Antibiotics: Mode of Action, Applications, Molecular Biology, and Epidemiology of Bacterial Resistance. *Microbiol. Mol. Biol. Rev.* **2001**, *65*, 232–260. [CrossRef] [PubMed]
3. Roberts, M.C. Update on acquired tetracycline resistance genes. *FEMS Microbiol. Lett.* **2005**, *245*, 195–203. [CrossRef] [PubMed]
4. Thaker, M.; Spanogiannopoulos, P.; Wright, G.D. The tetracycline resistome. *Cell. Mol. Life Sci.* **2010**, *67*, 419–431. [CrossRef] [PubMed]
5. Bryskier, A. *Antimicrobial Agents: Antibacterials and Antifungals*; Bryskier, A., Ed.; ASM Press: Washington, DC, USA, 2005.
6. Pruden, A.; Pei, R.; Storteboom, H.; Carlson, K.H. Antibiotic Resistance Genes as Emerging Contaminants: Studies in Northern Colorado. *Environ. Sci. Technol.* **2006**, *40*, 7445–7450. [CrossRef] [PubMed]
7. Zalewska, M.; Błażejewska, A.; Czapko, A.; Popowska, M. Antibiotics and Antibiotic Resistance Genes in Animal Manure—Consequences of Its Application in Agriculture. *Front. Microbiol.* **2021**, *12*, 640. [CrossRef] [PubMed]
8. Xiong, W.; Sun, Y.; Zeng, Z. Antimicrobial use and antimicrobial resistance in food animals. *Environ. Sci. Pollut. Res.* **2018**, *25*, 18377–18384. [CrossRef] [PubMed]
9. Zhao, X.; Wang, J.; Zhu, L.; Ge, W.; Wang, J. Environmental analysis of typical antibiotic-resistant bacteria and ARGs in farmland soil chronically fertilized with chicken manure. *Sci. Total Environ.* **2017**, *593*, 10–17. [CrossRef]
10. Torres, A.G. (Ed.) *Escherichia coli in the Americas*; Springer International Publishing: Cham, Switzerland, 2016. [CrossRef]
11. Agyare, C.; Boamah, V.E.; Zumbi, C.N.; Osei, F.B. Antibiotic Use in Poultry Production and Its Effects on Bacterial Resistance. In *Antimicrobial Resistance: A Global Threat*; Kumar, Y., Ed.; IntechOpen: London, UK, 2019; pp. 1–20. [CrossRef]
12. Moreno, M.; Dominguez, L.; Teshager, T.; Herrero, I.A.; Porrero, M.C. Antibiotic resistance monitoring: The Spanish programme. *Int. J. Antimicrob. Agents* **2000**, *14*, 285–290. [CrossRef]
13. Nguyen, M.C.; Woerther, P.L.; Bouvet, M.; Andremont, A.; Leclercq, R.; Canu, A. *Escherichia coli* as Reservoir for Macrolide Resistance Genes. *Emerg. Infect. Dis.* **2009**, *15*, 1648–1650. [CrossRef]
14. Burow, E.; Simoneit, C.; Tenhagen, B.A.; Käsbohrer, A. Oral antimicrobials increase antimicrobial resistance in porcine *E. coli*—A systematic review. *Prev. Vet. Med.* **2014**, *113*, 364–375. [CrossRef]
15. Chantziaras, I.; Boyen, F.; Callens, B.; Dewulf, J. Correlation between veterinary antimicrobial use and antimicrobial resistance in food-producing animals: A report on seven countries. *J. Antimicrob. Chemother.* **2014**, *69*, 827–834. [CrossRef]
16. Da Costa, P.M.; Bica, A.; Vaz-Pires, P.; Bernardo, F. Effects of Antimicrobial Treatment on Selection of Resistant *Escherichia coli* in Broiler Fecal Flora. *Microb. Drug Resist.* **2008**, *14*, 299–306. [CrossRef] [PubMed]
17. Herrero-Fresno, A.; Zachariasen, C.; Hansen, M.H.; Nielsen, A.; Hendriksen, R.S.; Nielsen, S.S.; Olsen, J.E. Apramycin treatment affects selection and spread of a multidrug-resistant *Escherichia coli* strain able to colonize the human gut in the intestinal microbiota of pigs. *Vet. Res.* **2016**, *47*, 12. [CrossRef] [PubMed]
18. Das, A.; Dhar, P.K.; Dutta, A.; Jalal, M.S.; Ghosh, P.; Das, T.; Barua, H.; Biswas, P.K. Circulation of oxytetracycline- and ciprofloxacin-resistant commensal *Escherichia coli* strains in broiler chickens and farm environments, Bangladesh. *Vet. World* **2020**, *13*, 2395–2400. [CrossRef]
19. Shah, S.H.; Sheikh, I.S.; Kakar, N.; Sumaira; Afzal, S.; Mehmood, K.; Rehman, H.U. In vivo analysis the effect of antibiotic growth promoters (AGPs), Oxytetracycline di-hydrate and Tylosin phosphate on the intestinal microflora in broiler chicken. *Braz. J. Biol.* **2022**, *84*, e258114. [CrossRef] [PubMed]
20. Pokrant, E.; Yévenes, K.; Trincado, L.; Terraza, G.; Galarce, N.; Maddaleno, A.; Martín, B.S.; Lapierre, L.; Cornejo, J. Evaluation of antibiotic dissemination into the environment and untreated animals, by analysis of oxytetracycline in poultry droppings and litter. *Animals* **2021**, *11*, 853. [CrossRef] [PubMed]
21. Martínez, J.L. Effect of antibiotics on bacterial populations: A multi-hierarchical selection process. *F1000Research* **2017**, *6*, 51. [CrossRef] [PubMed]
22. Berge, A.C.B.; Epperson, W.B.; Pritchard, R.H. Assessing the effect of a single dose florfenicol treatment in feedlot cattle on the antimicrobial resistance patterns in faecal *Escherichia coli*. *Vet. Res.* **2005**, *36*, 723–734. [CrossRef] [PubMed]
23. Fairchild, A.S.; Smith, J.L.; Idris, U.; Lu, J.; Sanchez, S.; Purvis, L.B.; Hofacre, C.; Lee, M.D. Effects of Orally Administered Tetracycline on the Intestinal Community Structure of Chickens and on *tet* Determinant Carriage by Commensal Bacteria and *Campylobacter jejuni*. *Appl. Environ. Microbiol.* **2005**, *71*, 5865–5872. [CrossRef]
24. Andersson, D.I.; Hughes, D. Microbiological effects of sublethal levels of Antibiotics. *Nat. Rev. Microbiol.* **2014**, *12*, 465–478. [CrossRef] [PubMed]
25. Gullberg, E.; Albrecht, L.M.; Karlsson, C.; Sandegren, L.; Andersson, D.I. Selection of a Multidrug Resistance Plasmid by Sublethal Levels of Antibiotics and Heavy Metals. *MBio* **2014**, *5*, e01918-14. [CrossRef] [PubMed]

26. Wistrand-Yuen, E.; Knopp, M.; Hjort, K.; Koskiniemi, S.; Berg, O.G.; Andersson, D.I. Evolution of high-level resistance during low-level antibiotic exposure. *Nat. Commun.* **2018**, *9*, 1599. [CrossRef] [PubMed]
27. Roth, N.; Käsbohrer, A.; Mayrhofer, S.; Zitz, U.; Hofacre, C.; Domig, K.J. The application of antibiotics in broiler production and the resulting antibiotic resistance in *Escherichia coli*: A global overview. *Poult. Sci.* **2019**, *98*, 1791–1804. [CrossRef] [PubMed]
28. Rapún Mas, L.; Angulo López, I.; Cecilio Irazola, Á.; Betrán Escartín, A.I. Discrepancy in the genotypic versus phenotypic testing for resistance to rifampicin in Mycobacterium tuberculosis. A case report. *Enf. Infecc. Microbiol. Clin. (Engl. ed.)* **2019**, *37*, 212–213. [CrossRef]
29. Davis, M.A.; Baker, K.N.K.; Orfe, L.H.; Shah, D.H.; Besser, T.E.; Call, D.R. Discovery of a Gene Conferring Multiple-Aminoglycoside Resistance in *Escherichia coli*. *Antimicrob. Agents Chemother.* **2010**, *54*, 2666–2669. [CrossRef] [PubMed]
30. Al-Bahry, S.; Al-Mashani, B.; Al-Ansari, A.; Elshafie, A.; Mahmoud, L. *Escherichia coli* tetracycline efflux determinants in relation to tetracycline residues in chicken. *Asian Pac. J. Trop. Med.* **2013**, *6*, 718–722. [CrossRef]
31. Seifi, S.; Khoshbakht, R. Prevalence of tetracycline resistance determinants in broiler isolated *Escherichia coli* in Iran. *Br. Poult. Sci.* **2016**, *57*, 729–733. [CrossRef]
32. Sreejith, S.; Shajahan, S.; Prathiush, P.R.; Anjana, V.M.; Viswanathan, A.; Chandran, V.; Ajith Kumar, G.S.; Jayachandran, R.; Mathew, J.; Radhakrishnan, E.K. Healthy broilers disseminate antibiotic resistance in response to tetracycline input in feed concentrates. *Microb. Pathog.* **2020**, *149*, 104562. [CrossRef]
33. Skočková, A.; Cupáková, Š.; Karpíšková, R.; Janštová, B. Detection of tetracycline resistance genes in *Escherichia coli* from raw cow's milk. *J. Microbiol. Biotechnol. Food Sci.* **2012**, *1*, 777–784.
34. Jurado-Rabadán, S.; De La Fuente, R.; Ruiz-Santa-Quiteria, J.A.; Orden, J.A.; De Vries, L.E.; Agersø, Y. Detection and linkage to mobile genetic elements of tetracycline resistance gene *tet*(M) in Escherichia coli isolates from pigs. *BMC Vet. Res.* **2014**, *10*, 155. [CrossRef]
35. Titilawo, Y.; Obi, L.; Okoh, A. Occurrence of virulence gene signatures associated with diarrhoeagenic and non-diarrhoeagenic pathovars of *Escherichia coli* isolates from some selected rivers in South-Western Nigeria. *BMC Microbiol.* **2015**, *15*, 204. [CrossRef]
36. Soufi, L.; Sáenz, Y.; Vinué, L.; Abbassi, M.S.; Ruiz, E.; Zarazaga, M.; Ben Hassen, A.; Hammami, S.; Torres, C. *Escherichia coli* of poultry food origin as reservoir of sulphonamide resistance genes and integrons. *Int. J. Food Microbiol.* **2011**, *144*, 497–502. [CrossRef]
37. Koo, H.J.; Woo, G.J. Distribution and transferability of tetracycline resistance determinants in *Escherichia coli* isolated from meat and meat products. *Int. J. Food Microbiol.* **2011**, *145*, 407–413. [CrossRef]
38. European Parliament and of the Council. Directive 2010/63/EU of the of 22 September 2010 on the protection of animals used for scientific purposes. *OJEU* **2010**, *L276*, 33–79.
39. American Veterinary Medical Association. *AVMA Guidelines for the Euthanasia of Animals: 2020 Edition*; AVMA: Schaumburg, IL, USA, 2020.
40. Comisión Nacional de Investigación Científica y Tecnológica. *Manual de Normas de Bioseguridad y Riesgos Asociados: Versión 2018*; CONICYT: Santiago, Chile, 2018.
41. Lupindu, A.M. Isolation and characterization of *Escherichia coli* from animals, humans, and environment. In *Escherichia Coli-Recent Advances on Physiology, Pathogenesis and Biotechnological Applications*; Samie, A., Ed.; InTechOpen: Rijeka, Croatia, 2017; pp. 187–206. [CrossRef]
42. Nkogwe, C.; Raletobana, J.; Stewart-Johnson, A.; Suepaul, S.; Adesiyun, A. Frequency of Detection of *Escherichia coli*, *Salmonella* spp., and *Campylobacter* spp. in the Faeces of Wild Rats (*Rattus* spp.) in Trinidad and Tobago. *Vet. Med. Int.* **2011**, *2011*, 686923. [CrossRef]
43. Toro, M.; Rivera, D.; Jiménez, M.F.; Díaz, L.; Navarrete, P.; Reyes-Jara, A. Isolation and characterization of non-O157 Shiga toxin-producing *Escherichia coli* (STEC) isolated from retail ground beef in Santiago, Chile. *Food Microbiol.* **2018**, *75*, 4847–4852. [CrossRef]
44. Clinical and Laboratory Standards Institute. *M100 Performance Standards for Antimicrobial Susceptibility Testing*, 29th ed.; Clinical and Laboratory Standards Institute: Wayne, PA, USA, 2019.
45. Noll, L.W.; Shridhar, P.B.; Dewsbury, D.M.; Shi, X.; Cernicchiaro, N.; Renter, D.G.; Nagaraja, T.G. A comparison of culture- and PCR- based methods to detect six major non-O157 serogroups of Shiga Toxin-producing *Escherichia coli* in cattle feces. *PLoS ONE* **2015**, *10*, e0135446. [CrossRef]
46. Ng, L.K.; Martin, I.; Alfa, M.; Mulvey, M. Multiplex PCR for the detection of tetracycline resistant genes. *Mol. Cell. Probes* **2001**, *15*, 209–215. [CrossRef]
47. Mamun, M.M.; Parvej, M.S.; Ahamed, S.; Hassan, J.; Nazir, K.H.M.N.H.; Nishikawa, Y.; Rahman, M.T. Prevalence and characterization of shigatoxigenic *Escherichia coli* in broiler birds in Mymensingh, Bangladesh. *J. Vet. Med.* **2016**, *14*, 5–8. [CrossRef]
48. Standard Operating Procedure for PulseNet PFGE of *Escherichia coli* O157:H7, *Escherichia coli* non-O157 (STEC), *Salmonella* serotypes, *Shigella sonnei* and *Shigella flexneri*. Available online: https://www.cdc.gov/pulsenet/pdf/ecoli-shigella-salmonella-pfge-protocol-508c.pdf (accessed on 24 May 2021).

Article

Exploring the Accessory Genome of Multidrug-Resistant *Rhodococcus equi* Clone 2287

Sonsiray Alvarez Narvaez * and Susan Sanchez *

Department of Infectious Diseases, College of Veterinary Medicine, University of Georgia, Athens, GA 30602, USA
* Correspondence: sonsiray.alvarez@uga.edu (S.A.N.); ssanchez@uga.edu (S.S.);
  Tel.: +1-706-410-3953 (S.A.N.); +1-706-542-5568 (S.S.)

Abstract: Decades of antimicrobial overuse to treat respiratory disease in foals have promoted the emergence and spread of zoonotic multidrug-resistant (MDR) *Rhodococcus equi* worldwide. Three main *R. equi* MDR clonal populations—2287, G2106, and G2017—have been identified so far. However, only clones 2287 and G2016 have been isolated from sick animals, with clone 2287 being the main MDR *R. equi* recovered. The genetic mechanisms that make this MDR clone superior to the others at infecting foals are still unknown. Here, we performed a deep genetic characterization of the accessory genomes of 207 *R. equi* isolates, and we describe IME2287, a novel genetic element in the accessory genome of clone 2287, potentially involved in the maintenance and spread of this MDR population over time. IME2287 is a putative self-replicative integrative mobilizable element (IME) carrying a DNA replication and partitioning operon and genes encoding its excision and integration from the *R. equi* genome via a serine recombinase. Additionally, IME2287 encodes a protein containing a Toll/interleukin-1 receptor (TIR) domain that may inhibit TLR-mediated NF-kB signaling in the host and a toxin–antitoxin (TA) system, whose orthologs have been associated with antibiotic resistance/tolerance, virulence, pathogenicity islands, bacterial persistence, and pathogen trafficking. This new set of genes may explain the success of clone 2287 over the other MDR *R. equi* clones.

Keywords: multidrug-resistant; *Rhodococcus equi*; clone 2287; IME2287

Citation: Alvarez Narvaez, S.; Sanchez, S. Exploring the Accessory Genome of Multidrug-Resistant *Rhodococcus equi* Clone 2287. *Antibiotics* **2023**, *12*, 1631. https://doi.org/10.3390/antibiotics12111631

Academic Editor: Carlos M. Franco

Received: 1 November 2023
Revised: 15 November 2023
Accepted: 16 November 2023
Published: 17 November 2023

1. Introduction

Rhodococcus equi is an animal and human pathogen mainly known for being the most common cause of severe pneumonia in foals [1,2]. *R. equi* is a soil saprotroph that becomes pathogenic upon acquiring a virulence plasmid (pVAP) that allows *R. equi* to infect and proliferate in macrophages [3]. So far, three different pVAPs have been reported, presenting different host tropisms: equine pVAPA, porcine pVAPB, and ruminant pVAPN [4–6]. *R. equi* is endemic on many horse-breeding farms [2,7,8]. It mainly produces disease in foals between 1 and 4 months of age, typically in the form of a multifocal purulent pneumonia [9,10]. The pulmonary infection is contracted during the neonatal period via inhalation of contaminated aerosolized dust [11–13]. No commercial vaccines or other consistently efficient immunoprophylactic strategies are available [14,15]. The elevated costs resulting from veterinary care, long-term therapy, and mortality of foals on endemic farms have forced many breeders to adopt practices such as chemoprophylaxis for all foals during the first weeks following birth [16] or to treat foals presenting ultrasonographic lung lesions with antimicrobials before the onset of clinical signs [17,18]. Only a few antibiotics have been seen to be clinically effective in horses to treat *R. equi* [19,20], and the combination of a macrolide and rifampin has been the treatment of choice since the 1980s [21–23]. Before the early 2000s, *R. equi* resistant to macrolides and rifampin were rarely found [24]. Then, the practice of thoracic screening plus subclinical treatment was implemented, resulting in a considerable increase in the use of macrolides and rifampin on endemic farms and the subsequent emergence of MDR *R. equi* isolates resistant to these two antibiotics [25].

Over time, the constant antimicrobial selective pressure exerted by the continuous prophylactic antimicrobial treatment has promoted the spread of multidrug-resistant (MDR) *R. equi* [24–26]. To date, isolates resistant to macrolides and/or rifampin have been reported in 16 US states [24], China [6,27], Ireland [28,29], France [30], and Poland [31]. Our previous work revealed that in the US, *R. equi* isolates resistant to macrolides and rifampin are mainly clustered in three clonal populations: clone 2287, clone G2016, and clone G2017 [26,32,33]. Clone 2287 harbors pRErm46, a 90 kb conjugative plasmid that carries antimicrobial resistance genes (ARGs) for macrolides, lincosamide, streptogramin B (MLS$_B$), tetracycline, and sulfamethoxazole [34] and a chromosomal *rpoB*^{S531F} mutation conferring resistance to rifampin. pRErm46 horizontal gene transfer (HGT) into a different *R. equi* genetic background gave rise to clone G2016, associated with a different rifampin resistance mutation (*rpoB*^{S531Y}). These two clones are the only resistant ones recovered so far, despite an intense search of sick horses, with clone 2287 accounting for most of the clinical cases [26]. Clone G2017, carrying MLS$_B$ resistance plasmids pRErm51 and pMobErm51, has been found exclusively in the environment [26,33]. Although the presence of pVAPA (a conjugative plasmid that carries the virulence factors required to colonize the equine host efficiently) has been reported in members of the three MDR clonal populations, clone 2287 is the most-recovered MDR isolate from sick animals [35,36]. Here, we investigated the accessory genome of 207 *R. equi* (environmental and clinical isolates) to elucidate the genetic mechanisms that make clone 2287 superior at infecting foals.

2. Materials and Methods

2.1. Bioinformatic Analysis

BLAST+ v2.9.0 [37] was used to align the contigs of 207 *R. equi* isolates characterized in our earlier studies [32,33,35] to known *R. equi* genetic elements (*R. equi* chromosome, pVAPA, pRErm46, and *erm*(51)) at >95% identity and >80% coverage. Similarly, BLASTn [37] was used to align the unknown contigs from sample 156 to the other novel contigs and the nucleotide NCBI database. IME2287 annotation was performed using Prokka V1.14.5 [38] and InterProScan V86.0 [39] and subsequently manually inspected and curated based on BLASTx [37] analysis.

2.2. Bacterial Strains and Culture Conditions

Two *R. equi* strains were used in this study: *R. equi* PAM2287 (NCBI BioSample database no. SAMN04880532), which is a macrolide- and rifampin-resistant clinical isolate carrying virulence plasmid pVAPA, macrolide resistance plasmid pRErm46, rifampin mutation *rpoB*^{S531}, and mobilizable element IME2287 [35], and *R. equi* 103$^-$Apra^R, which is a plasmidless derivative strain of reference strain *R. equi* 103 containing the *aac*(3)IV apramycin resistance cassette integrated on the chromosome [40]. *R. equi* isolates were routinely cultured in brain heart infusion medium (BHI; Difco Laboratories-BD, Franklin Lakes, NJ, USA) at 30 °C and 200 rpm unless otherwise stated. Agar media were prepared by adding 1.6% of bacteriological agar (Oxoid, Basingstoke, United Kingdom). Media were supplemented with antibiotics (erythromycin, 8 μg/mL; apramycin, 50 μg/mL; rifampin, 25 μg/mL; Sigma, Saint Louis, MO, USA) when required. All in vitro bacterial work (including the bacterial conjugation and the plasmid loss assays) were carried out in our laboratory at the Athens Veterinary Diagnostic Laboratory from the University of Georgia (Athens, GA, USA).

2.3. Bacterial Conjugation

Conjugation assays were carried out as described previously by Alvarez-Narvaez et al. [41]. Briefly, *R. equi* donor (MDR *R. equi* clone 2287) and recipient (macrolide-susceptible 103$^-$Apra^R) strains were grown overnight in BHI and in the presence of the corresponding antibiotic. Then, donor and recipient bacteria were mixed 1:1 in 5 μL of BHI and spotted in a thick drop onto a BHI plate. After 72 h of incubation at 30 °C, the bacterial mixture was scraped and resuspended in PBS. Serial dilutions were plated onto BHI agar supplemented with apramycin (recipient selection) or apramycin plus erythromycin (transconjugant selection). Conjugation ratios were calculated using the following formula: conjugation ratio = no. of transconjugant cells/no of recipient cells.

2.4. Plasmid Loss Assay

R. equi PAM2287 was inoculated into 10 mL of BHI, or donor horse serum (DHS), and grown at room temperature (RT, ~22 °C) or 37 °C with shaking (200 rpm) for 24 h. After that, bacteria were subcultured in 10 mL of fresh BHI or DHS at an initial optical density at 600 nm wavelength (OD_{600}) of 0.02. OD_{600} was measured using a Thermo Scientific™ Multiskan™ FC Microplate Photometer (Thermo Fisher Scientific, Waltham, MA, USA). Every subculture was considered a passage, and 45 passages were performed in this experiment. At passages 0, 15, 30, and 45, cultures were serially diluted 1:10 in PBS and plated onto BHI plates. In parallel, *R. equi* PAM2287 was inoculated into 5 mL test tubes (n = 10) containing 1 g of autoclaved soil (121 °C for 20 min; Ref. [42]) from the gardens of the University of Georgia School of Veterinary Medicine at a concentration of 5×10^4 CFU/g. Inoculated tubes were mixed well (shaking at ~1800 rpm in a vortex genie 2 bench mixer (Scientific Industries, Bohemia, NY, USA) for 30 min) and incubated statically at RT (n = 5) and 37 °C (n = 5) for up to 45 days. On days 0, 15, 30, and 45, the soil of 1 tube per temperature condition was resuspended in 2 mL phosphate-buffered saline (PBS) (by shaking at ~1800 rpm in a vortex genie 2 bench mixer (Scientific Industries) for 30 min) and quantitatively cultured through serial 10-fold dilutions in BHI. One hundred colonies per medium (BHI, DHS, and soil) and temperature (RT and 37 °C) were randomly selected and re-plated onto BHI, BHI-Ery, and BHI-Rmp plates. The presence of pRErm46 IME2287 and pVAPA was tested using PCR.

2.5. PCR

Transconjugants were confirmed using PCR for the macrolide-resistant gene *erm*(46) and apramycin-resistant gene *aac*(3)IV. pVAPA, pRErm46, and IME2287 mobilization were also tested using PCR. Table 1 contains the list of oligos used in this study. PCRs were carried out using a T100 thermocycler (Bio-Rad, Hercules, CA, USA) and GoTaq Flexi DNA polymerase (Promega, Madison, WI, USA) under the conditions specified in Alvarez-Narvaez et al. [41]. Briefly, we used an initial denaturation step for 5 min at 95 °C, followed by 30 amplification cycles (involving 30 s at 95 °C of denaturation, 30 s of oligonucleotide hybridization at the appropriate melting temperature, and 2 min of elongation at 72 °C) and a final elongation of 10 min at 72 °C.

Table 1. List of oligos used in this study.

Name	Sequence 5′–3′	Amplicon Size (bp)	Purpose	Source
IME2287_repAB_F IME2287_repAB_R	GGAGCACTACTACTGGACG GTTGACTGTGAACTCGGTGT	1746	IME2287 backbone marker	[43]
IME2287_sigma70_F IME2287_sigma70_R	CTTGCGAGTAGGACATGAAG GACCTTCGTCAGGGAGTAAG	1752	IME2287 backbone marker	[43]
IME2287_tnpR- helix_F IME2287_tnpR- helix_R	TCTACGTCGACAAGAAGTCC GTATGTGAACCGACCTTGTG	1745	IME2287 backbone marker	[43]
ChoE_F ChoE_R	AGTTGTCGATTCCCATCGTC AAGCGCAACTACTTCGAGGAG	672	*choE* gene, chromosomal marker	[5]
TraA-F1 TraA-R1	AGAGTTCATGCGTGACAACG GTCCACAGGTCACCGTTCTT	959	*traA* gene, pVAPA backbone marker	[5]
erm(46)F *erm*(46)R	TATGGAGTCGATCTGCAACG GAGATCGGACGAGTCTGACA	1098	macrolide resistance gene *erm*(46)	[34]
pRErm46_traG_F pRErm46_traG_R	ACCGTCGTAGCAGTAGCC CCTCAGCGAGTGTCTTCTC	1533	*traG* gene, pRErm46 backbone marker	[35]
ApraF ApraR	GGCCACTTGGACTGATCGAG GCATGACCGACTGGACCTTC	937	apramycin cassette *aac*(3)IV inserted in chromosome	[40]

3. Results

3.1. The Search for Genetic Elements in MDR R. equi Clone 2287 Selected and Maintained over Time

PacBio and Illumina whole-genome assemblies from 62 *R. equi* clinical isolates (40 MDR clone 2287 and 22 susceptible) collected between 2002 and 2017 in different US states and characterized in our earlier study [32] were used to identify potential new genetic elements associated exclusively with MDR *R. equi* clone 2287 (Table S1). Four thousand and one contigs from the 40 members of clonal population 2287 were first aligned to known *R. equi* genetic elements (chromosome, virulence plasmid pVAPA, and MDR plasmid pRErm46). In total, 2669, 85, and 168 contigs matched with the *R. equi* chromosome (accession number NC_014659), pVAPA1037 (accession number NC_011151), and pRErm46 (accession number KY494640), respectively, with >95% similarity and >80% coverage, and 1079 contigs were identified as unknown.

Strain 156 showed fewer unknown contigs, only two: contig_221 and contig_249. These two contigs were aligned to all the unknown contigs from the other resistant strains and to the genomes of the 18 susceptible isolates. Only contig_249 was common in all resistant strains and was not present in the assemblies of any susceptible isolates. Contig_249 was 22,578 bp long and comprised a novel 14,317 bp genetic element (Figure 1) that we noted, IME2287 (integrative mobilizable element from clone 2287), and a subsequent partial IME2287 duplication of 8260 bp in length. Further analysis of the contigs carrying this novel piece of DNA showed that in most of the cases, IME2287 was the only genetic element that appeared in the contigs, repeated (and sometimes inverted) in tandem (Figure 2). Further analysis showed that only one isolate (sample 169) presented with IME2287 randomly inserted in the bacterial chromosome (Figure 2).

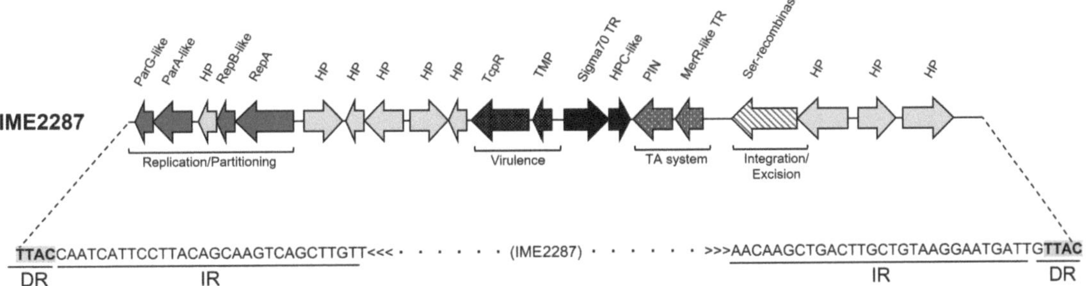

Figure 1. Genetic structure of IME2287. IME2287 carries its own replication and partitioning machinery (dark grey) comprised of ParG-like, ParA-like, RepB-like, and a RepA proteins in the same operon and excision/integration capacities via a serine recombinase (white with grey lines). Other exciting genes in IME2287 include the following: a putative virulence operon (black with white polka dots) composed of a Toll/interleukin-1 receptor protein (TcpR) and a transmembrane protein (TMP) and a putative toxin–antitoxin system containing a PIN domain-containing protein (toxin) and a MerR-like DNA binding protein (antitoxin). See text and Table 2 for other IME2287 components. DRs, direct repeats, (shaded) at the junction with genomic DNA and adjacent inverted repeats (IRs) comprise the TTAC sequence targeted by IME2287. This figure was produced using SnapGene Viewer v7.0.3 (www.snapgene.com) and manually edited with Microsoft PowerPoint for Mac v16.78.3.

Table 2. Annotation of IME2287 from *R. equi* sample 156.

IME2287 Locus Tag	Location (nt Position)		Size (bp)	Product (BlastX)	Coverage (AA Level)	Identity (AA Level)	Species
direct_repeat	1	4	3	-	-	-	-
inverted_repeat	5	34	29	-	-	-	-
IME2287_0010	46	303	257	ParG	>90%	>80%	Conserved in *Actinobacteria*
IME2287_0020	306	1067	761	ParA	>50%	>50%	Conserved in *Actinobacteria*
IME2287_0030	1165	1488	323	Conserved hypothetical protein	>90%	>80%	Conserved in *Rhodococcus* spp.
IME2287_0040	1478	1765	287	RepB	>70%	>80%	Conserved in *Rhodococcus* spp.
IME2287_0050	1765	2651	886	RepA	>70%	>70%	Conserved in *Rhodococcus* spp.
IME2287_0060	2873	3520	647	Conserved hypothetical protein	>70%	>50%	Conserved in *Actinobacteria*
IME2287_0070	3531	3854	323	Conserved hypothetical protein	>50%	>50%	Conserved in *Rhodococcus* spp.
IME2287_0080	3814	4413	599	Conserved hypothetical protein	>50%	>50%	Conserved in *Rhodococcus* spp.
IME2287_0090	4528	5208	680	Hypothetical protein	no match	no match	No match
IME2287_0100	5167	5409	242	Conserved hypothetical protein	>70%	>50%	Conserved in *Rhodococcus* spp.
IME2287_0110	5503	6471	968	TIR domain-containing protein	>90%	>40%	Conserved bacterial protein
IME2287_0120	6543	6731	188	Transmembrane protein	no match	no match	No match
IME2287_0130	7010	7903	893	Sigma70	>70%	>35%	Conserved in *Actinobacteria*
IME2287_0140	7900	8241	341	TPR-like domian-containing protein	no match	no match	No match
IME2287_0150	8392	9024	632	PIN domain-containing protein	>85%	>40%	Conserved in *Actinobacteria*
IME2287_0160	9073	9543	470	MerR-like DNA binding protein	>60%	60%	Conserved in *Actinobacteria*
IME2287_0170	10,014	11,267	1253	Serine recombinase	>40%	>70%	Conserved in *Rhodococcus* spp.
IME2287_0180	11,264	12,223	959	Conserved hypothetical protein	>85%	>25%	Conserved in *Actinobacteria*
IME2287_0190	12,470	12,886	416	Conserved hypothetical protein	>80%	>50%	Conserved in *Actinobacteria*
IME2287_0200	13,038	13,931	893	Conserved hypothetical protein	>90%	>40%	Conserved in *Actinobacteria*
inverted_repeat	14,284	14,313	29	-	-	-	-
direct_repeat	14,314	14,317	3	-	-	-	-

See Figure 1 for the genetic structure of the element.

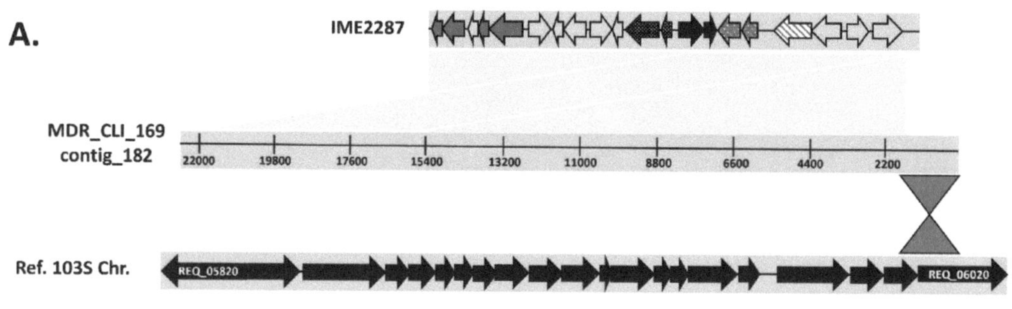

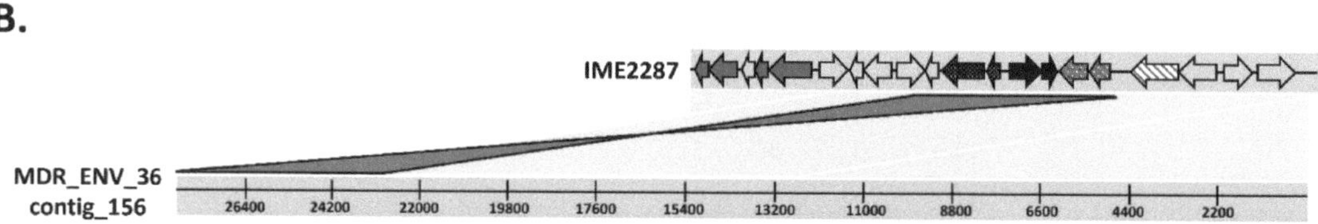

Figure 2. (**A**) IME2287 insertion in *R. equi* chromosome. Above, IME2287 with ORFs are represented by arrows. In the middle, contig 182 from MDR clinical isolate 169 (GenBank accession no. SAMN13392202). Below, a portion of the *Rhodococcus equi* 103S chromosome (ORFs are represented by black arrows; GenBank accession no. FN563149.1). (**B**) Example of IME2287 duplications. Above, IME2287 with ORFs represented by arrows. Below, the environmental MDR isolate 36 (GenBank accession no. WVCU00000000). Regions with significant similarity (nBlast, Score matrix Blosum62) are connected by colored lines (light grey, sequences in direct orientation; dark grey, sequences in reverse orientation). All connecting lines between sequences represent a percentage of identity > 97%. This figure was produced using the Artemis Comparison Tool (ACT) [44] and manually edited with Microsoft PowerPoint.

3.2. Molecular Characterization of IME2287, a Putative Mobilizable Element Associated with Macrolide Resistance in MDR R. equi

The in silico annotation of IME2287 identified 20 ORFs (Figure 1), of which 11 were predicted to have a biological function or a functional domain and 9 were classified as hypothetical proteins. Homology searches showed that the novel set of genes carried by IME2287 share high homology with genes previously found in the genus *Rhodococcus* and other *Actinobacteria*. Table 2 lists the 20 ORFs with the in silico functional predictions and corresponding homologies. Of interest, four genes are clustered together and code for proteins involved in DNA replication and partitioning. In this order, ORF IME2287_0010 was predicted to be a ParG-like DNA binding protein with a ribbon–helix–helix domain in its C-terminal region, ORF IME2287_0020 was identified as an AAA-ATPase ParA partitioning protein, ORF IME2287_0030 was classified as a RepB-like RNA polymerase with a DNA binding domain, and ORF IME2287_0050 was found to be a RepA plasmid replicase. These genes indicate that IME2287 is most likely a self-replicative genetic element.

ORF IME2287_0170 encodes a serine recombinase that would allow IME2287 to excise and reintegrate into the *R. equi* genome. A maximum identity value of 87% with its closest homolog unclassified ISBli29 from *Brevibacterium linens* (*Actinobacteria*) indicates that the serine recombinase is a novel orphan transposase, which we named ISRe2287. The presence of this gene and the replication and partitioning machinery suggested that IME2287 could be an integrative conjugative element (ICE). However, no conjugation genes were predicted in IME2287, and we found the inverted repeat (IR) sequence 5′-CAATCATTCCTTACAGCAAGTCAGCTTGTT-3′ invariably followed by the directly repeated (DR) tetranucleotide TTAC, supporting that IME2287 could be a transposable element. Additionally, IME2287 carries a gene that encodes a protein containing a Toll/interleukin-1 receptor (TIR) domain (TcpR, IME2287_0140) that acts by suppressing the innate immune system in other bacteria species [45–48] and a putative toxin–antitoxin system composed of a PIN domain-containing protein (toxin, IME2287_0150) and a MerR-like DNA binding protein (antitoxin, IME2287_0160).

3.3. IME2287 Moved from MDR R. equi Clone 2287 to other R. equi Genetic Backgrounds Associated with Macrolide Resistance

To explore if IME2287 was a mobile genetic element, the whole-genome assemblies of 10 additional MDR *R. equi* clinical isolates harboring pRErm46 but with genetic backgrounds different from clone 2287 were analyzed. Two isolates were members of the new MDR *R. equi* clonal population G2016 (Table 3), and eight were classified as singletons (an MDR *R. equi* isolate that shows a unique genetic background and therefore does not belong to any of the known *R. equi* clonal populations) in previous phylogenetic analysis [26,32]. We observed the presence of IME2287 in one of the *R. equi* G2016 clones and two of the singletons, indicating that this element moved from *R. equi* clone 2287 into other *R. equi* isolates through horizontal gene transfer (HGT).

Furthermore, we looked for evidence of IME2287 in the PacBio whole-genome assemblies from 135 *R. equi* environmental isolates collected from 100 farms in central Kentucky for a previous investigation [26,33]. The 45 environmental MDR *R. equi* that previous phylogenetic analysis [26,33] classified as part of clonal population 2287 were shown to harbor both pRErm46 and IME2287 in their genome, while none of the 38 susceptible isolates did (Table 3). *R. equi* clone G2016 (sample 52), the only MDR environmental isolate of this clone, carried pRErm46 but was not shown to have IME2287. Similarly, only 7 of the 39 MDR *R. equi* environmental isolates phylogenetically classified as members of the clonal population G2017 were shown to carry pRErm46/tnRErm46. Of these seven, three also carried IME2287, and four did not. Regarding the environmental singleton-resistant isolates, nine out of twelve were seen to carry pRErm46 or tnRErm46, and from those nine, seven isolates also carried IME2287 (Table 3).

Table 3. Association between genetic background, susceptibility, and the presence of pRErm46/tnErm46 and IME2287.

Source	Genetic Background	pRErm46/tnRErm46 and IME2287	pRErm46/ tnRErm46 Only	No pRErm46/tnRErm46 or IME2287	Total
Clinical	Clone 2287	40	0	0	40
	Clone G2016	1	1	0	2
	Clone G2017	0	0	0	0
	Singletons	2	6	0	8
	Susceptible	0	0	22	22
Environmental	Clone 2287	45	0	0	45
	Clone G2016	0	1	0	1
	Clone G2017	3	4	32	39
	Singletons	7	2	3	12
	Susceptible	0	0	38	38
	Total	98	14	95	207

Numbers correspond to the number of isolates found in each clonal population and carrying each genetic element combination.

As previously seen in the clinical isolates, IME2287 always appears in an individual contig that exclusively contains copies of these elements (Figure 2). Interestingly, we did not find any clinical or environmental isolates that presented IME2287 without any genetic elements (pRErm46 or tnRErm46) associated with the *erm*(46) resistance gene, suggesting that IME2287 could be hijacking pRErm46 conjugation machinery to move.

3.4. IME2287 Mobilization Is Independent of Macrolide Resistance Plasmid pRErm46

We previously reported that macrolide resistance plasmid pRErm46 was mobilizable at a high frequency through horizontal gene transfer via bacterial conjugation [35,41]. The findings above lead to the following question: is IME2287 mobilizable simultaneously with pRErm46? To answer this question, conjugation assays were carried out using MDR *R. equi* clone 2287 as a donor and susceptible (pRERrm46- negative) avirulent *R. equi* 103⁻ with an apramycin resistance *aac*(3)IV cassette (103⁻Apra^R) [40] as the recipient in a ratio 1:1. Transfer of pRErm46 to 103⁻Apra^R was observed at a frequency of $2.06 \pm 0.70 \times 10^{-3}$ transconjugants/recipient cells. In total, 90 CFUs were PCR-tested for virulence plasmid pVAPA and novel element IME2287. Five out of the ninety colonies tested were shown to have acquired pVAPA together with pRErm46, but no colonies were carrying IME2287, indicating that its mobilization is most likely independent from pRErm46 conjugal transfer.

Next, we investigated if IME2287 is lost over time in the same fashion pRErm46 is. We monitored changes in the antimicrobial resistance phenotype and genotype of MDR *R. equi* clone 2287 culture-passed daily in BHI and DHS at RT (~22 °C) or 37 °C for 45 passages. In parallel, we looked for changes in the antimicrobial resistance phenotype and genotype of MDR *R. equi* clone 2287 incubated in autoclaved soil at RT or 37 °C for 45 passages (Table 4). We observed that the percentage of macrolide- and rifampin-resistant colonies of clone 2287 was maintained for the entire experiment duration (45 passages) in all conditions, except for bacteria incubated in DHS at 37 °C, where macrolide resistance declined progressively over time. This indicates that, regardless of the temperature in BHI and soil, macrolide and rifampin resistance expression do not represent a fitness cost for *R. equi* clone 2287.

We PCR-tested the 46 CFUs that were shown to have lost macrolide resistance for the presence of pRErm46 and IME2287 (Table 5). As expected, all susceptible colonies had lost the macrolide resistance plasmid, but only half of those ($n = 23$) also lost IME2287, indicating again that the mobilization of these two elements is independent. Interestingly, looking at IME2287 loss over time, we realized that up to passage 30, the colonies that lost pRErm46 still maintained IME2287. In passage 45, 23 out of 36 (~64%) appeared to have lost both elements. This indicates that the loss of IME2287 is most likely subsequent to the loss of the pRErm46 plasmid.

Table 4. Maintenance of resistant phenotype to macrolides or rifampin in soil and during subculturing in DHS and BHI. The table exclusively presents conditions with a decrease in antimicrobial resistance phenotype.

| | CFU with Macrolide-Resistant Phenotype | | | |
| | DHS | | Soil | |
Passage	RT	37 °C	RT	37 °C
0	100%	100%	100%	100%
15	100%	99%	100%	100%
30	99%	97%	100%	100%
45	100%	64%	100%	95%

Numbers reflect the proportion of colony forming units (CFUs) that presented a macrolide-resistant phenotype after a particular number of passages under different temperature conditions (room temperature (RT) or 37 °C) and in different media (commercial donor horse serum (DHS) or autoclaved soil (Soil)). In bold are the conditions where we have seen a decrease in the resistance phenotype.

Table 5. pRErm46 and IME2287 stability in *R. equi* 2287 genetic background.

Media	Temp.	Passage	Loss of pRErm46 Only	Loss of pRErm46 and IME2287
Soil	37 °C	45	5	0
DHS	37 °C	30	1	0
	RT	15	1	0
		30	3	0
		45	13	23
	Total		23	23

In the plasmid loss experiment, 46 colonies lost the macrolide resistance phenotype. The numbers in the table reflect the number of macrolide-susceptible colonies that lost pRErm46 only or pRErm46 and IME2287 under different temperature conditions (room temperature (RT) or 37 °C) and in different media (brain heart infusion (BHI), commercial donor horse serum (DHS) or autoclaved soil (Soil)). The passage in which each colony was retrieved is also indicated in the table. PCR was used to determine the presence/absence of pRErm46 and IME2287 (see materials and methods).

4. Discussion

The emergence and spread of MDR bacterial clones represent a significant threat to animal and human health. The success of these dominant bacterial clades relies on the acquisition of virulence and antimicrobial resistance (AMR) genes associated with mobile genetic elements [49]. Although we currently know more about resistance mechanisms and their mobilization dynamics, there is still much to learn to help us prevent their selection and spread. Here, we explored the accessory genome of MDR *R. equi* clone 2287, looking for genetic elements that could explain its superiority over other MDR R. equi clones. Considering that bacterial genomes are subject to genetic drift that deletes superfluous sequences [50], we hypothesized that any genetic element that would provide an advantage in the adaptation of MDR clone 2287 would be exclusively present in all clone members and maintained over time. Based on this, we first performed a preliminary screening of clone 2287's accessory genome, trying to identify genetic elements present in all the genomes of the clonal population and not in susceptible strains. We only identified one genomic sequence under this threshold, and its genomic annotation indicated that we were dealing with a putative mobile genetic element. Consequently, we named the novel element IME2287.

The oldest member of clone 2287, in which we detected the presence of IME2287, is an isolate collected in Florida in 2002. Since then, IME2287 has been found in all the *R. equi* phylogenetically classified as clone 2287 that were analyzed ($n = 85$, over a period of 15 years in five different US states); the most recent isolate was found in Kentucky in 2017. It is not surprising to find the *R. equi* genome harboring new genetic elements because this bacteria species is a unique example of plasmid-driven full virulence, antimicrobial resistance, and host adaptation [10]. However, *R. equi* tends to lose its virulence host-adapted plasmids (pVAPs) when living as a saprotroph in the soil as they introduce a fitness cost due to the expression of the virulence genes [5,51]. The fact that IME2287 has been maintained for more than 15 years in this clonal population is intriguing and suggests that instead, IME2287 may not impact *R. equi* fitness, or if it does, the role that IME2287 plays for the bacterium overcomes that fitness cost.

We performed a deep genetic characterization of IME2287, looking for genes that could explain its persistence over time. We found that IME2287 carries a potential new virulence factor. The *tcpR* gene encodes a putative TIR domain protein that we designated as TcpR (Figure 1, Table 2). In eukaryotes, TIR domains are part of Toll-like receptors (TLRs), a family of proteins that recognize pathogens and initiate the innate immune response [52]. TIR domain proteins are also found in plants, where they mediate disease resistance, and in bacteria, associated with virulence and bacterial metabolic regulation [47]. TIR domain proteins have been reported to block TLR signaling, inhibiting innate immune responses in important animal and human pathogens such as *Brucella* spp. [53], uropathogenic *Escherichia coli* (UPEC) [54], nosocomial strains of *Enterococcus faecium* [55], and *Staphylococcus aureus* [48]. TIR domain proteins are frequently carried on mobile genetic elements and, in some cases, are associated with ARGs [48]. The innate immune system is the principal response against intracellular *R. equi* [56]. Therefore, an extra virulence factor, such as TcpR, would help *R. equi* circumvent innate host immunity during infection, which is extremely important for the bacteria. It would explain why IME2287 is still within all MDR *R. equi* clinical isolates 15 years after being first reported. Furthermore, this extra virulence factor could be the key to clone 2287's superiority at infecting animals over the other genomic backgrounds.

Another interesting operon also found in IME2287 consists of two genes that encode an MerR transcriptional regulator and a PIN domain ribonuclease (MerR-PIN). In silico work has classified MerR-PIN operons as putative Type 2 toxin–antitoxin (TA) systems [57], in which the PIN ribonuclease is toxic to the cell and stable. At the same time, the MerR antitoxin is unstable and requires continuous transcription to inhibit the produced toxin [58]. MerR-PIN and other Type II TA systems are often found in plasmids or, as appreciated in IME2287, inserted in chromosomes in association with mobile genetic elements [57,59]. TA systems have been associated with antibiotic resistance, virulence factors, and pathogenicity islands in pathogenic bacteria [58]. One of the first functions assigned to TA systems was plasmid stabilization through a process known as "post-segregational killing", a suicide mechanism for those cells that do not carry the TA plasmid after cell division [57]. More recently, Type II TA systems have been described as involved in bacterial permanence under stress conditions such as nutrient starvation or antibiotic exposure [60]. Under stress conditions, the antitoxins are selectively degraded, leaving toxins to promote growth arrest and dormancy. Once the stressor is removed, dormant persistent cells revert to the actively growing state and repopulate the original population [60]. Based on the current information regarding Type II TA systems, we hypothesized that the MerR-PIN system could have a double function for MDR *R. equi*, (i) avoiding the loss of IME2287 during cell division for over 15 years and (ii) mediating the permanence of MDR *R. equi* in the soil even in the absence of antimicrobial selective pressure. Future in vitro experimentation to determine MerR-PIN functionality is guaranteed.

IME2287 has been observed in several different genetic backgrounds, indicating that this genetic element can move through horizontal gene transfer. The IME2287 genetic characterization revealed that IME2287 carries a replication and partitioning operon (Figure 1, Table 2) and an integration/excision capability via a serine recombinase. This is consistent with the core modules of integrative conjugative elements (ICEs) previously identified in Actinomycetes [61]. However, IME2287 lacks the conjugation module, often consisting of a relaxase, a coupling protein, and a mating pair formation system [61]. We also identified IR and DR flanking IME2287, suggesting that this genetic element is likely a mobilizable transposon or a non-canonical IME [62]. The distinction between mobilizable transposons and IMEs is unclear and is considered irrelevant for some [62]. Various non-canonical IMEs (that do not encode any relaxase) have been identified in the Firmicutes and Proteobacteria phyla, sometimes with several copies integrated in tandem, like in the case of IME2287 [63]. Regardless of the nature of IME2287 (transposon, or IME), the in silico data obtained in this project reflect that at some point, IME2287 jumped from clone 2287, its native genomic background, to other *R. equi*. IME2287 was found in 50% (13/27) of pRErm46 containing non-clone 2287 *R. equi*, suggesting that its mobilization could be tied to the antimicrobial resistance plasmid mobilization. Hence, we hypothesized that IME2287 (lacking its conjugal machinery) could hijack the pRErm46 conjugation apparatus to spread. We explored this possibility using mating experiments, and we found that none of the transconjugants that acquired pRErm46 also received IME2287. This indicates that IME2287 mobilization is likely independent of pRErm46. However, we cannot discard that IME2287 could indeed be using pRErm46 mobilization machinery but with a very low mobilization rate, or maybe IME2287 transfer is not triggered or selected by exposure to macrolides.

Given the lack of success of our IME2287 transfer attempt, we decided to test if the loss of this novel genetic element was tied to the loss of pRErm46. Our experiments subculturing clone 2287 in the absence of antimicrobials and under different media and temperature conditions showed that the loss of IME2287 is independent of pRErm46 and that the recently discovered genetic element would get lost, most probably after the loss of the macrolide-resistant plasmid. Additionally, this last experiment allowed us to study the persistence of clone 2287 resistant phenotypes without selective pressure. We observed that clone 2287 maintained a rifampin resistance phenotype during the entire duration of the experiment, independent of the incubation conditions. Willingham-Lane and collaborators studied the stability of three lab-generated rifampin-resistant *rpoB* mutations in the *R. equi* 103S genetic background by passing bacteria in BHI at 37 °C [64]. They reported that two of the three *rpoB* mutations reverted to the wild-type form after 20 passages (>90% revert) and that these mutations resulted in decreased growth in vitro and in soil. This, together with the fact that the mutation present in MDR clone 2287 ($rpoB^{S531F}$) has been impossible to recreate in the laboratory (Willingham-Lane personal communication), suggests that $rpoB^{S531F}$ has been naturally selected over other potential *rpoB* mutations because it does not impact *R. equi* fitness either in vitro (soil and BHI at RT) or in vivo (DHS and 37 °C). In our experiment, clone 2287 maintained its macrolide resistance phenotype when incubated at room temperature independently of the medium. We only found a significant reduction (of ~35%) in macrolide resistance when clone 2287 was subcultured in horse serum at 37 °C. Willingham-Lane et al. also investigated the stability of macrolide resistance in *R. equi* 103S subcultured in BHI at 37 °C and the performance of the macrolide-resistant strain in vitro and soil [64]. Under their conditions, macrolide resistance was lost over time (decrease ~50% after 40 passages), and the growth of the macrolide-resistant strains was affected in vitro and in soil. We did not observe a decrease in the macrolide-resistant phenotype in BHI over time, but the duration of our experiment was shorter. Additionally, our previous work did not detect a fitness cost associated with the presence of pRErm46 in *R. equi* 103S [35]. We decided to subculture bacteria in DHS and soil at different temperatures to recreate the closest conditions *R. equi* will face inside the host (DHS at 37 °C) and in the environment (soil at RT). Taking this into consideration, our results indicate that in the environment (i.e., bacteria in soil), the presence of pRErm46 does not have an impact on the growth of clone 2287, which will be able to keep the macrolide-resistant plasmid for long periods in the absence of antimicrobial selective pressure. However, inside the host, when antimicrobials are not administered (DHS at 37 °C), the expression of the virulence genes is essential for *R. equi* survival at the expense of reduced growth [3], which will prompt the bacteria to lose the non-essential MGEs such as pRErm46 and IME2287.

5. Conclusions

The work presented herein studied a clonal competitive event in the zoonotic human pathogen *R. equi*. We characterized an accessory genetic element, IME2287, in the genome of MDR *R. equi* clone 2287 that could explain its superiority over other MDR *R. equi* clones. More specifically, this study elucidated at least two molecular traits in IME2287, a toxin–antitoxin (TA) system and a host NF-kB signaling inhibitor, that successful MDR *R. equi* clones may acquire to their advantage when infecting and colonizing the host. Unfortunately, the factors that trigger IME2287 mobilization and the genetic mechanisms required for this task are still unknown.

Supplementary Materials: The following supporting information can be downloaded at: https://www.mdpi.com/article/10.3390/antibiotics12111631/s1, Table S1: Genomes of the 72 *R. equi* clinical isolates used in this study.

Author Contributions: Conceptualization, S.A.N. and S.S.; methodology, S.A.N.; software, S.A.N.; formal analysis, S.A.N.; investigation, S.A.N. and S.S.; resources, S.S.; data curation, S.A.N.; writing—original draft preparation, S.A.N.; writing—review and editing, S.A.N. and S.S.; visualization, S.A.N.; supervision, S.S.; project administration, S.A.N. and S.S.; funding acquisition, S.S. All authors have read and agreed to the published version of the manuscript.

Funding: This research was funded by the US Department of Agriculture (USDA), grant number 1U18FD006157-04.

Institutional Review Board Statement: Not applicable.

Informed Consent Statement: Not applicable.

Data Availability Statement: All genomic data used in this study have been previously used in other studies and are available in the GenBank database (https://www.ncbi.nlm.nih.gov/genbank/).

Acknowledgments: We thank L. Berghaus, K. Hart, and the Large Animal Medicine Department from the University of Georgia for facilitating the *R. equi* strains PAM2287 and 103⁻Apra^R.

Conflicts of Interest: The authors declare no conflict of interest.

References

1. Prescott, J.F. *Rhodococcus equi*: An animal and human pathogen. *Clin. Microbiol. Rev.* **1991**, *4*, 20–34. [CrossRef] [PubMed]
2. Bordin, A.I.; Huber, L.; Sanz, M.G.; Cohen, N.D. *Rhodococcus equi* foal pneumonia: Update on epidemiology, immunity, treatment and prevention. *Equine Vet. J.* **2022**, *54*, 481–494. [CrossRef] [PubMed]
3. Giguere, S.; Hondalus, M.K.; Yager, J.A.; Darrah, P.; Mosser, D.M.; Prescott, J.F. Role of the 85-kilobase plasmid and plasmid-encoded virulence-associated protein A in intracellular survival and virulence of *Rhodococcus equi*. *Infect. Immun.* **1999**, *67*, 3548–3557. [CrossRef] [PubMed]
4. Valero-Rello, A.; Hapeshi, A.; Anastasi, E.; Alvarez, S.; Scortti, M.; Meijer, W.G.; MacArthur, I.; Vazquez-Boland, J.A. An Invertron-Like Linear Plasmid Mediates Intracellular Survival and Virulence in Bovine Isolates of *Rhodococcus equi*. *Infect. Immun.* **2015**, *83*, 2725–2737. [CrossRef] [PubMed]
5. Ocampo-Sosa, A.A.; Lewis, D.A.; Navas, J.; Quigley, F.; Callejo, R.; Scortti, M.; Leadon, D.P.; Fogarty, U.; Vazquez-Boland, J.A. Molecular epidemiology of *Rhodococcus equi* based on traA, vapA, and vapB virulence plasmid markers. *J. Infect. Dis.* **2007**, *196*, 763–769. [CrossRef]
6. Song, Y.; Xu, X.; Huang, Z.; Xiao, Y.; Yu, K.; Jiang, M.; Yin, S.; Zheng, M.; Meng, H.; Han, Y.; et al. Genomic Characteristics Revealed Plasmid-Mediated Pathogenicity and Ubiquitous Rifamycin Resistance of *Rhodococcus equi*. *Front. Cell. Infect. Microbiol.* **2022**, *12*, 807610.
7. Chaffin, M.K.; Cohen, N.D.; Martens, R.J.; Edwards, R.F.; Nevill, M. Foal-related risk factors associated with development of *Rhodococcus equi* pneumonia on farms with endemic infection. *J. Am. Vet. Med. Assoc.* **2003**, *223*, 1791–1799. [CrossRef]
8. Chaffin, M.K.; Cohen, N.D.; Martens, R.J.; Edwards, R.F.; Nevill, M.; Smith, R., 3rd. Hematologic and immunophenotypic factors associated with development of *Rhodococcus equi* pneumonia of foals at equine breeding farms with endemic infection. *Vet. Immunol. Immunopathol.* **2004**, *100*, 33–48. [CrossRef]
9. Giguere, S.; Prescott, J.F. Clinical manifestations, diagnosis, treatment, and prevention of *Rhodococcus equi* infections in foals. *Vet. Microbiol.* **1997**, *56*, 313–334. [CrossRef]
10. Vazquez-Boland, J.A.; Giguere, S.; Hapeshi, A.; MacArthur, I.; Anastasi, E.; Valero-Rello, A. *Rhodococcus equi*: The many facets of a pathogenic actinomycete. *Vet. Microbiol.* **2013**, *167*, 9–33. [CrossRef]
11. Muscatello, G.; Gilkerson, J.R.; Browning, G.F. Detection of virulent *Rhodococcus equi* in exhaled air samples from naturally infected foals. *J. Clin. Microbiol.* **2009**, *47*, 734–737. [CrossRef] [PubMed]
12. Muscatello, G.; Leadon, D.P.; Klayt, M.; Ocampo-Sosa, A.; Lewis, D.A.; Fogarty, U.; Buckley, T.; Gilkerson, J.R.; Meijer, W.G.; Vazquez-Boland, J.A. *Rhodococcus equi* infection in foals: The science of 'rattles'. *Equine Vet. J.* **2007**, *39*, 470–478. [CrossRef] [PubMed]

13. Takai, S.; Yoda, A.; Sasaki, Y.; Kakuda, T.; Suzuki, Y.; Oikawa, M. Birth month associated with tracheal colonization of *Rhodococcus equi* in newborn foals on horse-breeding farms with sporadic rhodococcosis in Japan. *Vet. Microbiol.* **2022**, *267*, 109373. [CrossRef] [PubMed]

14. Giles, C.; Vanniasinkam, T.; Ndi, S.; Barton, M.D. *Rhodococcus equi* (*Prescottella equi*) vaccines; the future of vaccine development. *Equine Vet. J.* **2015**, *47*, 510–518. [CrossRef]

15. Kahn, S.K.; Cohen, N.D.; Bordin, A.I.; Coleman, M.C.; Heird, J.C.; Welsh, T.H., Jr. Transfusion of hyperimmune plasma for protecting foals against *Rhodococcus equi* pneumonia. *Equine Vet. J.* **2022**, *55*, 376–388. [CrossRef]

16. Perkins, G.A.; Yeager, A.; Erb, H.N.; Nydam, D.V.; Divers, T.J.; Bowman, J.L. Survival of foals with experimentally induced *Rhodococcus equi* infection given either hyperimmune plasma containing *R. equi* antibody or normal equine plasma. *Vet. Ther.* **2002**, *3*, 334–346.

17. Higuchi, T.; Arakawa, T.; Hashikura, S.; Inui, T.; Senba, H.; Takai, S. Effect of prophylactic administration of hyperimmune plasma to prevent *Rhodococcus equi* infection on foals from endemically affected farms. *Zentralbl Vet. B* **1999**, *46*, 641–648. [CrossRef]

18. Cohen, N. Control and prevention of *Rhodococcus equi* foal pneumonia. *AAEP Proc.* **2010**, *56*, 129–131.

19. Giguere, S. Treatment of Infections Caused by *Rhodococcus equi*. *Vet. Clin. N. Am. Equine Pract.* **2017**, *33*, 67–85. [CrossRef]

20. Erol, E.; Scortti, M.; Fortner, J.; Patel, M.; Vazquez-Boland, J.A. Antimicrobial Resistance Spectrum Conferred by pRErm46 of Emerging Macrolide (Multidrug)-Resistant *Rhodococcus equi*. *J. Clin. Microbiol.* **2021**, *59*, e0114921. [CrossRef]

21. Hillidge, C.J. Use of erythromycin-rifampin combination in treatment of *Rhodococcus equi* pneumonia. *Vet. Microbiol.* **1987**, *14*, 337–342. [CrossRef] [PubMed]

22. Sweeney, C.R.; Sweeney, R.W.; Divers, T.J. *Rhodococcus equi* pneumonia in 48 foals: Response to antimicrobial therapy. *Vet. Microbiol.* **1987**, *14*, 329–336. [CrossRef]

23. Erol, E.; Shaffer, C.L.; Lubbers, B.V. Synergistic combinations of clarithromycin with doxycycline or minocycline reduce the emergence of antimicrobial resistance in *Rhodococcus equi*. *Equine Vet. J.* **2021**, *54*, 799–806. [CrossRef]

24. Giguere, S.; Lee, E.; Williams, E.; Cohen, N.D.; Chaffin, M.K.; Halbert, N.; Martens, R.J.; Franklin, R.P.; Clark, C.C.; Slovis, N.M. Determination of the prevalence of antimicrobial resistance to macrolide antimicrobials or rifampin in *Rhodococcus equi* isolates and treatment outcome in foals infected with antimicrobial-resistant isolates of *R. equi*. *J. Am. Vet. Med. Assoc.* **2010**, *237*, 74–81. [CrossRef] [PubMed]

25. Huber, L.; Giguere, S.; Slovis, N.M.; Carter, C.N.; Barr, B.S.; Cohen, N.D.; Elam, J.; Erol, E.; Locke, S.J.; Phillips, E.D.; et al. Emergence of resistance to macrolides and rifampicin in clinical isolates of *Rhodococcus equi* from foals in central Kentucky, USA: 1995 to 2017. *Antimicrob. Agents Chemother.* **2018**, *63*, e01714-18. [PubMed]

26. Alvarez-Narvaez, S.; Huber, L.; Giguere, S.; Hart, K.A.; Berghaus, R.D.; Sanchez, S.; Cohen, N.D. Epidemiology and Molecular Basis of Multidrug Resistance in *Rhodococcus equi*. *Microbiol. Mol. Biol. Rev.* **2021**, *85*, e00011-21. [CrossRef]

27. Liu, H.; Wang, Y.; Yan, J.; Wang, C.; He, H. Appearance of multidrug-resistant virulent *Rhodococcus equi* clinical isolates obtained in China. *J. Clin. Microbiol.* **2014**, *52*, 703. [CrossRef]

28. Buckley, T.; McManamon, E.; Stanbridge, S. Resistance studies of erythromycin and rifampin for *Rhodococcus equi* over a 10-year period. *Ir. Vet. J.* **2007**, *60*, 728–731. [CrossRef]

29. Val-Calvo, J.; Darcy, J.; Gibbons, J.; Creighton, A.; Egan, C.; Buckley, T.; Schmalenberger, A.; Fogarty, U.; Scortti, M.; Vazquez-Boland, J.A. International Spread of Multidrug-Resistant *Rhodococcus equi*. *Emerg. Infect. Dis.* **2022**, *28*, 1899–1903. [CrossRef]

30. Duchesne, R.; Castagnet, S.; Maillard, K.; Petry, S.; Cattoir, V.; Giard, J.C.; Leon, A. In vitro antimicrobial susceptibility of equine clinical isolates from France, 2006–2016. *J. Glob. Antimicrob. Resist.* **2019**, *19*, 144–153. [CrossRef]

31. Kalinowski, M.; Jarosz, L.; Gradzki, Z. Assessment of Antimicrobial Susceptibility of Virulent Strains of *Rhodococcus equi* Isolated From Foals and Soil of Horse Breeding Farms With and Without Endemic Infections. *J. Equine Vet. Sci.* **2020**, *91*, 103114. [CrossRef]

32. Alvarez-Narvaez, S.; Giguère, S.; Cohen, N.; Slovis, N.M.; Vazquez-Boland, J. Spread of emerging multidrug-resistant *Rhodococcus equi* in USA. *Emerg. Infect. Dis.* **2021**, *27*, 529. [CrossRef] [PubMed]

33. Huber, L.; Giguere, S.; Slovis, N.M.; Alvarez-Narvaez, S.; Hart, K.A.; Greiter, M.; Morris, E.R.A.; Cohen, N.D. The novel and transferable erm(51) gene confers macrolides, lincosamides and streptogramins B (MLSB) resistance to clonal *Rhodococcus equi* in the environment. *Environ. Microbiol.* **2020**, *22*, 2858–2869. [CrossRef]

34. Anastasi, E.; Giguere, S.; Berghaus, L.J.; Hondalus, M.K.; Willingham-Lane, J.M.; MacArthur, I.; Cohen, N.D.; Roberts, M.C.; Vazquez-Boland, J.A. Novel transferable erm(46) determinant responsible for emerging macrolide resistance in *Rhodococcus equi*. *J. Antimicrob. Chemother.* **2015**, *70*, 3184–3190. [PubMed]

35. Alvarez-Narvaez, S.; Giguere, S.; Anastasi, E.; Hearn, J.; Scortti, M.; Vazquez-Boland, J.A. Clonal Confinement of a Highly Mobile Resistance Element Driven by Combination Therapy in *Rhodococcus equi*. *mBio* **2019**, *10*, e02260-19. [CrossRef] [PubMed]

36. Alvarez-Narvaez, S.; Berghaus, L.J.; Morris, E.R.A.; Willingham-Lane, J.M.; Slovis, N.M.; Giguere, S.; Cohen, N.D. A Common Practice of Widespread Antimicrobial Use in Horse Production Promotes Multi-Drug Resistance. *Sci. Rep.* **2020**, *10*, 911. [CrossRef]

37. Altschul, S.F.; Gish, W.; Miller, W.; Myers, E.W.; Lipman, D.J. Basic local alignment search tool. *J. Mol. Biol.* **1990**, *215*, 403–410. [CrossRef]

38. Seemann, T. Prokka: Rapid prokaryotic genome annotation. *Bioinformatics* **2014**, *30*, 2068–2069. [CrossRef]

39. Blum, M.; Chang, H.Y.; Chuguransky, S.; Grego, T.; Kandasaamy, S.; Mitchell, A.; Nuka, G.; Paysan-Lafosse, T.; Qureshi, M.; Raj, S.; et al. The InterPro protein families and domains database: 20 years on. *Nucleic Acids Res.* **2021**, *49*, D344–D354. [CrossRef]

40. Tripathi, V.N.; Harding, W.C.; Willingham-Lane, J.M.; Hondalus, M.K. Conjugal transfer of a virulence plasmid in the opportunistic intracellular actinomycete *Rhodococcus equi*. *J. Bacteriol.* **2012**, *194*, 6790–6801. [CrossRef]

41. Alvarez-Narvaez, S.; Giguere, S.; Berghaus, L.J.; Dailey, C.; Vazquez-Boland, J.A. Horizontal Spread of *Rhodococcus equi* Macrolide Resistance Plasmid pRErm46 across Environmental Actinobacteria. *Appl. Environ. Microbiol.* **2020**, *86*, e00108-20. [CrossRef]

42. Berns, A.E.; Philipp, H.; Narres, H.D.; Burauel, P.; Vereecken, H.; Tappe, W. Effect of gamma-sterilization and autoclaving on soil organic matter structure as studied by solid state NMR, UV and fluorescence spectroscopy. *Eur. J. Soil. Sci.* **2008**, *59*, 540–550. [CrossRef]

43. Alvarez-Narvaez, S. Conjugal Transfer of Host-Adaptive Determinants in the Pathogenic Actinobacterium *Rhodococcus equi*. Ph.D. Thesis, Edinburgh Research Archive, University of Edinburgh, Edinburgh, Scotland, 2017.

44. Carver, T.J.; Rutherford, K.M.; Berriman, M.; Rajandream, M.-A.; Barrell, B.G.; Parkhill, J. ACT: The Artemis comparison tool. *Bioinformatics* **2005**, *21*, 3422–3423. [CrossRef]

45. Askarian, F.; van Sorge, N.M.; Sangvik, M.; Beasley, F.C.; Henriksen, J.R.; Sollid, J.U.; van Strijp, J.A.; Nizet, V.; Johannessen, M. A *Staphylococcus aureus* TIR domain protein virulence factor blocks TLR2-mediated NF-kappaB signaling. *J. Innate Immun.* **2014**, *6*, 485–498. [CrossRef]

46. Cirl, C.; Wieser, A.; Yadav, M.; Duerr, S.; Schubert, S.; Fischer, H.; Stappert, D.; Wantia, N.; Rodriguez, N.; Wagner, H.; et al. Subversion of Toll-like receptor signaling by a unique family of bacterial Toll/interleukin-1 receptor domain-containing proteins. *Nat. Med.* **2008**, *14*, 399–406. [CrossRef]

47. Essuman, K.; Summers, D.W.; Sasaki, Y.; Mao, X.; Yim, A.K.Y.; DiAntonio, A.; Milbrandt, J. TIR Domain Proteins Are an Ancient Family of NAD(+)-Consuming Enzymes. *Curr. Biol.* **2018**, *28*, 421–430 e424. [CrossRef]

48. Patot, S.; Imbert, P.R.; Baude, J.; Martins Simoes, P.; Campergue, J.B.; Louche, A.; Nijland, R.; Bes, M.; Tristan, A.; Laurent, F.; et al. The TIR Homologue Lies near Resistance Genes in *Staphylococcus aureus*, Coupling Modulation of Virulence and Antimicrobial Susceptibility. *PLoS Pathog.* **2017**, *13*, e1006092.

49. Klemm, E.J.; Wong, V.K.; Dougan, G. Emergence of dominant multidrug-resistant bacterial clades: Lessons from history and whole-genome sequencing. *Proc. Natl. Acad. Sci. USA* **2018**, *115*, 12872–12877. [CrossRef]

50. Bobay, L.M.; Ochman, H. The Evolution of Bacterial Genome Architecture. *Front. Genet.* **2017**, *8*, 72. [CrossRef]

51. Takai, S.; Sugawara, T.; Watanabe, Y.; Sasaki, Y.; Tsubaki, S.; Sekizaki, T. Effect of growth temperature on maintenance of virulent *Rhodococcus equi*. *Vet. Microbiol.* **1994**, *39*, 187–192. [CrossRef]

52. Kawasaki, T.; Kawai, T. Toll-like receptor signaling pathways. *Front. Immunol.* **2014**, *5*, 461. [CrossRef] [PubMed]

53. Coronas-Serna, J.M.; Louche, A.; Rodriguez-Escudero, M.; Roussin, M.; Imbert, P.R.C.; Rodriguez-Escudero, I.; Terradot, L.; Molina, M.; Gorvel, J.P.; Cid, V.J.; et al. The TIR-domain containing effectors BtpA and BtpB from Brucella abortus impact NAD metabolism. *PLoS Pathog.* **2020**, *16*, e1007979. [CrossRef] [PubMed]

54. Yadav, M.; Zhang, J.; Fischer, H.; Huang, W.; Lutay, N.; Cirl, C.; Lum, J.; Miethke, T.; Svanborg, C. Inhibition of TIR domain signaling by TcpC: MyD88-dependent and independent effects on *Escherichia coli* virulence. *PLoS Pathog.* **2010**, *6*, e1001120. [CrossRef] [PubMed]

55. Wagner, T.M.; Janice, J.; Paganelli, F.L.; Willems, R.J.; Askarian, F.; Pedersen, T.; Top, J.; de Haas, C.; van Strijp, J.A.; Johannessen, M.; et al. *Enterococcus faecium* TIR-Domain Genes Are Part of a Gene Cluster Which Promotes Bacterial Survival in Blood. *Int. J. Microbiol.* **2018**, *2018*, 1435820. [CrossRef] [PubMed]

56. Dawson, T.; Horohov, D.W.; Meijer, W.G.; Muscatello, G. Current understanding of the equine immune response to *Rhodococcus equi*. An immunological review of *R. equi* pneumonia. *Vet. Immunol. Immunopathol.* **2010**, *135*, 1–11. [CrossRef]

57. Makarova, K.S.; Wolf, Y.I.; Koonin, E.V. Comprehensive comparative-genomic analysis of type 2 toxin-antitoxin systems and related mobile stress response systems in prokaryotes. *Biol. Direct* **2009**, *4*, 19. [CrossRef]

58. Arcus, V.L.; McKenzie, J.L.; Robson, J.; Cook, G.M. The PIN-domain ribonucleases and the prokaryotic VapBC toxin-antitoxin array. *Protein Eng. Des. Sel.* **2011**, *24*, 33–40. [CrossRef]

59. Fraikin, N.; Goormaghtigh, F.; Van Melderen, L. Type II Toxin-Antitoxin Systems: Evolution and Revolutions. *J. Bacteriol.* **2020**, *202*, e00763-19. [CrossRef]

60. Page, R.; Peti, W. Toxin-antitoxin systems in bacterial growth arrest and persistence. *Nat. Chem. Biol.* **2016**, *12*, 208–214. [CrossRef]

61. Ghinet, M.G.; Bordeleau, E.; Beaudin, J.; Brzezinski, R.; Roy, S.; Burrus, V. Uncovering the prevalence and diversity of integrating conjugative elements in actinobacteria. *PLoS ONE* **2011**, *6*, e27846. [CrossRef]

62. Guedon, G.; Libante, V.; Coluzzi, C.; Payot, S.; Leblond-Bourget, N. The Obscure World of Integrative and Mobilizable Elements, Highly Widespread Elements that Pirate Bacterial Conjugative Systems. *Genes* **2017**, *8*, 337. [CrossRef] [PubMed]

63. Puymege, A.; Bertin, S.; Guedon, G.; Payot, S. Analysis of *Streptococcus agalactiae* pan-genome for prevalence, diversity and functionality of integrative and conjugative or mobilizable elements integrated in the tRNA(Lys CTT) gene. *Mol. Genet. Genom.* **2015**, *290*, 1727–1740. [CrossRef] [PubMed]

64. Willingham-Lane, J.M.; Berghaus, L.J.; Berghaus, R.D.; Hart, K.A.; Giguere, S. Effect of Macrolide and Rifampin Resistance on the Fitness of *Rhodococcus equi*. *Appl. Environ. Microbiol.* **2019**, *85*, e02665-18. [CrossRef] [PubMed]

Article

Detection and Phylogenetic Analysis of Extended-Spectrum β-Lactamase (ESBL)-Genetic Determinants in Gram-Negative Fecal-Microbiota of Wild Birds and Chicken Originated at Trimmu Barrage

Muhammad Adnan Saeed [1,*], Aman Ullah Khan [1], Syed Ehtisham-ul-Haque [1], Usman Waheed [1], Muhammad Fiaz Qamar [1], Aziz ur Rehman [1], Amar Nasir [2], Muhammad Arfan Zaman [1], Muhammad Kashif [2], Jean-Paul Gonzalez [3] and Hosny El-Adawy [4,5,*]

[1] Department of Pathobiology, University of Veterinary and Animal Sciences, Lahore, CVAS Campus, 12-Km Chiniot Road, Jhang 35200, Pakistan; amanullah.khan@uvas.edu.pk (A.U.K.); ehtishamsyed@uvas.edu.pk (S.E.-u.-H.); usman.waheed@uvas.edu.pk (U.W.); fiaz.qamar@uvas.edu.pk (M.F.Q.); aziz.rehman@uvas.edu.pk (A.u.R.); arfan.zaman@uvas.edu.pk (M.A.Z.)

[2] Department of Clinical Sciences, University of Veterinary and Animal Sciences, Lahore, CVAS Campus, 12-Km Chiniot Road, Jhang 35200, Pakistan; amar.nasir@uvas.edu.pk (A.N.); muhammad.kashif@uvas.edu.pk (M.K.)

[3] Department of Microbiology & Immunology, School of Medicine, Georgetown University, Washington, DC 20057, USA; jean.paul.gonzalez@georgetown.edu

[4] Institute of Bacterial Infections and Zoonoses, Friedrich-Loeffler-Institut, 07743 Jena, Germany

[5] Faculty of Veterinary Medicine, Kafrelsheikh University, Kafr El-Sheikh 35516, Egypt

[*] Correspondence: adnan.saeed@uvas.edu.pk (M.A.S.); hosny.eladawy@fli.de (H.E.-A.); Tel.: +49-3641-804-2249 (H.E.-A.)

check for updates

Citation: Saeed, M.A.; Khan, A.U.; Ehtisham-ul-Haque, S.; Waheed, U.; Qamar, M.F.; Rehman, A.u.; Nasir, A.; Zaman, M.A.; Kashif, M.; Gonzalez, J.-P.; et al. Detection and Phylogenetic Analysis of Extended-Spectrum β-Lactamase (ESBL)-Genetic Determinants in Gram-Negative Fecal-Microbiota of Wild Birds and Chicken Originated at Trimmu Barrage. *Antibiotics* **2023**, *12*, 1376. https://doi.org/10.3390/antibiotics12091376

Academic Editor: Nahla O. Eltai

Received: 21 July 2023
Revised: 21 August 2023
Accepted: 24 August 2023
Published: 28 August 2023

Abstract: Extended-spectrum β-lactamases (ESBL) give rise to resistance against penicillin and cephalosporin antibiotics in multiple bacterial species. The present study was conducted to map genetic determinants and related attributes of ESBL-producing bacteria in three wild aquatic bird species and chickens at the "Trimmu Barrage" in district Jhang, Punjab province, Pakistan. To study the prevalence of ESBL-producing bacteria, a total of 280 representative samples were collected from wild bird species; cattle egrets (*Bubulcus ibis*), little egrets (*Egretta garzetta*) and common teals (*Anas crecca*) as well as from indigenous chickens (*Gallus gallus domesticus*) originating from a local wet market. The isolates were confirmed as ESBL producers using a double disc synergy test (DDST) and bacterial species were identified using API-20E and 20NE strips. A polymerase chain reaction (PCR) was used to detect ESBL genetic determinants and for genus identification via 16S rRNA gene amplification. A phenotypic antimicrobial susceptibility test was performed for ESBL-producing isolates against 12 clinically relevant antibiotics using the Kirby–Bauer disk diffusion susceptibility test. A phylogenetic tree was constructed for the sequence data obtained in this study and comparative sequence data obtained from GenBank. The overall prevalence of ESBL-producing bacteria was 34.64% (97/280). The highest percentage (44.28%; 31/70) of ESBL-producing bacteria was recovered from chickens (*Gallus gallus domesticus*), followed by little egrets (*Egretta garzetta*) (41.43%; 29/70), common teal (*Anas crecca*) (28.57%; 20/70) and cattle egrets (*Bubulcus ibis*) (24.28%; 17/70). Five different ESBL-producing bacteria were identified biochemically and confirmed via 16S rRNA gene sequencing, which included *Escherichia coli* (72; 74.23%), *Enterobacter cloacae* (11; 11.34%), *Klebsiella pneumoniae* (8; 8.25%), *Salmonella enterica* (4; 4.12%) and *Pseudomonas aeruginosa* (2; 2.06%). Based on PCR, the frequency of obtained ESBL genes in 97 isolates was bla_{CTX-M} (51.55%), bla_{TEM} (20.62%), bla_{OXA} (6.18%) and bla_{SHV} (2.06%). In addition, gene combinations bla_{CTX-M} + bla_{TEM}, bla_{TEM} + bla_{OXA} and bla_{CTX-M} + bla_{SHV} were also detected in 16.49%, 2.06% and 1.03% of isolates, respectively. The ESBL gene variation was significant ($p = 0.02$) in different bacterial species while non-significant in relation to different bird species ($p = 0.85$). Phylogenetic analysis of amino acid sequence data confirmed the existence of CTX-M-15 and TEM betalactamases. The average susceptibility of the antibiotics panel used was lowest for both *Klebsiella pneumoniae* (62.5% ± 24.42) and *Salmonella enterica* (62.5% ± 31.08) as compared to *Enterobacter cloacae* (65.90% ± 21.62), *Pseudomonas aeruginosa* (70.83% ± 33.42) and *Escherichia coli* (73.83% ± 26.19). This study provides insight into the role of aquatic wild birds as

---

reservoirs of ESBL-producing bacteria at Trimmu Barrage, Punjab, Pakistan. Hence, active bio-surveillance and environment preservation actions are necessitated to curb antimicrobial resistance.

Keywords: ESBL; microbiota; wild birds; aquatic birds; Trimmu Barrage; phylogenetics

1. Introduction

Antimicrobial resistance (AMR) has been categorized among the top ten lead threats of public health significance by the World Health Organization (WHO) [1]. AMR-associated global human mortality is expected to rise to the hefty figure of 10 million per annum by 2050 as compared to 0.7 million in 2019 [2]. Antibiotics, mainly cephalosporin and penicillin, are vulnerable to structural degradation through a wide variety of bacteria-origin hydrolyzing enzymes known as extended-spectrum β-lactamase (ESBL) [3]. Hundreds of individual strains of bacteria belonging mainly to several species of Enterobacterales and a few non-Enterobacterales orders are known to produce ESBL enzymes [4]. β-lactamases include a wide variety of hydrolytic enzymes that tend to destabilize the structural integrity of β-lactam antibiotics through degradation of the amide bond of the β-lactam ring [4]. ESBL-producing bacteria are used as bio-surveillance markers to map the emergence and extent of antimicrobial resistance (AMR) under one-health research paradigms [5]. Classically, the ESBL-producing bacteria species were thought to be associated with clinical settings which would transfer mainly via nosocomial, and community-acquired channels [6]. However, the larger picture revealed by recent studies shows a complicated and multi-origin spread of resistant bacteria [7,8]. Trans-border dissemination of resistant bacteria includes wild migratory birds, human traffic, farm animals, arthropod-borne vectors, fomites, water, and food chains, although it is not limited to these factors [7,9,10]. Due to multiple factors, including habitat destruction, annual migration, and the search for feeding sites or breeding needs, wild birds often move into urban or peri-urban areas where they may encounter humans and other animals [11]. Many species of bacteria colonize wild aquatic birds as a part of gut microbiota [12]. In this way, resistant bacteria thrive, are carried away to far-furlong areas and are dispensed to new geographical locations and environments as soon as the birds migrate seasonally or interact with other species [13,14]. To feed and breed, the aquatic birds often contact stagnant water sources, streams, garbage landfills, farmed animals, other bird species and even human settlements [13]. For bacteria, AMR is a key component for adaptation in a changing environment under evolutionary process [15]. Two of the main driving forces behind AMR include first the mutations (intrinsic factors), which in most cases decreases the permeability of antibiotics to enter inside the bacterial cells, and second the acquired genetic elements of resistance (e.g., plasmids, integrons, transposons) which are transmissible horizontally [15].

In the Jhang district, Punjab province, Pakistan, Trimmu Barrage controls water flow at the downstream site of the junction between the Chenab and Jhelum rivers. The Trimmu Barrage represents a typical wetland that provides a balanced ecosystem consisting of both terrestrial and aquatic environments. As many as eighty-nine different avian species, including year-round inhabitants and seasonal migratory birds, were recorded at Trimmu Barrage in a previous study [16]. The present study aimed to determine the nature of ESBL genetic determinants in different species of ESBL-producing bacteria isolated from three different wild bird species, including the cattle egret (*Bubulcus ibis*), little egret (*Egretta garzetta*), and common teal (*Anas crecca*) along with indigenous chickens (*Gallus gallus domesticus*) sourced from the local wet-market, by using molecular characterization and phylogenetic analysis. This project was carried out to provide insight into the potential of wild birds to harbor resistant bacteria.

2. Results

2.1. Prevalence of ESBL-Producing Bacteria and Genetic Determinants in Bird Species

Out of 280 collected samples, 97 (34.6%) were found to contain ESBL-producing bacteria based on the double disc synergy test (DDST) (Table 1). The highest prevalence of ESBL-producing bacteria was recorded in chickens (44.3%). However, among the three studied wild bird species, ESBL-producing bacteria were found to be highest in the little egrets (41.4%), followed by the common teals (28.6%) and cattle egrets (24.3%) (Table 1). PCR detected all ESBL genetic determinants, including bla_{CTX-M} (593 bp), bla_{TEM} (445 bp), bla_{SHV} (1016 bp) and bla_{OXA} (296 bp) with varying percentages and combinations in different samples (Table 1). Only bla_{CTX-M} and bla_{TEM} were detected consistently in all species of birds, while bla_{SHV} was detected only in cattle egrets (5.9%) and little egrets (3.5%). Similarly, bla_{OXA} was found only in little egrets (13.8%) and chickens (6.5%) (Table 1).

Table 1. ESBL genes detected in bacteria isolated from different bird species at Trimmu Barrage.

Bird Species	Total ESBL n (%)	ESBL Gene/Gene Combinations Identified						
		bla_{CTX-M} n (%)	bla_{TEM} n (%)	bla_{SHV} n (%)	bla_{OXA} n (%)	$bla_{CTX-M} + bla_{TEM}$ n (%)	$bla_{CTX-M} + bla_{SHV}$ n (%)	$bla_{TEM} + bla_{OXA}$ n (%)
Cattle egret (*Bubulcus ibis*)	17 (24.3)	11 (64.7)	3 (17.7)	1 (5.9)	-	2 (11.8)	-	-
Little egret (*Egretta garzetta*)	29 (41.4)	18 (62.1)	5 (17.2)	1 (3.5)	4 (13.8)	-	1 (3.5)	-
Common teal (*Anas crecca*)	20 (28.6)	5 (25)	8 (4)	-	-	5 (25)	-	2 (10)
Chicken (*Gallus gallus domesticus*)	31 (44.3)	16 (51.6)	4 (12.9)	-	2 (6.5)	9 (29)	-	-
Total number	97 (34.6)	50 (51.6)	20 (20.6)	2 (2.1)	6 (6.2)	16 (16.5)	1 (1)	2 (2.1)

2.2. Diversity of ESBL-Producing Bacteria

Five different bacteria species of ESBL producers were identified using biochemical commercial kits, API 20E, and 20NE, (bioMérieux, Craponne, France). Ninety-seven isolates were recovered, including *Escherichia coli* (72; 74.2%), *Enterobacter cloacae* (11; 11.3%), *Klebsiella pneumoniae* (8; 8.3%), *Salmonella enterica* (4; 4.1%) and *Pseudomonas aeruginosa* (2; 2.1%). The distribution of the isolated ESBL-producing bacteria concerning bird species was demonstrated in Table 2. Selected isolates belonging to each species were further confirmed by 16S rRNA gene amplification and sequencing. The sequence data was submitted to the GenBank as *Enterobacter cloacae* (OP744530), *Klebsiella pneumoniae* (OP744528, OP744534), *Salmonella enterica* (OP744581), and *Pseudomonas aeruginosa* (OP745421).

2.3. Genetic Determinants Detected in Different ESBL-Producing Bacteria

The One-way ANOVA was used to determine the statistical significance of the ESBL gene variation (types and percentage distribution) in relation to different bird species and bacterial species. The ESBL gene variation was significant ($p = 0.02$) in different bacterial species while non-significant in relation to different bird species ($p = 0.85$). bla_{CTX-M} remained the most frequent yet least variable determinant, with a prevalence ranging from 37.5% to 54.5%. bla_{TEM} was detected in all species except *K. pneumoniae*. bla_{SHV} was rare and appeared in just 2.8% of *E. coli*. Only one *E. cloacae* (9.1%) and five *E. coli* (6.94%) harbored the bla_{OXA} gene. A total of three paired ESBL gene combinations were detected; however, bla_{CTX-M} and bla_{SHV} combinations were found exclusive to *K. pneumoniae* (Table 3).

Table 2. Prevalence of ESBL-producing bacterial species recovered from wild birds and chickens at Trimmu Barrage.

Bird Species	No. of Samples	ESBL Bacterian (%)	*E. coli* n (%)	*E. cloacae* n (%)	*K. pneumoniae* n (%)	*S. enterica* n (%)	*P. aeruginosa* n (%)
Cattle egret (*Bubulcus ibis*)	70	17 (24.3%)	12 (12.37%)	4 (4.12%)	0	1 (1.03%)	0
Little egret (*Egretta garzetta)*	70	29 (41.4%)	24 (24.74%)	1 (1.03%)	2 (2.06%)	1 (1.03%)	1 (1.03%)
Common teal (*Anas crecca)*	70	20 (28.6%)	16 (16.49%)	3 (3.09%)	1 (1.03%)	0	0
Chicken (*Gallus gallus domesticus)*	70	31 (44.3%)	20 (20.62%)	3 (3.09%)	5 (5.15%)	2 (2.06%)	1 (1.03%)
Total number	280	97 (34.6%)	72 (74.2%)	11 (11.3%)	8 (8.3%)	4(4.1%)	2 (2.1%)

Table 3. Frequency and diversity of ESBL genetic determinants associated with different bacterial species recovered from wild birds and chickens at Trimmu Barrage.

Bacterial Species	Total ESBL n (%)	ESBL Gene/Gene Combinations Identified						
		bla_{CTX-M} n (%)	bla_{TEM} n (%)	bla_{SHV} n (%)	bla_{OXA} n (%)	$bla_{CTX-M} + bla_{TEM}$ n (%)	$bla_{CTX-M} + bla_{SHV}$ n (%)	$bla_{TEM} + bla_{OXA}$ n (%)
Escherichia coli	72 (74.2%)	38 (52.8%)	14 (19.5%)	2 (2.8%)	5 (6.9%)	13 (18.1%)	0	0
Enterobacter cloacae	11 (11.3%)	6 (54.5%)	4 (36.4%)	0	1 (9.1%)	0	0	0
Klebsiella pneumoniae	8 (8.3%)	3 (37.5%)	0	0	0	3 (37.5%)	1 (12.5%)	1 (12.5%)
Salmonella enterica	4 (4.1%)	2 (50%)	1 (25%)	0	0	0	0	1 (25%)
Pseudomonas aeruginosa	2 (2.1%)	1 (50%)	1 (50%)	0	0	0	0	0
Total number	97 (34.6%)	50 (51.55%)	20 (20.6%)	2 (2.06%)	6 (6.2%)	16 (16.5%)	1 (1.03%)	2 (2.1%)

2.4. Phylogenetic Analysis

The phylogenetic tree constructed showed three clades and included an outgroup (catalase enzyme). CTX-M partial protein sequences obtained in this study (UZZ47306.1, UZZ47307.1, and UZZ47308.1) were clustered in clade I and consisted of different groups of CTX-M enzyme. Out of the five groups of CTX-M enzymes (CTX-M-1, CTX-M-2, CTX-M-8, CTX-M-9, and CTX-M-25), the present study sequences were grouped next to the CTX-M-1 and CTX-M-15 (a subtype of CTX-M-1 group). TEM enzyme sequence (UZZ47309.1) appeared in clade II, consisting of TEM sequences from India and Croatia (Figure 1).

2.5. Antibiotic Susceptibility Profile

A complete antibiogram of the ESBL-producing bacteria isolated in this study against 12 antimicrobial agents is given in Table 4. All isolated ESBL-producing bacteria in this study were highly resistant to ceftiofur and ampicillin. The resistance of *S. enterica*, *K. pneumoniae*, and *E. cloacae* against tylosin was 50%, 50% and 45.5%, respectively, while *E. coli* and *P. aeruginosa* were sensitive (Table 4). All isolated ESBL-producing bacteria were highly sensitive to the other antimicrobial agents tested in this study.

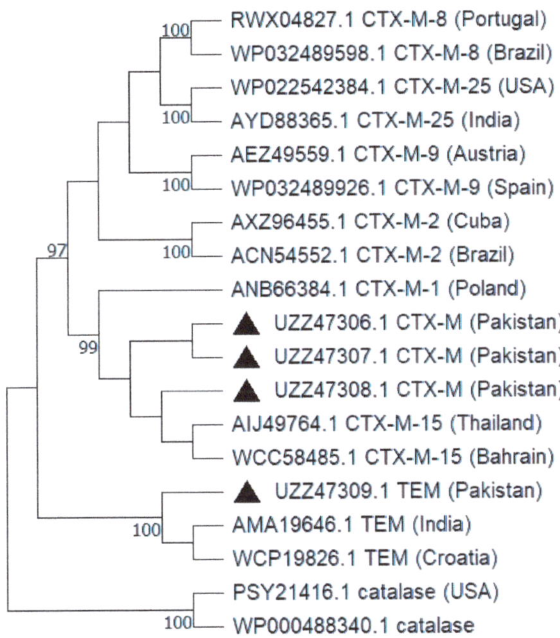

Figure 1. Phylogenetic relationship of amino acid sequence data where present study sequences has been marked with black triangles. The phylogenetic tree was constructed using the maximum likelihood method and JTT matrix-based model. Numeric values below the nodes represent the bootstrap frequency obtained via 1000 replicates.

Table 4. Antibiogram of ESBL-producing bacteria isolated from wild birds and chickens at Trimmu Barrage.

Antibiotic	E. cloacae (n = 11)			K. pneumoniae (n = 8)			S. enterica (n = 4)			P. aeruginosa (n = 2)			E. coli (n = 72)		
	R	S	I	R	S	I	R	S	I	R	S	I	R	S	I
									n (%)						
Streptomycin	1 (9)	10 (90.9)	0	2 (25)	6 (75)	0	2 (50)	2 (50)	0	1 (50)	1 (50)	0	3 (4.16)	69 (95.8)	0
Neomycin	2 (18.2)	8 (72.7)	1 (9)	1 (12.5)	6 (75)	1 (12.5)	1 (25)	3 (75)	0	0	1 (50)	1 (50)	21 (29.2)	46 (63.9)	5 (6.9)
Gentamicin	0	11 (100)	0	3 (37.5)	3 (37.5)	2 (25)	1 (25)	3 (75)	0	0	2 (100)	0	12 (16.7)	55 (76.4)	5 (6.9)
Florfenicol	3 (27.3)	8 (72.7)	0	2 (25)	6 (75)	0	0	4 (100)	0	0	2 (100)	0	7 (9.7)	61 (84.7)	4 (5.6)
Ceftiofur	6 (54.5)	5 (45.5)	0	4 (50)	3 (37.5)	1 (12.5)	2 (50)	2 (50)	0	1 (50)	1 (50)	0	64 (88.9)	8 (11.1)	0
Enrofloxacin	3 (27.3)	6 (54.5)	2 (18.2)	3 (37.5)	5 (62.5)	0	1 (25)	3 (75)	0	0	2 (100)	0	6 (8.3)	65 (90.3)	1 (1.4)
Norfloxacin	4 (36.4)	7 (63.4)	0	2 (25)	6 (75)	0	0	4 (100)	0	0	2 (100)	0	0	70 (97.2)	2 (2.8)
Tylosin	5 (45.5)	5 (45.5)	1 (9)	4 (50)	4 (50)	0	2 (50)	1 (25)	1 (25)	0	1 (50)	1 (50)	11 (15.3)	61 (84.7)	0
Ampicillin	7 (63.6)	4 (36.4)	0	6 (75)	1 (12.5)	1 (12.5)	3 (75)	0	1 (25)	2 (100)	0	0	41 (56.9)	27 (37.5)	4 (5.5)
Doxycycline	3 (27.3)	6 (54.5)	2 (18.2)	1 (12.5)	5 (62.5)	2 (25)	1 (25)	2 (50)	1 (25)	1 (50)	1 (50)	0	14 (19.4)	47 (65.3)	11 (15.3)
Colistin	4 (36.4)	6 (54.5)	1 (9)	0	8 (100)	0	2 (50)	2 (50)	0	0	2 (100)	0	9 (12.5)	60 (83.3)	3 (4.2)
Imipenem	0	11 (100)	0	1 (12.5)	7 (87.5)	0	0	4 (100)	0	0	2 (100)	0	3 (4.2)	69 (95.8)	0

R: Resistant: S: Sensitive; I: Intermediate.

The overall susceptibility to the panel of antibiotics was lowest for both *K. pneumoniae* (62.5% ± 24.42) and *S. enterica* (62.5% ± 31.08) as compared to *E. cloacae* (65.9% ± 21.62), *P. aeruginosa* (70.8% ± 33.42) and *E. coli* (73.8% ± 26.19) (Table 5). The mean susceptibility of ESBL-producing bacteria was found to be highest for imipenem (96.7%) followed by norfloxacin (87.2%), florfenicol (86.5%), gentamicin (77.7%), colistin (77.5%), enrofloxacin (76.5%), streptomycin (72.3%), neomycin (67.3%), doxycycline (56.5%), tylosin (51%), ceftiofur (38.8%) and ampicillin (17.3%).

Table 5. The overall susceptibility antibiogram of ESBL-producing bacteria isolated from wild birds and chickens at Trimmu Barrage.

Antimicrobials	*E. cloacae*	*K. pneumoniae*	*S. enterica*	*P. aeruginosa*	*E. coli*
Streptomycin	90.9	75	50	50	95.8
Neomycin	72.7	75	75	50	63.9
Gentamicin	100	37.5	75	100	76.4
Florfenicol	72.7	75	100	100	84.7
Ceftiofur	45.5	37.5	50	50	11.1
Enrofloxacin	54.5	62.5	75	100	90.3
Norfloxacin	63.6	75	100	100	97.2
Tylosin	45.5	50	25	50	84.7
Ampicillin	36.4	12.5	0	0	37.5
Doxycycline	54.5	62.5	50	50	65.3
Colistin	54.5	100	50	100	83.3
Imipenem	100	87.5	100	100	95.8
Mean	65.90416667	62.5	62.5	70.83333	73.83833
Standard Deviation	21.62692533	24.42521052	31.07907803	33.4279	26.19029
Standard Error	6.243155581	7.050950935	8.971757032	9.649802	7.560486
Minimum	36.36	12.5	0	0	11.11
Maximum	100	100	100	100	97.22
Count	12	12	12	12	12

3. Discussion

Antimicrobial resistance (AMR) is a phenomenon wherein microbes attain the ability to survive a particular concentration of a given antimicrobial substance where they would otherwise be inhibited or killed [17]. Wild birds are reservoirs of pathogens and bacteria resistant to highly and critically important antimicrobials, i.e., ESBL-producing bacteria and MDR strains. Wild birds can occupy different ecological niches and adapt to many urban, suburban and rural environments, representing true environmental sentinels. Due to their ability to move and through the deposition of droppings, birds can play an important role as vectors in the environmental circulation and spread of zoonotic agents, antimicrobial-resistant bacteria and resistance genes [18]. The bio-surveillance system based on monitoring the frequency rates of ESBL-positive bacteria, especially *E. coli*, has become a primary indicator for tracking AMR globally [19]. This system goes beyond a singular focus by embracing the One Health approach, which considers the interconnectedness of animal, human and environmental health in tackling AMR across these sectors [1,5]. Additionally, it facilitates direct comparisons of AMR prevalence and patterns across various regions worldwide.

In Pakistan, most of the selected antibiotics are used in veterinary prescriptions and food animal production, including in the poultry sector [20]. However, some of the selected antibiotics are exclusively used in human medicine (e.g., imipenem and cefotaxime). The antibiotics panel was finally selected under the One Health approach to represent antibiotics of both human and veterinary importance [21]. The emergence of resistant bacteria in wild birds is attributed to environmental contamination with antibiotic residues. The use of tylosin, specifically, was found in 100% (30/30 flocks) of the studied broiler population in Pakistan [21] and vigorous use of tylosin in broiler production was reported previously [20,21]. Although tylosin is primarily active against Gram-positive bacteria, there are reports of a correlation between exposure to tylosin and a change in susceptibility to other antibacterial drugs and even the emergence of multidrug-resistant strains of enterobacteria [22,23]. Such reports suggest that tylosin susceptibility of resistant strains is variable.

There are reports of a high prevalence of ESBL-producing bacteria from clinical and non-clinical sources (mainly poultry and water bodies) in Pakistan [24,25]. The present study included three targeted and commonly found aquatic wild bird species, including cattle egrets (*Bubulcus ibis*), little egrets (*Egretta garzetta*) and common teals (*Anas crecca*) as well as the commercial chickens (*Gallus gallus domesticus*) originated at the diverse aquatic and land environment of the Trimmu Barrage in the district Jhang, Punjab province, Pakistan. Of 280 samples, 97 (34.6%) contained ESBL-positive bacteria based on DDST. The highest prevalence of ESBL-producing bacteria was recorded in chickens (44.3%). Previously, in Pakistan, various studies have reported the existence of ESBL-positive bacteria in commercial chickens with variable prevalence in different regions. Similar to our findings, a study conducted in three districts of the province of Khyber Pakhunkhwa (KPK) (Peshawar, Kohat, and Nowshera) reported an overall prevalence of 43.5% among samples from chickens (liver, spleen and meat) [26]. A study conducted in the Faisalabad metropolitan area (Punjab, Pakistan) reported a relatively low prevalence of ESBL-positive bacteria in poultry (13.7%) as compared to cattle (31%) and humans (55%) (28). From Pakistan, a meta-analysis study on ESBL-producing bacteria revealed an overall prevalence of 40% [24]. Recent studies in Pakistan have reported a much higher prevalence; in a study conducted in Islamabad, ESBL-positive bacteria in chicken was 62.4% [27] and 82% in KPK [28]. Overall, the prevalence of ESBL-positive bacteria in commercial chicken is quite high in Pakistan, as suggested by the current study and previous studies on this subject. This situation is partly explained due to the practice of high antimicrobial use (AMU) for treating bacterial infections, prophylaxis and growth promotion [21]. In Pakistan, the usage of antimicrobials in commercial broiler farming is very high; 22 antibiotics belonging to different classes have been reported to be used on poultry farms, 60% of which have been categorized as critically important antimicrobial classes (CIA) for human medicine by the World Health Organization [20]. Antimicrobials are known to induce selective pressure among normal microflora as well as the avian pathogenic species for evolution into the resistance types [9].

Beta-lactamases (*bla*) are the largest class of antibiotic-resistance genes, which cause resistance to beta-lactam antibiotics (Penicillins and cephalosporins). ESBL enzymes TEM and SHV are known penicillinases. Both have been detected in the present study. According to Beta-Lactamase DataBase (BLDB), 7996 types of beta-lactamases have been identified (http://www.bldb.eu/ accessed on 16 August 2023) [29]. The present study was limited to extended-spectrum beta-lactamases (ESBL) reporting. Amoxicillin was used in the present study as a co-amoxiclav disc while performing a double disc synergy test (DDST). Therefore, ampicillin was used for antibiogram data. Ampicillin was mainly included as a representative of a medically important antibacterial drug; however, it is also used in broiler production in Pakistan. A recent study reported that 91.7% of *Salmonella* spp. recovered from chicken meat in Pakistan were found resistant to ampicillin [30]. A comparable prevalence of 38.18% ESBL-producing *E. coli* has been reported in Bangladesh in migratory wild bird species where various genetic determinants related to ESBL production mechanisms were reported as bla_{TEM}, bla_{CMY}, bla_{CTX-M} and bla_{SHV} were detected in 95.2%, 90.5%, 85.7% and 42.9% of isolates, respectively [31]. In this study, ESBL-producing bacteria were highly prevalent in the little egret (*Egretta garzetta*) (41.4%), followed by the common teal (*Anas crecca*) (28.57%) and cattle egret (*Bubulcus ibis*) (24.28%) at the Trimmu Barrage, Jhang, Punjab, Pakistan. Previous studies conducted in Africa, reported very high prevalence (92.9%) of ESBL-producing bacteria in cattle egrets (*Bubulcus ibis*) as compared to our findings (24.3%), however, consistent to our findings the $bla_{CTX-M-15}$ (83.3%) was reported as the major resistance determinant gene along with $bla_{CTX-M-9}$ (11.8%) and carbapenemase resistance genes (bla_{KPC-2} and bla_{KPC-3}) [32,33].

To our knowledge, no study has been conducted to report the prevalence of ESBL-producing bacteria originating from the little egret (*Egretta garzetta*) in Pakistan. The present study spotlights the emergence of ESBL *E. coli* in little egrets as the highest (41.43%) among all the studied bird species. A study conducted in Greece reported ESBL genes in 11 out of 12 *E. coli* isolates in fecal samples of different wild and feral bird species (n = 362) but did not find any positive samples from little egrets [34]. This might be attributed to the lower sample size, consisting of only eight samples. A study conducted in China in little egrets reported the ESBL gene prevalence as 46.7% (7/15), with ESBL genes bla_{OXY} and bla_{TEM} as dominant types [35]. In the present study, a comparable prevalence of ESBL bacteria (41.43%) has been noted in little egrets; however, with a different set of genetic determinants, including bla_{CTX-M}, bla_{OXA}, bla_{SHV} and bla_{TEM}. The little egret is considered a successful aquatic bird species colonizing and thriving in wetland environments in the Punjab and Sindh provinces of Pakistan. Sewage, as well as untreated industrial and municipal waste, causes pollution of the wetlands in Pakistan [36]. Heavy metals have been previously detected in the eggshells of little egrets at the Trimmu Barrage area [37]. Given that, it is evident that the environmental contamination of wetlands impacts the types of gut microbiota colonizing the aquatic bird species, especially the little egret. This exposure leads to the colonization of antimicrobial-resistant bacteria in aquatic wildlife [37].

The present study detected 28.6% ESBL *E. coli* in common teals (*Anas crecca*). Previous studies indicated a variable prevalence of ESBL-producing bacteria in ducks: 0% in mallards (*Anas platyrhynchos*) in the Czech Republic [38] and 33.3% (1/3) in mallards in Poland [39]; 31% ESBL *E. coli* were reported from multiple migratory bird species, including common teals (*Anas crecca*) in Pakistan with bla_{TEM} dominant gene type [40] and 47% in ducks in Sweden with $bla_{CTX-M-15}$ as the dominant ARG [41]. In 2017, the first preliminary report was published that provided evidence of the existence of ESBL-producing *Klebsiella pneumoniae* in wild migratory birds in wetlands in Pakistan [42].

Similar to these findings, the present study reported CTX-M-1 as the predominant group with 88.8% of bla_{CTX-M}-producing isolates, whereas $bla_{CTX-M-15}$ remained the most dominant genotype as per DNA sequencing analysis in Eurasian coot (*Fulica atra*), rosy starling (*Pastor roseus*) and northern shoveler (*Anas clypeata*) [42]. However, our study also reported the existence of combinations of genetic determinants related to ESBL genes (bla_{CTX-M} and bla_{TEM}; 37%), (bla_{CTX-M} and bla_{SHV}; 12.5%) and (bla_{TEM} and bla_{OXA}; 12.5%) in *Klebsiella pneumoniae* isolates in different species of wild birds. The "CTX-M pandemic" is the term used to describe the global spread of ESBL-containing isolates. Since the early 2000s, the CTX-M-type enzymes have been recognized as the most common ESBL category, deposing TEM and SHV as the leading ESBL enzyme types [43]. Previously, CTX-M enzyme genetic determinants have been largely reported in clinical settings, communities, food products, pet animals, environments and farm animals [44]. However, our study reports the existence and dissemination of bla_{CTX-M} genes in the fecal microbiota of wild birds and domestic poultry. The $bla_{CTX-M-15}$ type gene was identified in the present study, as demonstrated by phylogenetic analysis.

A previous study reported an overall prevalence of ESBL-producing *E. coli* at 17.3% in fecal samples of wild migratory birds in Pakistan [45]. Of these, 88.4% of isolates exhibited multidrug resistance (MDR) characteristics and were found resistant to antibiotics, including cefotaxime, ceftazidime, ampicillin, doxycycline, tetracycline and sulfamethoxazole/trimethoprim, whereas $bla_{CTX-M-15}$ was found the most common ESBL gene [45]. Similar to these findings, the ESBL-producing bacteria isolated in the present study were found resistant to multiple antibiotics, and the mean resistance percentage obtained as ampicillin (74.1%), ceftiofur (58.7%), tylosin (32.1%), streptomycin (27.6%), doxycycline (26.8%), colistin (19.8%), enrofloxacin (19.6%), neomycin (17%), gentamicin (15.8%), florfenicol (12.4%), norfloxacin (12.3%) and imipenem (3.3%). These findings indicate that the multidrug resistance phenomenon is common in ESBL-producing bacteria.

Five different species of ESBL-producing bacteria have been identified in our study, including *E. coli*, 74.2%, *E. cloacae*, 11.3%, *K. pneumoniae*, 8.3%, *S. enterica*, 4.1% and *P. aeruginosa*, 2.1%. High prevalence and diversity of ESBL-producing bacteria were noted in Catalonia, including ESBL prevalence (11.5%) in wild birds with bla_{CMY-2} (50%) and $bla_{CTX-M-15}$ (18%) and various species of bacteria, including *K. pneumoniae* (20%), *C. freundii* (15%), *E. cloacae* (5%), *P. mirabilis* (5%), *Providencia* spp. (5%) and *Serratia marcescens* (2.5%) were also isolated [46]. The nature and diversity of gut microflora in wild birds are dependent on multiple intrinsic (genetic makeup, age, sex, breed, etc.) and extrinsic factors (diet, environment, behavior, social contact, etc.) [47]. Although different species of ESBL-producing bacteria can be isolated by using modified (antibiotics-supplemented) selective culture media, mostly the members of the family *Enterobacteriaceae* are resistant to β-lactam antibiotics. The prevalence of ESBL-producing *E. coli* (74.2%) was notably higher than other co-species of bacteria in the present study. Nevertheless, this finding is inconsistent and varies in different geographical regions and wild bird species. However, it can partially be explained by the remarkable ability of *E. coli* to sustain a wide range of regulatory and metabolic pathways that allow it to survive in different habitats and the alimentary canals of a diverse range of host species [48].

As employed in the present study, 16S rRNA gene sequencing provides an effective tool for bio-surveillance of antimicrobial resistance in wild birds. It can be utilized as a relatively economical, reliable, less labor-intensive and user-friendly alternative to biochemical and serological identification techniques for bacteria, especially in low and middle-income countries. The occurrence of antimicrobial-resistant bacteria in wild bird species could pose a significant challenge for public health in the future. Therefore, it becomes increasingly important to periodically monitor the wild bird species as silent spreaders of resistant bacteria.

4. Materials and Methods

4.1. Collection of Pooled Fecal Material

The sampling period spanned November 2021 to December 2022. The selection of the three target species was based on relative abundance, convenience sampling and available resources. Selected bird species included two indigenous flying species, cattle egrets (*Bubulcus ibis*) and little egrets (*Egretta garzetta*), along with one winter visitor species, common teals (*Anas crecca*). The birds were observed at their nesting, feeding and breeding hotspots at Trimmu Barrage (district Jhang), Punjab, Pakistan. Sterile aluminum foil sheets (1 m^2) were spread on the ground at different hotspots to collect fresh droppings. Fresh droppings (3–4) obtained from individual birds of the same species were pooled to prepare a representative sample. Similarly, fresh fecal material was collected from the indigenous chickens (*Gallus gallus domesticus*) originating from a local wet market. A total of 280 representative samples, consisting of seventy samples from each species, were collected in vials of sterile transport media and shipped to the Microbiology laboratory of the College of Veterinary and Animal Sciences, Jhang. Samples were stored at 4 °C until processed further.

4.2. Isolation and Confirmation of ESBL-Producing Bacteria

The representative specimens (n = 280) were pre-enriched in 10 mL of tryptone soy broth (CM0129, Oxoid, Hampshire, UK) supplemented with 4 mg/L cefotaxime (Caisson C032-100G, Smithfield, UT, USA) and incubated at 37 °C for 24 h. Pre-enriched broth (100 µL) was streaked on MacConkey agar (Oxoid CM0007, Hampshire, UK) plates supplemented with 4 mg/L cefotaxime and incubated at 37 °C for 24 h [49]. Individual colonies selected from MacConkey agar were sub-cultured in tryptone soy broth to obtain a pure culture. A double disc synergy test (DDST) was used to confirm pure cultures as ESBL producers by following the protocol described in the M100 performance standards for antimicrobial susceptibility testing by the Clinical and Laboratory Standards Institute (CLSI) [50]. Briefly, the 0.5 McFarland standard equivalent test culture was swabbed onto the Mueller-Hinton (CM0337B, Oxoid, Hampshire, UK) agar plates and two antibiotic discs cefotaxime (CTX-30) and amoxicillin/clavulanic acid (AMC-30) were placed at a center-center distance of 20 mm and plates were incubated at 37 °C for 24 h. Expansion of the zone of inhibition of CTX-30 toward the AMC-30 disc was marked as a positive DDST. The species of ESBL bacteria were identified by using API-20E and 20NE strips (bioMérieux, Craponne, France) according to the manufacturer's instructions.

4.3. Molecular Identification of ESBL-Producing Bacteria and Detection of Genetic Determinants of β-Lactamase Associated Genes

The primer sets utilized for the amplification of ESBL genetic determinants (bla_{CTX-M}, bla_{TEM}, bla_{SHV} and bla_{OXA}) as well as the 16S rRNA gene sequencing-based genus confirmation of selected isolates have been summarized in Table 6. All the isolates, pre-confirmed as ESBL producers by DDST, were sub-cultured on Tryptone Soy Broth and incubated overnight at 37 °C for 18 h. Fresh broth cultures (1 mL) were processed for extraction of genomic DNA by using WizPrep™ gDNA Kit (Wizbiosolutions, Gyeonggi-do, Republic of Korea) as per manufacturer's instructions to obtain a DNA concentration of minimally 10 ng/μL. PCR-positive controls for genetic determinants were maintained at the Microbiology laboratory of the College of Veterinary and Animal Sciences, Jhang, Pakistan, through preserved stains, including *E. coli* strain MASMS_A3 (GenBank: ON736876.1) for bla_{CTX-M} and bla_{TEM} genes. In contrast, *K. pneumoniae* strain MASJG8 (GenBank: OP744534.1) was used for bla_{SHV} and bla_{OXA} genes. Positive control amplicons were verified through nucleotide sequencing and the BLAST tool of the NCBI. Nuclease-free distilled water was used as a negative control template. Mono-plex PCRs for ESBL genetic determinants were carried out by mixing 25 μL PCR master mix (Dream Taq Green 2x, ThermoFisher, Waltham, MA, USA), template DNA (4 μL), forward and reverse primers (10 pM/μL) as 2 μL per reaction for each primer and completed to total volume 50 μL with nuclease-free water (AM9932, Invitrogen™ Waltham, MA, USA). All PCR tubes were kept in a thermal cycler (Biorad, T100, Hercules, CA, USA) for amplification of the targets under these conditions: pre-denaturation monocycle at 94 °C for 15 min, 30× (denaturation at 94 °C for 30 s, annealing temperatures as per Table 6, for 30 s and extension at 72 °C for 2 min) followed by a 10 min final extension at 72 °C.

Table 6. Primers used for detection of ESBL genetic determinants and 16S rRNA gene.

Target Gene	Primers	Primer Sequences (5′-3′)	Annealing Temp. (°C)	Amplicon Size	Reference
bla_{CTX-M}	F	ATGTGCAGYACCAGTAARGTKATGGC	58	593 bp	[51]
	R	TGGGTRAARTARGTSACCAGAAYCAGCGG			
bla_{TEM}	F	TCGCCGCATACACTATTCTCAGAATGA	50	445 bp	
	R	ACGCTCACCGGCTCCAGATTTAT			
bla_{SHV}	F	CGCCGGGTTATTCTTATTTGTCGC	68	1016 bp	[52]
	R	TCTTTCCGATGCCGCCGCCAGTCA			
bla_{OXA}	F	ATTATCTACAGCAGCGCCAGTG	56	296 bp	[53]
	R	TGCATCCACGTCTTTGGTG			
16S rRNA	fD1	AGAGTTTGATCCTGGCTCAG	52	1500 bp	[54]
	rP2	ACGGCTACCTTGTTACGACTT			

The reaction mixture for amplification of 16S rRNA gene was prepared by using 12.5 μL PCR master mix (Dream Taq Green 2x, Thermo Fisher Scientific, Waltham, MA, USA), gDNA (2 μL), primers (fD_1 and rP_2) at concentration 10 pM as 1 μL each and completed to total volume 25 μL with nuclease-free water. The thermocycling parameters for amplification of the 16S rRNA gene included initial denaturation at 94 °C for 10 min, 36x (denaturation at 94 °C for 60 s, annealing at 52 °C for 30 s, and extension at 72 °C for 2 min) followed by a single final extension step at 72 °C for 10 min. The PCR products were electrophoresed using the Mupid One Electrophoresis System (NIPPON Genetics, Tokyo, Japan) at 90 volts for 45 min by using agarose gel (1.3%) stained with ethidium bromide at 0.5 μg/mL of the gel. Stained gels were examined with a UV-transilluminator (Fisher Scientific, Hampton, NH, USA) to visualize and capture gel images.

4.4. Sequencing of PCR Amplicons and Phylogenetic Analysis

Selected PCR amplicons were tested for DNA concentration (>30 ng/μL) and DNA purity (A_{260}/A_{280} ~1.8) via Nanodrop-1000 Spectrophotometer (ThermoScientific, Greenville, SC, USA) and processed for sanger sequencing by commercial service provider, Beijing Genomics Institute (BGI), Shenzhen (518083), China. For phylogenetic analysis, raw sequence reads were trimmed using BioEdit 7.0 software. Selected sequence data were uploaded to the GenBank database. Sequence data obtained in this study was processed by using the BLAST-p program from the National Center for Biotechnology Information (NCBI) to obtain similar sequence data available at the NCBI. For phylogenetic analysis protein sequence data included present study GenBank accession numbers CTX-M (UZZ47306.1, UZZ47307.1 and UZZ47308.1) and TEM (UZZ47309.1) while comparative sequence data included CTX-M-1 (ANB66384.1), CTX-M-2 (AXZ96455.1, ANB66384.1), CTX-M-8 (RWX04827.1, WP032489598.1), CTX-M-9 (AEZ49559.1, WP032489926.1), CTX-M-15 (AIJ49764.1, WCC58485.1), CTX-M-25 (WP022542384.1, AYD88365.1), TEM (AMA19646.1, WCP19826.1) and catalase as outgroup sequences (PSY21416.1 and WP000488340.1). A phylogenetic tree was constructed using the maximum likelihood method with bootstraps (1000) and the Jones–Taylor–Thornton (JTT) model using MEGA 11 software (64-bit) [55].

4.5. Antibiotic Susceptibility Testing

The phenotypic antimicrobial susceptibility test was performed using the Kirby–Bauer disk diffusion susceptibility test for ESBL-producing isolates, including *E. cloacae* (n = 11), *K. pneumoniae* (n = 8), *S. enterica* (n = 4), *P. aeruginosa* (n = 2) and *E. coli* (n = 72) against the panel of 12 antibiotics discs (Oxoid, UK; Condalab, Spain) included streptomycin (10 μg), neomycin (30 μg), gentamicin (10 μg), florfenicol (30 μg), ceftiofur (30 μg), enrofloxacin (5 μg), norfloxacin (10 μg), tylosin (30 μg), ampicillin (10 μg), doxycycline (30 μg), colistin (10 μg) and imipenem (10 μg). Testing and result interpretation protocols were followed as described in the M100s manual of the Clinical Laboratory Standards Institute (CLSI), 2020 [50].

4.6. Statistical Analysis

Descriptive statistical analysis, including prevalence, percentages and analysis of variance using One-way ANOVA, was performed using IBM-SPSS (Version-25) software.

5. Conclusions

This study described the prevalence, diversity, genotypic characterization, phylogeny and antibiotic susceptibility of the extended-spectrum β-lactamase (ESBL) producing bacteria of wild bird and commercial chicken origin at the Trimmu Barrage (district Jhang), Punjab, Pakistan. The study found that several member species of *Enterobacteriaceae* are ESBL producers, possess a wide range of genetic determinants ($bla_{CTX-M-15}$, bla_{TEM}, bla_{SHV}, and bla_{OXA}), and are resistant to multiple antibiotics of clinical relevance. The presence of ESBL genes in bacteria connects to the co-emergence of resistance against non-beta lactam antibiotics as well. Conclusively, the wild aquatic birds often found in urban and peri-urban areas have the potential to disseminate resistant microflora through their movement and interaction with biotic and abiotic subjects in the environment. The findings of the present study may be helpful for health officials, veterinary professionals and public policymakers to chalk out an effective control and prevention plan by creating public awareness, preserving wildlife, and environmental interventions to slow down the pace of emerging antimicrobial resistance.

Author Contributions: Conceptualization, M.A.S., A.U.K. and U.W.; Data curation, M.A.S., A.U.K., S.E.-u.-H., U.W. and A.u.R.; Formal analysis, M.A.S. and J.-P.G.; Investigation, M.A.S., A.U.K., S.E.-u.-H., U.W., A.u.R., A.N., M.A.Z. and M.K.; Methodology, M.A.S. and S.E.-u.-H.; Resources, M.A.S., A.U.K. and U.W.; Supervision, M.F.Q. and H.E.-A.; Writing—original draft, M.A.S. and A.U.K.; Writing—review & editing, S.E.-u.-H., U.W., M.F.Q., A.u.R., A.N., M.A.Z., J.-P.G., M.K. and H.E.-A. All authors have read and agreed to the published version of the manuscript.

Funding: This research received no external funding.

Institutional Review Board Statement: The study was conducted in accordance with the Declaration of Helsinki and approved by the Institutional Review Board and standard recommendations of the Ethical Review internal committee of the Department of Pathobiology, College of Veterinary and Animal Sciences (CVAS), Jhang, Approval Code: MICR/LEC/21/11, Approval Date: 26 October 2021.

Informed Consent Statement: Not applicable.

Data Availability Statement: The data presented in this study are available on request from the corresponding author.

Acknowledgments: The authors would like to acknowledge the field staff and zoologists who helped us in the collection of specimens and identification of wild bird species.

Conflicts of Interest: The authors declare no conflict of interest.

References

1. Perestrelo, S.; Amaro, A.; Brouwer, M.S.M.; Clemente, L.; Duarte, A.S.R.; Kaesbohrer, A.; Karpiskova, R.; Lopez-Chavarrias, V.; Morris, D.; Prendergast, D.; et al. Building an international One Health strain level database to characterise the epidemiology of AMR threats: ESBL-AmpC producing *E. coli* as an example-challenges and perspectives. *Antibiotics* **2023**, *12*, 552. [CrossRef]
2. Murray, C.J.L.; Ikuta, K.S.; Sharara, F.; Swetschinski, L.; Aguilar, G.R.; Gray, A.; Han, C.; Bisignano, C.; Rao, P.; Wool, E.; et al. Global burden of bacterial antimicrobial resistance in 2019: A systematic analysis. *Lancet* **2022**, *399*, 629–655. [CrossRef]
3. Mora-Ochomogo, M.; Lohans, C.T. beta-Lactam antibiotic targets and resistance mechanisms: From covalent inhibitors to substrates. *RSC Med. Chem.* **2021**, *12*, 1623–1639. [CrossRef]
4. Castanheira, M.; Simner, P.J.; Bradford, P.A. Extended-spectrum β-lactamases: An update on their characteristics, epidemiology and detection. *JAC-Antimicrob. Resist.* **2021**, *3*, dlab092. [CrossRef]
5. Hashim, R.; Husin, S.A.; Ahmad, N.; Bahari, N.; Abu, N.; Ali, R.M.; Hashim, N.H.F.; Ramasamy, P.; Saberi, M.R.H.; Laine, L.Y.; et al. Tricycle Project—One Health approach: Whole genome sequencing(WGS) of Extended-spectrum beta-lactamase (ESBL) producing *Eschericia (E.) coli* derived from human, food chain and environment. *Int. J. Infect. Dis.* **2022**, *116*, S105–S106. [CrossRef]
6. Chong, Y.; Shimoda, S.; Shimono, N. Current epidemiology, genetic evolution and clinical impact of extended-spectrum beta-lactamase-producing *Escherichia coli* and *Klebsiella pneumoniae*. *Infect. Genet. Evol.* **2018**, *61*, 185–188. [CrossRef]
7. Dame-Korevaar, A.; Fischer, E.A.J.; van der Goot, J.; Stegeman, A.; Mevius, D. Transmission routes of ESBL/pAmpC producing bacteria in the broiler production pyramid, a literature review. *Prev. Vet. Med.* **2019**, *162*, 136–150. [CrossRef]
8. Castronovo, C.; Agozzino, V.; Schirò, G.; Mira, F.; Di Bella, S.; Lastra, A.; Antoci, F.; Pennisi, M.; Giudice, E.; Guercio, A. Evaluation of the antimicrobial resistance of different serotypes of *Salmonella enterica* from livestock farms in Southern Italy. *Appl. Sci.* **2023**, *13*, 442. [CrossRef]
9. Hernando-Amado, S.; Coque, T.M.; Baquero, F.; Martinez, J.L. Defining and combating antibiotic resistance from One Health and Global Health perspectives. *Nat. Microbiol.* **2019**, *4*, 1432–1442. [CrossRef]
10. Alcala, L.; Alonso, C.A.; Simon, C.; Gonzalez-Esteban, C.; Oros, J.; Rezusta, A.; Ortega, C.; Torres, C. Wild birds, Frequent carriers of extended-spectrum beta-Lactamase (ESBL) producing *Escherichia coli* of CTX-M and SHV-12 types. *Microb. Ecol.* **2016**, *72*, 861–869. [CrossRef]
11. Navarro, J.; Gremillet, D.; Afan, I.; Miranda, F.; Bouten, W.; Forero, M.G.; Figuerola, J. Pathogen transmission risk by opportunistic gulls moving across human landscapes. *Sci. Rep.* **2019**, *9*, 10659. [CrossRef]
12. Baez, J.; Hernandez-Garcia, M.; Guamparito, C.; Diaz, S.; Olave, A.; Guerrero, K.; Canton, R.; Baquero, F.; Gahona, J.; Valenzuela, N.; et al. Molecular characterization and genetic diversity of ESBL-producing *Escherichia coli* colonizing the migratory Franklin's gulls (*Leucophaeus pipixcan*) in Antofagasta, North of Chile. *Microb. Drug Resist.* **2015**, *21*, 111–116. [CrossRef]
13. Zeballos-Gross, D.; Rojas-Sereno, Z.; Salgado-Caxito, M.; Poeta, P.; Torres, C.; Benavides, J.A. The role of Gulls as reservoirs of antibiotic resistance in aquatic environments: A scoping review. *Front. Microbiol.* **2021**, *12*, 703886. [CrossRef]
14. Wu, J.; Huang, Y.; Rao, D.; Zhang, Y.; Yang, K. Evidence for environmental dissemination of antibiotic resistance mediated by wild birds. *Front. Microbiol.* **2018**, *9*, 745. [CrossRef]
15. Sultan, I.; Rahman, S.; Jan, A.T.; Siddiqui, M.T.; Mondal, A.H.; Haq, Q.M.R. Antibiotics, resistome and resistance mechanisms: A bacterial perspective. *Front. Microbiol.* **2018**, *9*, 2066. [CrossRef]
16. Mahboob, S.; Zaib-un-Nisa. Diversity of Avifauna of Trimmu Barrage, District Jhang, Punjab, Pakistan. *Pak. J. Zool.* **2009**, *41*, 43–49.
17. Prestinaci, F.; Pezzotti, P.; Pantosti, A. Antimicrobial resistance: A global multifaceted phenomenon. *Pathog. Glob. Health* **2015**, *109*, 309–318. [CrossRef]
18. Russo, T.P.; Minichino, A.; Gargiulo, A.; Varriale, L.; Borrelli, L.; Pace, A.; Santaniello, A.; Pompameo, M.; Fioretti, A.; Dipineto, L. Prevalence and phenotypic antimicrobial resistance among ESKAPE bacteria and Enterobacterales strains in wild birds. *Antibiotics* **2022**, *11*, 1825. [CrossRef]

19. Anjum, M.F.; Schmitt, H.; Borjesson, S.; Berendonk, T.U.; Stehling, E.G.; Boerlin, P.; Topp, E.; Jardine, C.; Li, X.W.; Li, B.; et al. The potential of using *E. coli* as an indicator for the surveillance of antimicrobial resistance (AMR) in the environment. *Curr. Opin. Microbiol.* **2021**, *64*, 152–158. [CrossRef]

20. Umair, M.; Tahir, M.F.; Ullah, R.W.; Ali, J.; Siddique, N.; Rasheed, A.; Akram, M.; Zaheer, M.U.; Mohsin, M. Quantification and trends of antimicrobial use in commercial broiler chicken production in Pakistan. *Antibiotics* **2021**, *10*, 598. [CrossRef]

21. Mohsin, M.; Van Boeckel, T.P.; Saleemi, M.K.; Umair, M.; Naseem, M.N.; He, C.; Khan, A.; Laxminarayan, R. Excessive use of medically important antimicrobials in food animals in Pakistan: A five-year surveillance survey. *Glob. Health Action* **2019**, *12*, 1697541. [CrossRef]

22. do V. Barroso, M.; da Silva, J.S.; Moreira, S.M.; Sabino, Y.N.; Rocha, G.C.; Moreira, M.A.; Bazzolli, D.M.; Mantovani, H.C. Selection of multidrug-resistant *Enterobacteria* in weaned pigs and its association with in-feed subtherapeutic combination of colistin and tylosin. *Curr. Microbiol.* **2022**, *79*, 349. [CrossRef]

23. Mechesso, A.F.; Park, S. Tylosin exposure reduces the susceptibility of *Salmonella Typhimurium* to florfenicol and tetracycline. *BMC Vet. Res.* **2020**, *16*, 22. [CrossRef]

24. Abrar, S.; Hussain, S.; Khan, R.A.; Ul Ain, N.; Haider, H.; Riaz, S. Prevalence of extended-spectrum-β-lactamase-producing *Enterobacteriaceae*: First systematic meta-analysis report from Pakistan. *Antimicrob. Resist. Infect. Control* **2018**, *7*, 26. [CrossRef]

25. Ahsan, A.; Rehman, T.A.U.; Irshad, H.; Shahzad, M.A.; Siddique, A.; Jamil, A.; Ali, A. Antibiotic resistance pattern and molecular detection of ESBL-associated genes in *E. coli* from surface and wastewater of Islamabad capital territory, Pakistan. *J. Water Health* **2022**, *20*, 601–609. [CrossRef]

26. Zainab, L.; Ibrar, K.; Sadiq, A.; Hamid, A.K.; Ullah, M.; Noor, R. Extended spectrum beta lactamases-producing *Escherichia coli* in retail chicken meat from Khyber Pakhtunkhwa, Pakistan. *Saudi J. Biol. Sci.* **2022**, *29*, 103280. [CrossRef]

27. Ilyas, S.; Rasool, M.H.; Arshed, M.J.; Qamar, M.U.; Aslam, B.; Almatroudi, A.; Khurshid, M. The *Escherichia coli* sequence type 131 harboring extended-spectrum beta-lactamases and carbapenemases genes from poultry birds. *Infect. Drug Resist.* **2021**, *14*, 805–813. [CrossRef]

28. Liaqat, Z.; Khan, I.; Azam, S.; Anwar, Y.; Althubaiti, E.H.; Maroof, L. Isolation and molecular characterization of extended spectrum beta lactamase producing *Escherichia coli* from chicken meat in Pakistan. *PLoS ONE* **2022**, *17*, e0269194. [CrossRef]

29. Naas, T.; Oueslati, S.; Bonnin, R.A.; Dabos, M.L.; Zavala, A.; Dortet, L.; Retailleau, P.; Iorga, B.I. Beta-lactamase database (BLDB)—structure and function. *J. Enzym. Inhib. Med. Chem.* **2017**, *32*, 917–919. [CrossRef]

30. Tagar, S.; Qambrani, N.A. Bacteriological quality assessment of poultry chicken meat and meat contact surfaces for the presence of targeted bacteria and determination of antibiotic resistance of *Salmonella* spp. in Pakistan. *Food Control* **2023**, *151*, 109786. [CrossRef]

31. Islam, M.S.; Sobur, M.A.; Rahman, S.; Ballah, F.M.; Ievy, S.; Siddique, M.P.; Rahman, M.; Kafi, M.A.; Rahman, M.T. Detection of bla(TEM), bla(CTX-M), bla(CMY), and bla(SHV) genes among extended-spectrum beta-lactamase-producing *Escherichia coli* isolated from migratory birds travelling to Bangladesh. *Microb. Ecol.* **2022**, *83*, 942–950. [CrossRef] [PubMed]

32. Fashae, K.; Engelmann, I.; Monecke, S.; Braun, S.D.; Ehricht, R. Molecular characterisation of extended-spectrum ss-lactamase producing *Escherichia coli* in wild birds and cattle, Ibadan, Nigeria. *BMC Vet. Res.* **2021**, *17*, 33. [CrossRef] [PubMed]

33. Ben Yahia, H.; Chairat, S.; Gharsa, H.; Alonso, C.A.; Ben Sallem, R.; Porres-Osante, N.; Hamdi, N.; Torres, C.; Ben Slama, K. First report of KPC-2 and KPC-3-producing *Enterobacteriaceae* in wild birds in Africa. *Microb. Ecol.* **2020**, *79*, 30–37. [CrossRef] [PubMed]

34. Athanasakopoulou, Z.; Diezel, C.; Braun, S.D.; Sofia, M.; Giannakopoulos, A.; Monecke, S.; Gary, D.; Krahmer, D.; Chatzopoulos, D.C.; Touloudi, A.; et al. Occurrence and characteristics of ESBL- and carbapenemase- producing *Escherichia coli* from wild and feral birds in Greece. *Microorganisms* **2022**, *10*, 1217. [CrossRef] [PubMed]

35. Zhang, S.H.; Lv, X.; Han, B.; Gu, X.; Wang, P.F.; Wang, C.; He, Z. Prevalence of antibiotic resistance genes in antibiotic-resistant *Escherichia coli* isolates in surface water of Taihu Lake Basin, China. *Environ. Sci. Pollut. Res. Int.* **2015**, *22*, 11412–11421. [CrossRef] [PubMed]

36. Noor, R.; Maqsood, A.; Baig, A.; Pande, C.B.; Zahra, S.M.; Saad, A.; Anwar, M.; Singh, S.K. A comprehensive review on water pollution, South Asia Region: Pakistan. *Urban. Clim.* **2023**, *48*, 101413. [CrossRef]

37. Hashmi, M.Z.; Malik, R.N.; Shahbaz, M. Heavy metals in eggshells of cattle egret (*Bubulcus ibis*) and little egret (*Egretta garzetta*) from the Punjab province, Pakistan. *Ecotoxicol. Environ. Saf.* **2013**, *89*, 158–165. [CrossRef]

38. Tausova, D.; Dolejska, M.; Cizek, A.; Hanusova, L.; Hrusakova, J.; Svoboda, O.; Camlik, G.; Literak, I. *Escherichia coli* with extended-spectrum beta-lactamase and plasmid-mediated quinolone resistance genes in great cormorants and mallards in Central Europe. *J. Antimicrob. Chemother.* **2012**, *67*, 1103–1107. [CrossRef]

39. Osińska, M.; Nowakiewicz, A.; Zięba, P.; Gnat, S.; Łagowski, D.; Trościańczyk, A. A rich mosaic of resistance in extended-spectrum β-lactamase-producing *Escherichia coli* isolated from red foxes (*Vulpes vulpes*) in Poland as a potential effect of increasing synanthropization. *Sci. Total Environ.* **2022**, *818*, 151834. [CrossRef]

40. Shah, A.; Alam, S.; Kabir, M.; Fazal, S.; Khurshid, A.; Iqbal, A.; Khan, M.M.; Khan, W.; Qayyum, A.; Hussain, M. Migratory birds as the vehicle of transmission of multi drug resistant extended spectrum β lactamase producing *Escherichia fergusonii*, an emerging zoonotic pathogen. *Saudi J. Biol. Sci.* **2022**, *29*, 3167–3176. [CrossRef]

41. Hessman, J.; Atterby, C.; Olsen, B.; Jarhult, J.D. High prevalence and temporal variation of extended spectrum beta-lactamase-producing bacteria in urban Swedish Mallards. *Microb. Drug Resist.* **2018**, *24*, 822–829. [CrossRef] [PubMed]

42. Raza, S.; Mohsin, M.; Madni, W.A.; Sarwar, F.; Saqib, M.; Aslam, B. First Report of bla (CTX-M-15)-Type ESBL-producing *Klebsiella pneumoniae* in wild migratory birds in Pakistan. *Ecohealth* **2017**, *14*, 182–186. [CrossRef] [PubMed]

43. Picão, R.C.; Poirel, L.; Gales, A.C.; Nordmann, P. Further identification of CTX-M-2 extended-spectrum β-lactamase in *Pseudomonas aeruginosa*. *Antimicrob. Agents Chemother.* **2009**, *53*, 2225. [CrossRef] [PubMed]

44. Liu, C.M.; Stegger, M.; Aziz, M.; Johnson, T.J.; Waits, K.; Nordstrom, L.; Gauld, L.; Weaver, B.; Rolland, D.; Statham, S. *Escherichia coli* ST131-H 22 as a Foodborne Uropathogen. *MBio* **2018**, *9*, e00470-18. [CrossRef] [PubMed]

45. Mohsin, M.; Raza, S.; Schaufler, K.; Roschanski, N.; Sarwar, F.; Semmler, T.; Schierack, P.; Guenther, S. High prevalence of CTX-M-15-type ESBL-producing *E. coli* from migratory avian species in Pakistan. *Front. Microbiol.* **2017**, *8*, 2476. [CrossRef]

46. Darwich, L.; Vidal, A.; Seminati, C.; Albamonte, A.; Casado, A.; Lopez, F.; Molina-Lopez, R.A.; Migura-Garcia, L. High prevalence and diversity of extended-spectrum beta-lactamase and emergence of OXA-48 producing Enterobacterales in wildlife in Catalonia. *PLoS ONE* **2019**, *14*, e0210686. [CrossRef]

47. Sun, F.; Chen, J.; Liu, K.; Tang, M.; Yang, Y. The avian gut microbiota: Diversity, influencing factors, and future directions. *Front. Microbiol.* **2022**, *13*, 934272. [CrossRef]

48. Leimbach, A.; Hacker, J.; Dobrindt, U. *E. coli* as an All-Rounder: The thin line between commensalism and pathogenicity. *Curr. Top. Microbiol. Immunol.* **2013**, *358*, 3–32. [CrossRef]

49. He, D.D.; Zhao, S.Y.; Wu, H.; Hu, G.Z.; Zhao, J.F.; Zong, Z.Y.; Pan, Y.S. Antimicrobial resistance-encoding plasmid clusters with heterogeneous MDR regions driven by IS 26 in a single *Escherichia coli* isolate. *J. Antimicrob. Chemother.* **2019**, *74*, 1511–1516. [CrossRef]

50. CLSI. *Performance Standards for Antimicrobial Susceptibility Testing*; CLSI Supplement M100-S25; Clinical and Laboratory Standards Institute: Wayne, PA, USA, 2020.

51. Monstein, H.J.; Ostholm-Balkhed, A.; Nilsson, M.V.; Nilsson, M.; Dornbusch, K.; Nilsson, L.E. Multiplex PCR amplification assay for the detection of blaSHV, blaTEM and blaCTX-M genes in *Enterobacteriaceae*. *APMIS* **2007**, *115*, 1400–1408. [CrossRef]

52. Coculescu, B.; Palade, A.; Delcaru, C.; Coculescu, E.C. Genetic analysis of multidrug-resistant *Salmonella enterica* serovar Typhimurium strains producing extended-spectrum B-lactamases (ESBL) associated with diarrhea in romanian pediatric patients. *Rom. Biotechnol. Lett.* **2016**, *21*, 11393–11403.

53. Uysal, A.; Gunes, E.; Arslan, E.; Durak, Y. Characterization of uropathogenic *Escherichia coli* stains: Antibiotic resistance patterns, detection of Esbl genes and interactions by lytic phages. *Fresenius Environ. Bullet.* **2018**, *27*, 402–414.

54. Weisburg, W.G.; Barns, S.M.; Pelletier, D.A.; Lane, D.J. 16S ribosomal DNA amplification for phylogenetic study. *J. Bacteriol.* **1991**, *173*, 697–703. [CrossRef]

55. Wollenberg, K.R.; Atchley, W.R. Separation of phylogenetic and functional associations in biological sequences by using the parametric bootstrap. *Proc. Natl. Acad. Sci. USA* **2000**, *97*, 3288–3291. [CrossRef] [PubMed]

Article

Antimicrobial Resistance in *Escherichia coli* and Its Correlation with Antimicrobial Use on Commercial Poultry Farms in Bangladesh

Nelima Ibrahim [1,2,*], Filip Boyen [3], Md. Abu Shoieb Mohsin [4], Moniek Ringenier [1], Anna Catharina Berge [1], Ilias Chantziaras [1], Guillaume Fournié [5,6,7], Dirk Pfeiffer [7,8] and Jeroen Dewulf [1]

1   Faculty of Veterinary Medicine, Department of Internal Medicine, Reproduction and Population Medicine, Ghent University, Salisburylaan 133, 9820 Merelbeke, Belgium
2   Department of Livestock Services, Dhaka 1215, Bangladesh
3   Faculty of Veterinary Medicine, Department of Pathobiology, Pharmacology and Zoological Medicine, Ghent University, Salisburylaan 133, 9820 Merelbeke, Belgium
4   Department of Medicine and Surgery, Chattogram Veterinary and Animal Sciences University, Chattogram 4202, Bangladesh; shoieb.tox@gmail.com
5   Université de Lyon, INRAE, VetAgro Sup, UMR EPIA, 69280 Marcy l'Etoile, France
6   Université Clermont Auvergne, INRAE, VetAgro Sup, UMR EPIA, 63122 Saint Genes Champanelle, France
7   Department of Pathobiology and Population Sciences, Royal Veterinary College, Royal College Street, London NW1 0TU, UK
8   Department of Infectious Diseases and Public Health, Jockey Club College of Veterinary Medicine and Life Sciences, City University of Hong Kong, Hong Kong SAR, China
*   Correspondence: nelima.ibrahim@ugent.be or nibrahim421@gmail.com; Tel.: +32-484-88-09-78

Abstract: Antimicrobial resistance is a global concern, posing risks to human and animal health. This research quantified antimicrobial resistance (AMR) in *E. coli* isolates from poultry fecal and environmental samples in Bangladesh and explored their association with antimicrobial use (AMU). We screened 725 fecal and 250 environmental samples from 94 conventional broilers and 51 Sonali farms for *E. coli* presence using MALDI-TOF mass spectrometry. AMU data were collected at flock levels, expressed as treatment incidence (TI), while minimum inhibitory concentrations (MIC) for 14 antibiotics were determined on five fecal *E. coli* isolates per farm and on all environmental isolates. MIC results were interpreted using human clinical breakpoints and EUCAST epidemiological cut-off values (ECOFFs). Acquired resistance against commonly used antimicrobial agents such as ciprofloxacin, tetracycline and ampicillin, was extremely high and predominantly clinically relevant. There was a moderate correlation between fecal and environmental antibiotic resistance index (ARI), but there was no significant correlation between AMU and AMR, suggesting that the observed AMR prevalence is unrelated to current AMU in poultry, but may be due to high historical AMU. A high level of multidrug resistance, including against critically important antimicrobials, was found in both farm types. Therefore, an AMR/AMU surveillance program is urgently needed in the poultry production sector of Bangladesh.

Keywords: MIC; fecal and environmental samples; ECOFF value; CLSI; broiler; Sonali; multidrug resistance (MDR); *E. coli*; Bangladesh

Citation: Ibrahim, N.; Boyen, F.; Mohsin, M.A.S.; Ringenier, M.; Berge, A.C.; Chantziaras, I.; Fournié, G.; Pfeiffer, D.; Dewulf, J. Antimicrobial Resistance in *Escherichia coli* and Its Correlation with Antimicrobial Use on Commercial Poultry Farms in Bangladesh. *Antibiotics* **2023**, *12*, 1361. https://doi.org/10.3390/antibiotics12091361

Academic Editor: Nahla O. Eltai

Received: 21 July 2023
Revised: 11 August 2023
Accepted: 16 August 2023
Published: 24 August 2023

1. Introduction

Antimicrobial resistance (AMR) has been identified by the World Health Organization (WHO) as one of the major threats to public health in the present and near future [1]. The WHO launched a Global Action Plan (GAP) in 2015 to tackle this developing global issue in a comprehensive manner, based on a 'One Health' strategy that emphasizes the interdependence of human, animal and environmental health [2,3]. South Asia is considered to be exposed to the highest risk of AMR among all WHO regions due to its large human population and high level of antimicrobial use (AMU) in both humans and animals [4]. The escalation of evolutionary trends leading to AMR poses a significant threat to the health of both humans and animals [2,5,6]. Apart from their crucial role in treating and preventing human infections, antibiotics have been extensively used in food-producing animals. However, this practice poses a significant concern as it creates a reservoir of antibiotic-resistant bacteria and AMR genes, which have the potential to be transferred to humans [7]. This transfer of antibiotic resistance from animals to humans further exacerbates the global challenge of antimicrobial resistance and brings attention to the effectiveness of antibiotics in human healthcare [6,8,9]. The World Organization for Animal Health (WOAH) recommends monitoring AMR in commensal *E. coli* sampled from animals [10]. Indeed, although commensal *E. coli* is known not to be harmful to their host, several studies have shown that *E. coli* can develop resistance and serve as a reservoir for multidrug resistance (MDR) both in animal populations and environment, making it a useful indicator organism for measuring antimicrobial resistance (AMR) [11–13]. According to the Bangladesh Fish Feed and Animal Feed Act, 2010, the government banned the use of antimicrobials in animal feed [14]. In accordance with WHO GAP standards, Bangladesh has approved a National Action Plan (NAP) to combat AMR [15]. Population density, easy access and inappropriate use of antimicrobials, contamination of environments with animal manure, all contribute to the occurrence of drug-resistant community-acquired infections [16–20]. In Bangladesh, Sonali chicken consumption holds the second position after broiler meat in terms of popularity [18]. Sonali chicken, which is a cross-breed of Rhode Island Red cocks and Fayoumi hens, exhibits a similar appearance and taste to native chickens. In Bangladesh, several studies have been conducted on the prevalence of MDR *E. coli* in broilers [21–25]. However, none of these studies have investigated the link between AMR and AMU. Additionally, there is a lack of available data on AMR of *E. coli* in Sonali chickens. Furthermore, there is a notable absence of studies that use epidemiological cut-off values (ECOFFs) to interpret the results. ECOFF values provide a standardized framework for interpreting antimicrobial susceptibility data, allowing for more reliable comparisons and informed decision-making. To address these gaps, this study aimed to quantify antimicrobial resistance patterns in *E. coli* isolates obtained from chicken fecal and environmental samples collected from farms raising conventional broiler and Sonali chickens, the most produced and consumed poultry types in Bangladesh [18], using appropriate testing protocols and interpretation criteria. In addition, the associations between antimicrobial use (AMU) and resistance (AMR), as well as any differences in resistance among broiler and Sonali chicken isolates, were explored.

2. Results

In total, 725 fresh fecal and 250 environmental samples were collected. *E. coli* was recovered in 98% and 78% of fecal and environmental samples, respectively. The number of isolates obtained per farm ranged from three to seven. Minimum inhibitory concentrations (MICs) testing was conducted on 691 fecal and 191 environmental *E. coli* isolates, corresponding to 3–5 isolates randomly selected per farm. The MIC values for QC strains were within the acceptable ranges as described by CLSI [26].

2.1. AMR Results

Results of the antimicrobial susceptibility testing (MIC distributions) are summarized in Table 1 for fecal and in Table 2 for environmental isolates of both broiler and Sonali farms. The Supplementary Tables provide separate results for broiler and Sonali farms, including findings for both fecal and environmental samples (Supplementary Tables S1–S4). All fecal and environmental isolates showed MDR towards three to seven antibiotic classes. Based on ECOFFs, the proportion of isolates resistant (i.e., non-wild type) to ciprofloxacin, tetracycline, ampicillin, nalidixic acid, trimethoprim or sulfamethaxole ranged from 93% to 99%. About two-thirds of fecal isolates were resistant to chloramphenicol or azithromycin. In contrast, about two-thirds of isolates were wild types with respect to gentamicin or tigecycline and the proportion of resistant isolates was ≤6% for ceftazidime and cefotaxime, whereas for colistin, it was 12%. The antimicrobial compounds associated with the highest levels of resistance in the environmental isolates were the same as in fecal samples, despite slightly lower proportion of resistant isolates. The main difference was for chloramphenicol, with the proportion of resistant isolates dropping to a third of all tested environmental isolates. No significant difference was observed in the resistance prevalence between broiler and Sonali fecal and environmental isolates.

Acquired resistance in non-wild-type isolates was predominantly clinically relevant as assessed using CLSI clinical breakpoints (Tables 1 and 2).

As shown in Figure 1, a significant ($p < 0.001$) correlation ($R^2 = 0.35$) was found between the antimicrobial resistance index (ARI) of fecal and environmental isolates in both broiler and Sonali.

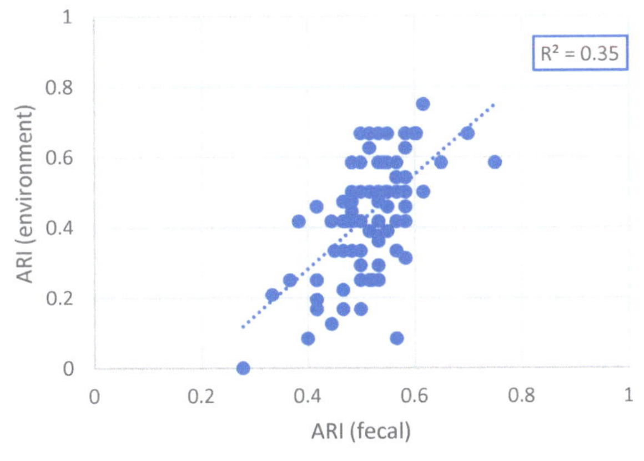

Figure 1. The correlation between antimicrobial resistance (ARI) of fecal and environmental isolates of all sampled farms.

Table 1. Minimum inhibitory concentration distribution for *E. coli* in fecal isolates.

Antimicrobial Agent	Number of Strains with MIC (mg/mL)														Susceptible		Intermediate		Resistant		Wild Type		Non-Wild Type		No. of Isolates
	≤0.03	0.06	0.12	0.25	0.5	1	2	4	8	16	32	64	128	256	[n]	[%]	[n]	[%]	[n]	[%]	[n]	[%]	[n]	[%]	
Gentamicin					*67	231	200	38	3	1	5	146			498	72	41	6	152	22	498	72	193	28	691
Tetracycline							*2	6	4		6	69	604		2	0	10	1	679	98	12	2	679	98	691
Ceftazidime					*656	5	4	3	11	12					665	96	14	2	12	2	661	96	30	4	691
Trimethoprim				*18	11	6	1	2			1	652			38	5	0	0	653	95	36	5	655	95	691
Ampicillin						*1	21	25	2	1		3	638		47	7	3	0	641	93	49	7	642	93	691
Chloramphenicol									*109	138	73	89	78	204	0	0	247	36	444	64	247	36	444	64	691
Azithromycin									1	37	201	104	77	271	38	5	201	29	452	65	239	35	452	65	691
Nalidixic acid								*10	17	40	9	8	607		27	4	40	6	624	90	27	4	664	96	691
Colistin						*604	7	34	44	1	1				604	87	7	1	80	12	611	88	80	12	691
Cefotaxime				*655	8	2	7	1	18						663	96	9	1	19	3	655	95	36	5	691
Meropenem	*691														691	100	0	0	0	0	691	100	0	0	691
Ciprofloxacin	*8	1	2	2	38	29	23	27	39	522					11	1	40	6	640	93	9	1	682	99	691
Tigecycline					*274	230	147	25	4	4	7				-	-	-	-	-	-	504	73	187	27	691

Antimicrobial Agent	Number of Strains with MIC (mg/mL)									Susceptible		Intermediate		Resistant		Wild Type		Non-Wild Type		No. of Isolates
	≤8	16	32	64	128	256	512	1024	2048	[n]	[%]	[n]	[%]	[n]	[%]	[n]	[%]	[n]	[%]	
Sulfamethoxazole	*10	11	11	5				1	653	37	5	0	0	654	95	37	5	654	95	691

* No visible growth at this concentration, meaning MIC is equal to or below this concentration. ⌐CLSI breakpoint between susceptible and intermediate, ⌐CLSI breakpoint between intermediate and resistant and ▌ECOFF value between wild type and non-wild type.

Table 2. Minimum inhibitory concentration distribution for *E. coli* in environmental isolates.

Antimicrobial Agent	≤0.03	0.06	0.12	0.25	0.5	1	2	4	8	16	32	64	128	256	Susceptible [n]	[%]	Intermediate [n]	[%]	Resistant [n]	[%]	Wild Type [n]	[%]	Non-Wild Type [n]	[%]	No. of Isolates
Gentamicin					*12	72	48	8	1		2	20			132	81	9	6	22	13	132	81	31	19	163
Tetracycline							*4	6	2	3	1	19	128		4	2	8	5	151	93	12	7	151	93	163
Ceftazidime				*152	5	5				1					162	99	0	0	1	1	157	96	6	4	163
Trimethoprim			*21	13	4						125				38	23	0	0	125	77	38	23	125	77	163
Ampicillin						12	13	1	3	2	4	128			25	15	4	2	134	82	26	16	137	84	163
Chloramphenicol								*50	59	4	15	7	28		0	0	109	67	54	33	109	67	54	33	163
Azithromycin								5	49	28	28	53			5	3	49	30	109	67	54	33	109	67	163
Nalidixic acid								*23	22	12	6		2	98	45	28	12	7	106	65	45	28	118	72	163
Colistin						*143	4		2	6	2	6			143	88	4	2	16	10	147	90	16	10	163
Cefotaxime			*152		7	1	2		1						159	98	3	2	1	0	152	93	11	7	163
Meropenem	*163														163	100	0	0	0	0	163	100	0	0	163
Ciprofloxacin	*13	3	2	8	27	8	4	10	6	82					18	11	35	21	110	67	16	10	147	90	163
Tigecycline				*45	73	32	8	4	1						-	-	-	-	-	-	118	72	45	28	163

Antimicrobial Agent	≤8	16	32	64	128	256	512	1024	2048	Susceptible [n]	[%]	Intermediate [n]	[%]	Resistant [n]	[%]	Wild Type [n]	[%]	Non-wild Type [n]	[%]	No. of Isolates
Sulfamethoxazole	*9	7	9	5	2				131	32	20	0	0	131	80	32	20	131	80	163

* No visible growth at this concentration, meaning MIC is equal to or below this concentration. └─CLSI breakpoint between susceptible and intermediate, └─CLSI breakpoint between intermediate and resistant and ▮─ ECOFF value between wild type and non-wild type.

2.2. AMU in Flock Level

Antimicrobials were used on all farms, but there was a huge variation in the amount of AMU in the different farms. The quantification of AMU (TI$_{DDDvet}$) and the detailed results were described by Ibrahim et al., 2023 [27]. The median of treatment incidence which expresses the number of Defined Daily Dose (TI$_{DDDvet}$) was 60.0 (range 18.3–188.2) for conventional broilers and 58.3 (range 31.1–212.6) for Sonali chickens. This indicates that conventional broilers and Sonali birds were treated with antimicrobials for approximately 60% and 58% of their lifetime, respectively [27].

2.3. Linking Antimicrobial Use and Antimicrobial Resistance

Prior to exploring the association between antimicrobial use and antimicrobial resistance, Figure 2 depicts the frequency distribution of the ARI.

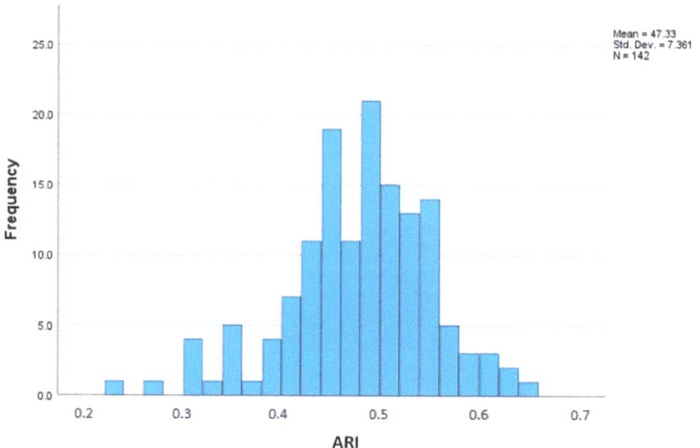

Figure 2. Distribution of antimicrobial resistance index of farms.

There was no significant ($p = 0.73$) correlation between flock level of total antimicrobial use (TI$_{DDDvet}$) and antimicrobial resistance (ARI), as shown in Figure 3.

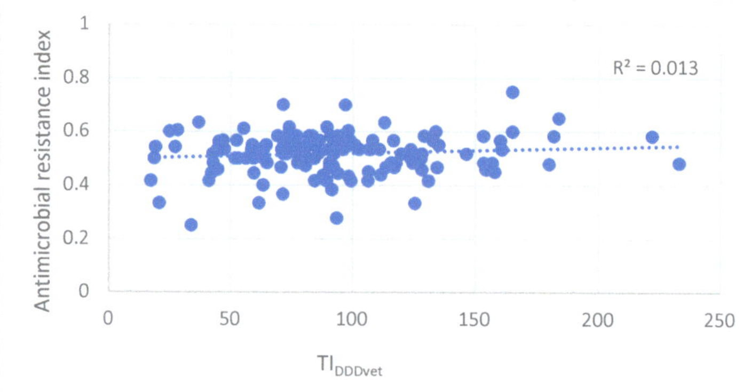

Figure 3. The correlation between antimicrobial use (TI$_{DDDvet}$) and antimicrobial resistance index (ARI) on all sampled farms.

In both broiler and Sonali farms, ciprofloxacin, ampicillin, tetracycline, colistin, gentamicin and trimethoprim were reported as being most commonly used. In Table 3, the association between the use of these molecules and the observed resistance prevalence is shown. Only for tetracycline ($p = 0.01$) and gentamicin ($p = 0.02$) a significant association was found between the use of the compound in the farm and the resistance prevalence.

Table 3. Proportion of farms that used most common six (6) antimicrobials and two antimicrobial class.

Name of Antibiotics		Number of Farms	No. of Susceptible Isolates	No. of Resistant Isolates	Prevalence of Resistance (%)	p Value
Ciprofloxacin	Use	64	4	311	99	0.399
	No use	78	8	368	98	
Ampicillin	Use	91	33	422	93	0.983
	No use	51	17	219	93	
Tetracycline	Use	75	2	361	99	0.01
	No use	67	10	318	97	
Trimethoprim	Use	22	7	101	94	0.760
	No use	120	32	551	95	
Colistin	Use	80	378	30	7	0.570
	No use	62	266	17	6	
Gentamicin	Use	8	25	15	38	0.02
	No use	134	511	140	22	
Fluoroquinolones class	Use	89	9	485	98	0.812
	No use	53	3	194	98	
Sulfonamides class	Use	101	26	475	95	0.829
	No use	41	11	179	94	

3. Discussion

The emergence of antimicrobial resistance has become a major animal and public health threat. The current findings reveal a high level of antimicrobial resistance in fecal and environmental isolates towards the most commonly used antimicrobials in both broiler and Sonali farms. Furthermore, in all fecal and environmental isolates, 100% of the investigated poultry *E. coli* strains showed MDR towards three to seven antibiotic classes. Although this is not the first report of MDR *E. coli* in poultry in Bangladesh [21–25,28,29] this manuscript's results have added value over previous reports both in terms of the methods used and interpretation of the results. For example, we have used MIC testing, compared to disk diffusion in the past, which is generally regarded as less reliable than MIC testing and is not reliable at all for certain antibiotics, such as colistin [30] and we interpreted susceptibility testing results using both wild-type cut-off value and clinical breakpoints. Additionally, this manuscript describes for the first time, the association between antimicrobial usage (AMU) and antimicrobial resistance (AMR), as well as the correlation between ARI of fecal and environmental isolates based on Bangladesh data.

The *E. coli* isolates obtained both from fecal and environmental samples from broiler and Sonali chicken in the current study show acquired resistance in varying degrees to different antimicrobial agents. All the isolates collected from fecal samples in this study had similar to higher resistance levels (higher percentage of resistant bacteria) to commonly used antimicrobials on the farm compared to some of the previous studies in Bangladesh [21–25]. Based on most recent publications [28,31], *E. coli* in poultry and poultry environments were found to have varying but generally high degrees of ampicillin, ciprofloxacin, tetracycline, sulfamethoxazole and trimethoprim resistance in Bangladesh, reaching up to 100%, similar to this study. These high levels of acquired resistance are likely due to the long-term use of these antimicrobials in poultry in Bangladesh.

In the current study, colistin resistance was found in 12% of fecal isolates, which may more closely estimate colistin resistance prevalence in poultry in Bangladesh, compared with previous investigations using the disk diffusion method, which reported higher resistance rates [21,22,28], that is not reliable for colistin susceptibility testing [21,30]. Since many laboratories still rely on the cheaper disk diffusion test, the emergence of colistin resistance may be misjudged and needs to be monitored closely using the appropriate test methods [30]. Nevertheless, it has been shown using molecular methods that mobilized colistin resistance (mcr) genes, which are associated with colistin resistance, has been detected in up to 25% of *E. coli* isolates obtained from poultry in Bangladesh [32]. It is important to mention that during the time of this study, antimicrobial combinations including colistin, but excluding the liquid single oral solution in bottles of at least one liter, were prohibited by the Directorate General of Drug Administration in 2019 [33]. However, despite this ban, farmers continued to use colistin due to its availability in liquid form on the market. Finally, colistin was fully prohibited in all forms in 2022 [34]. It is unclear to what extent the observed lower resistance prevalence for colistin in the current study, compared to previous studies in Bangladesh, is due to the use of a different methodology of sampling and/or resistance detection or due to the (partial) ban of colistin use. Further molecular characterization of current isolates may confirm observed phenotypic results but were out of the scope of the current investigations.

In case of cefotaxime, this study found 5% and 7% resistance in fecal and environmental isolates, respectively, whereas high resistance has been reported in isolates from poultry cloacal swabs and farm sewage samples in Bangladesh, though in a different region and using a different susceptibility testing protocol [22]. In this study, *E. coli* isolates obtained from both fecal and environmental samples were 100% wild type for meropenem, though other studies showed meropenem resistance in cloacal samples, sewerage and hand washes samples in Bangladesh [22,23,35]. Considering the fact that meropenem is probably not used in poultry in Bangladesh, the rare occasion of isolating meropenem-resistant *E. coli* in poultry, likely indicates human to poultry transmission of such isolates.

In this study, *E. coli* isolates demonstrated high levels of acquired resistance (98% and 89% in fecal and environmental isolates, respectively) to the quinolone-class antibiotic ciprofloxacin [36]. Ciprofloxacin has been widely used in commercial poultry farms in Bangladesh over the last decade [33,35] though its use in poultry is strictly regulated in the European Union (EU) or even forbidden in the USA and in large parts of the world [37,38]. According to a recent study in Bangladesh, fluoroquinolones were the most frequently used antimicrobial class in broiler chickens [27]. In this context, combinations of ciprofloxacin with trimethoprim were banned by the Directorate General of Drug Administration in Bangladesh in 2019 [33]. Furthermore, ciprofloxacin use can cause cross-resistance to other members of the quinolone class [36]. Consequently, resistance to nalidixic acid was found to be 96% and 72% for fecal and environmental isolates, respectively, despite the fact that this antibiotic was not used in the farms of Bangladesh.

Azithromycin is a commonly used macrolide for the treatment of invasive *E. coli* infections in humans in Bangladesh [28]. The fact that 65% fecal and 67% environmental isolates obtained in the current study had acquired resistance against this critically important agent indicates a potential serious human health issue. The high percentage of azithromycin resistance in *E. coli* isolates found in poultry and poultry environments in the current study is somewhat unexpected because this antibiotic is not commonly used in the poultry farms of Bangladesh [27]. One explanation might be the fact that farmers often raise other animals such as cattle or goats on the same farm and azithromycin is a commonly used antibiotic for large animal treatment in Bangladesh [29]. In a recent study, Amin et al. (2020) found that *E. coli* resistance against azithromycin was 100% in cattle and their environment [29]. In this context, azithromycin was recently banned for veterinary use by the Directorate General of Drug Administration, Bangladesh (2022) [34].

We also interpreted the MIC results using human clinical breakpoint showing that the acquired resistance was very often also clinically relevant. For example, in Bangladesh, fluoroquinolones, which are considered as a first-line antibiotic therapy for *E. coli* infections [28], are widely used to treat bacterial infections in humans, poultry and other animals.

The fecal AMR index and environmental AMR index had a moderate correlation of 0.54. This correlation indicates that flocks contaminated the farm environment and/or vice versa. This could be attributed to the lack of farm biosecurity and the fact that poultry litter is often dried on the farm premises before sale, or directly used as a fertilizer in vegetable fields. This correlation, however, was only moderate, implying that there could be other (fecal) sources of *E. coli* isolates with a different antimicrobial resistance pattern in the farm environment. Possible sources might be the proximity of other farms or animal species (cattle, goat, etc.), or even the presence of humans and related wastewater. Resistance in environmental isolates may also be indirectly related to AMU in chickens and other host species through resistance selection induced by antibiotic residues present in both manure and wastewater. Contact with these bacteria has the potential to spread AMR in humans.

All the *E. coli* isolates (100%) of the current study showed MDR against at least three, but up to seven antimicrobial classes which is consistent with previous studies [25,39–41]. According to a recent comprehensive analysis, food animals and particularly poultry, are probably responsible for a proportion of *E. coli* infections in humans with extra-intestinal, extended-spectrum cephalosporin resistance [41,42]. Interventions that restrict antibiotic use in food-producing animals have been linked to a decrease in the presence of antibiotic-resistant bacteria in these animals [43]. There is a limited but indicative set of evidence that suggests a similar link in the human populations studied, particularly those with direct exposure to food-producing animals [43].

In the current study, the observed associations between AMU and AMR were very weak to absent. Stuart Levy introduced the threshold theory, which suggests that a certain level of antimicrobial drug consumption is required to trigger the emergence of resistance in a particular environment [44]. Austin et al. (1999) supported this theory by describing the sigmoidal rise in resistance over time in the presence of a constant rate of antimicrobial consumption [45]. This suggests that small changes in the amount of antimicrobials used in a population with a low level of AMR may lead to much larger changes in resistance than the effect of comparable changes in use in a population where already a (very) high level of resistance is present [46]. This might explain why we could not find significant associations. It also highlights the importance of reacting on emerging resistance at the earliest possible phase. Furthermore, when studying the link between use and resistance in bacteria using field data, the observed levels of resistance are a reflection of current and historical use, whereas measured use often only reflects recent use or, at best, a retrospect of only a short period [36,47–49].

The Bangladesh government has developed a National Action Plan (NAP) for the period 2017–2022 to combat AMR [15]. The main goals were to identify and restrict the sale of critical antimicrobials used in food animals, to stop the "over-the-counter drug sale", to monitor and assess compliance with withdrawal periods, by providing training to farmers and poultry workers to raise awareness. However, the findings of this study clearly indicate that these goals have not been achieved so far. Establishing a regulatory framework (e.g., antimicrobial use law) to control AMU is of utmost importance in Bangladesh to effectively combat AMR. Given that the current timeframe of the NAP has expired, it is essential to extend it and identify the reasons for the NAP's shortcomings in order to take appropriate measures for its successful implementation. The findings of this study are expected to provide valuable insights for policymakers and practitioners, aiding them in revising the NAP to address the present circumstances and expedite its practical execution. To comprehend the resistance mechanisms and relatedness of the *E. coli* strains, future research should involve conducting genotypic resistance and phylogenetic analysis due to the high level of resistance observed.

4. Materials and Methods

4.1. Study Design

A cross-sectional study was conducted in seven (7) districts of Bangladesh to collect AMU data and samples for isolating *E. coli*. A total of 145 small-scale (range: 500–2500 number of birds) commercial conventional broiler (94) and Sonali (51) poultry farms were recruited. The recruitment criteria of the farms were described in Ibrahim et al., 2023 [27]. Eligible farms adopted an all-in all-out production system and had a farm size of >1000 birds per batch. In cases where multiple sheds were present, only one shed was considered randomly. Each farm was visited twice, upon the delivery of day-old chicks and within two days prior to the chicken being sold [27].

On each farm, samples were collected on the second visit. Fresh feces were collected from 5 healthy-appearing chickens and environmental samples (swab from soil) were collected in the area between the farm gate and the shed and near the vegetable field due to the practice of using poultry litter as manure for vegetables. Two environmental samples were collected from each farm located in the northern districts (n = 105) and one environmental sample from each farm located in the southeast districts (n = 40). The swabs were transported in ice-pack cooled boxes to the Central Disease Investigation Laboratory (CDIL), Bangladesh where they were stored at –80 °C until further use.

Fecal samples were inoculated on MacConkey III agar (Oxoid company, Dhaka, Bangladesh) and incubated aerobically at 37 °C for 18 to 24 h. Environmental samples were first resuscitated in 10 mL Brain Heart Infusion (BHI) broth and incubated aerobically at 37 °C for 18 to 24 h before being inoculated on MacConkey agar and incubated aerobically at 37 °C for another 18 to 24 h. Before sending all lactose positive Enterobacteriaceae isolates to the laboratory of Ghent University, Belgium, these isolates were passaged three times on MacConkey III agar, as required by Belgian law to minimize the chance of importing Newcastle Disease virus and highly pathogenic Avian Influenza virus.

Upon arrival at the laboratory, isolates were purified on Columbia agar with 5% sheep blood and subsequently confirmed to be *E. coli* using MALDI-TOF mass spectrometry, as described previously [50]. Antimicrobial susceptibility testing of the *E. coli* isolates was performed using a Sensititre EU Surveillance *E. coli* EUVSEC Plate (Trek Diagnostic Systems, Thermofisher Scientific, Merelbeke, Belgium) according to the manufacturer's guidelines. In short, 1 to 3 colonies were suspended in sterile saline to an optical density of 0.5 McFarland. Fifty microliter of this suspension was inoculated in 10 mL sterile Mueller Hinton broth. Again, fifty microliter of the Mueller Hinton broth with bacteria was transferred to each of the wells in the Sensititre micro-titer plate with the lyophilized antimicrobials using a multichannel (final concentration of 2.5×10^4 CFU/well).

After incubation at 35 °C for 18–24 h, the plates were examined and growth endpoints were established for each antimicrobial to provide MICs. The Minimum inhibitory concentration (MIC) was defined as the lowest concentration by which no visible growth could be detected. Quality control (QC) strains, *E. coli* ATCC 25922 and *E. coli* NCTC 13846 (for QC colistin resistance), were used throughout the study [26]. MIC values were interpreted based on (1) the epidemiological cut-off values (ECOFFs) published by the European Committee on Antimicrobial Susceptibility Testing (EUCAST) [51] and (2) the human clinical breakpoints published by CLSI [26] as no clinical breakpoints for poultry were available (Table 4).

Isolates having MIC values greater or equal than the ECOFF were considered to have acquired resistance and classified as non-wild type. Isolates having MIC values greater than the clinical breakpoints for susceptibility or resistance were considered intermediate or resistant, respectively. Note that no clinical breakpoints were available for tigecycline and while no ECOFF value was available for sulfamethoxazole, non-wild type was assumed if a bi- or multimodal MIC distribution was observed.

Table 4. Panel of antimicrobial substances and concentration ranges included in antimicrobial susceptibility testing and applied epidemiological cut-offs (ECOFFs) and clinical breakpoints.

Antimicrobial (Abbreviation)	Concentration Range Tested (mg/L)	Non-Wild-Type Population * (mg/L)	Clinical Breakpoint for Susceptibility [#] (mg/L)	Clinical Breakpoint for Resistance [#] (mg/L)
Ampicillin (AMP)	1–64	≥ 8	≤ 8	≥ 32
Cefotaxime (FOT)	0.25–4	≥ 0.25	≤ 1	≥ 4
Ceftazidime (TAZ)	0.5–8	≥ 1	≤ 8	≥ 16
Meropenem (MERO)	0.03–16	≥ 0.06	≤ 1	≥ 4
Nalidixic acid (NAL)	4–128	≥ 8	-	≥ 32
Ciprofloxacin (CIP)	0.015–8	≥ 0.06	≤ 0.25	≥ 1
Tetracycline (TET)	2–64	≥ 8	≤ 4	≥ 16
Colistin (COL)	1–16	≥ 2	≤ 1	≥ 4
Gentamicin (GEN)	0.5–32	≥ 2	≤ 4	≥ 16
Trimethoprim (TMP)	0.25–32	≥ 2	-	≥ 16
Sulfamethoxazole (SMX)	8–1024	-	-	≥ 512
Chloramphenicol (CHL)	8–128	≥ 16	≤ 8	≥ 32
Azithromycin (AZI)	2–64	≥ 16	-	≥ 32
Tigecycline (TGC)	0.25–8	≥ 0.5	-	-

* According to EUCAST (https://mic.eucast.org/search/, most recently viewed on 15 May 2023); # According to CLSI M100-ED32:2022 [26].

The antimicrobial susceptibility results were primarily interpreted using the EUCAST epidemiological cut-off values [52], which identify whether an isolate has acquired resistance against a certain antibiotic compared to the wild-type population [53,54]. On the other hand, human clinical breakpoints provides insights into whether the observed acquired resistance patterns are clinically relevant. We therefore chose to report both interpretations.

The European Food Safety Authority (EFSA) defined antimicrobial resistance level as the percentage of tested isolates of a given microorganism that were resistant to a given antimicrobial. These levels are described as rare (<0.1%), very low (0.1% to 1%), low (>1% to 10%), moderate (>10% to 20%), high (>20% to 50%), very high (>50% to 70%) and extremely high (>70%) [55].

An isolate was defined as multi-drug resistant if it was resistant to antimicrobial compounds belonging to at least three different antimicrobial classes. The antimicrobial resistance index (ARI) of an isolate was calculated as the proportion of tested antimicrobial compounds against which resistance was observed. It was computed based on 13 rather than 14 compounds because cefotaxime and ceftazidime belong to the same antibiotic class and exhibited resistance simultaneously. The average antimicrobial resistance index (ARI) for each farm was calculated by determining the ARI of all isolates from that farm and taking the mean of those values.

4.2. AMU

The quantification of AMU was described in detail in Ibrahim et. al, 2023 [27]. In brief, AMU was quantified by computing the treatment incidence (TI) which expresses the number of Defined Daily Dose (DDDvet) administered per 100 animal days at risk. It reflects the percentage of the lifetime of a bird for which it was treated with antimicrobials.

$$TI_{DDDvet} = \frac{total\ amount\ of\ AS\ administered\ or\ purchased}{DDDvet\ (mg\,/kg/\,day) \times no.\ of\ days\ at\ risk \times kg\ of\ AAR} \times 100\ AAR$$

AS—active substance; AAR—animal at risk.

4.3. Data Analysis

Descriptive statistics related to AMU (TI_{DDDvet}) and AMR were computed using Excel® 2016. Correlations between all variables were explored by means of Pearson's correlation test. Additionally, R^2 was computed to assess the proportion of variance in antimicrobial resistance explained by the TI. The assumptions of normality and homogeneity of variance were assessed by examining histograms and normal probability plots of residuals. A t-test was conducted to compare the resistance of *E. coli* between broiler and Sonali poultry. To test whether the use of antibiotics was associated with the presence of resistant *E. coli* strains, we used the generalized linear mixed model (GLMM) with a binomial logistic probability function. Farm was included as subject, isolate as within-subject factor and their working correlation were set as independent. The use or no use of antibiotic was included as fixed factor. Associations were considered significant when p-values were ≤ 0.05. The data were analyzed using SPSS software (SPSS version 27®; IBM, Armonk, NY, USA).

5. Conclusions

In Bangladesh, *E. coli* isolates obtained from poultry feces and the environment of broiler and Sonali chicken farms exhibit high levels of multidrug resistance to commonly used antibiotic classes including fluoroquinolones, which are classified as a "high priority critically important antibiotic" for humans. Unexpectedly, high antimicrobial use was not associated with the level of AMR, probably due to an overall (very) high level of resistance. The fecal and environmental AMR indexes were moderately correlated, which may indicate a lack of biosecurity on Bangladesh poultry farms. A comprehensive and multisectoral approach is necessary to address these factors and combat the spread of resistance in the poultry industry.

Supplementary Materials: The following supporting information can be downloaded at: https://www.mdpi.com/article/10.3390/antibiotics12091361/s1, Table S1: Minimum inhibitory concentration distribution for *E. coli* in fecal samples of broiler; Table S2: Minimum inhibitory concentration distribution for *E. coli* in fecal samples of Sonali; Table S3: Minimum inhibitory concentration distribution for *E. coli* in environmental samples of broiler; Table S4: Minimum inhibitory concentration distribution for *E. coli* in environmental samples of Sonali.

Author Contributions: Conceptualization: J.D., F.B. and N.I.; methodology: J.D., F.B. and N.I.; software: I.C. and N.I.; validation: J.D., F.B., I.C. and N.I.; formal analysis: J.D., F.B., I.C., M.R. and N.I.; investigation: N.I. and M.A.S.M.; data curation: J.D., F.B., I.C. and N.I.; writing—original draft preparation: N.I.; writing—review and editing: J.D., F.B., A.C.B., M.R., G.F. and N.I.; supervision: J.D. and F.B.; project administration: J.D., F.B. and N.I.; funding acquisition: D.P. All authors have read and agreed to the published version of the manuscript.

Funding: This work was supported by the 'Strengthening the Capacity of Public Administration for Achieving SDGs (SCPA-SDG)' project of the Prime Minister's Office, Bangladesh, City University, Hongkong and the UKRI GCRF One Health Poultry Hub (BB/S011269/1), 1 of 12 interdisciplinary research hubs funded under the UK government's Grand Challenge Research Fund Interdisciplinary Research Hub initiative. The MALDI-TOF mass spectrometer was financed by the Research Foundation Flanders (FWO-Vlaanderen, Brussels, Belgium) as Hercules project G0H2516N (AUGE/15/05).

Institutional Review Board Statement: Not applicable.

Informed Consent Statement: Not applicable.

Data Availability Statement: Data were provided in the manuscript.

Acknowledgments: We would like to thank all participating farmers and all field veterinarians participating in the sampling. Furthermore, we would like to express our sincere gratitude and appreciation to the Department of Livestock Services (DLS), Bangladesh.

References

1. World Health Organization (WHO). Antibiotic Resistance. 2020. Available online: https://www.who.int/news-room/factsheets/detail/antibiotic-resistance (accessed on 17 May 2023).
2. World Health Organization (WHO). Global Action Plan on Antimicrobial Resistance. 2015. Available online: https://www.amcra.be/swfiles/files/WHO%20actieplan_90.pdf (accessed on 16 May 2023).

3. World Health Organization (WHO). Tripartite and UNEP Support OHHLEP's Definition of 'One Health'. 2021. Available online: https://www.who.int/news/item/01-12-2021-tripartite-and-unep-support-ohhlep-s-definition-of-one-health (accessed on 17 May 2023).

4. Chereau, F.; Opatowski, L.; Tourdjman, M.; Vong, S. Risk assessment for antibiotic resistance in South East Asia. *BMJ* **2017**, *358*, j3393. [CrossRef] [PubMed]

5. Harbarth, S.; Participants, F.T.W.H.-A.I.R.F.; Balkhy, H.H.; Goossens, H.; Jarlier, V.; Kluytmans, J.; Laxminarayan, R.; Saam, M.; Van Belkum, A.; Pittet, D. Antimicrobial resistance: one world, one fight! *Antimicrob. Resist. Infect. Control.* **2015**, *4*, 49.

6. European Food Safety Authority and European Centre for Disease Prevention Control. EU Summary Report on antimicrobial resistance in zoonotic and indicator bacteria from humans, animals and food in 2013. *EFSA J.* **2015**, *13*, 4036.

7. Marshall, B.M.; Levy, S.B. Food Animals and Antimicrobials: Impacts on Human Health. *Clin. Microbiol. Rev.* **2011**, *24*, 718–733. [CrossRef]

8. Aarestrup, F.M.; Wegener, H.C.; Collignon, P. Resistance in bacteria of the food chain: epidemiology and control strategies. *Expert Rev. Anti-Infective Ther.* **2008**, *6*, 733–750. [CrossRef]

9. Mellata, M. Human and avian extraintestinal pathogenic Escherichia coli: Infections, zoonotic risks, and antibiotic resistance trends. *Foodborne Pathog. Dis.* **2013**, *10*, 916–932. [CrossRef]

10. World Organisation for Animal Health (WOAH). Terrestrial Animal Health Code. Available online: https://www.woah.org/app/uploads/2022/06/a-sixth-annual-report-amu-final.pdf (accessed on 21 May 2023).

11. Katakweba, A.A.; Muhairwa, A.P.; Lupindu, A.M.; Damborg, P.; Rosenkrantz, J.T.; Minga, U.M.; Mtambo, M.M.; Olsen, J.E. First report on a randomized investigation of antimicrobial resistance in fecal indicator bacteria from livestock, poultry, and humans in Tanzania. *Microb. Drug Resist.* **2018**, *24*, 260–268. [CrossRef]

12. Odonkor, S.T.; Addo, K.K. Prevalence of multidrug-resistant Escherichia coli isolated from drinking water sources. *Int. J. Microbiol.* **2018**, *2018*, 7204013. [CrossRef]

13. Arsène-Ploetze, F.; Chiboub, O.; Lièvremont, D.; Farasin, J.; Freel, K.C.; Fouteau, S.; Barbe, V. Adaptation in toxic environments: comparative genomics of loci carrying antibiotic resistance genes derived from acid mine drainage waters. *Environ. Sci. Pollut. Res.* **2017**, *25*, 1470–1483. [CrossRef]

14. Department of Livestock Services of Bangladesh. Fish Feed and Animal Feed Act 2010. Bangladesh Government. 2010. Available online: http://extwprlegs1.fao.org/docs/pdf/bgd165024.pdf (accessed on 23 May 2023).

15. Directorate General of Health Services (DGHS). National Action Plan: Antimicrobial Resistance Containment in Bangladesh 2017–2022. In: Ministry of Health & Family Welfare, ed. 2017. Available online: https://cdn.who.int/media/docs/default-source/antimicrobial-resistance/amr-spc-npm/nap-library/antimicrobial-resistance-containment-in-bangladesh-2017-2022.pdf?sfvrsn=bfa46b_3&download=true (accessed on 23 May 2023).

16. Goutard, F.L.; Bordier, M.; Calba, C.; Erlacher-Vindel, E.; Góchez, D.; de Balogh, K.; Benigno, C.; Kalpravidh, W.; Roger, F.; Vong, S. Antimicrobial policy interventions in food animal production in South East Asia. *BMJ* **2017**, *358*, j3544. [CrossRef]

17. Imam, T.; Gibson, J.S.; Foysal, M.; Das, S.B.; Gupta, S.D.; Fournié, G.; Hoque, M.A.; Henning, J. A Cross-Sectional Study of Antimicrobial Usage on Commercial Broiler and Layer Chicken Farms in Bangladesh. *Front. Vet. Sci.* **2020**, *7*, 576113.

18. Netherlands Enterprise Agency. Poultry sector study Bangladesh. 2020. Available online: https://www.rvo.nl/sites/default/files/2020/12/Poultry%20sector%20study%20Bangladesh.pdf (accessed on 22 May 2023).

19. Antunes, P.; Novais, C.; Peixe, L. Food-to-Humans Bacterial Transmission. *Microbiol. Spectr.* **2020**, *8*, 161–193.

20. Tasmim, S.T.; Hasan, M.; Talukder, S.; Mandal, A.K.; Parvin, M.S.; Ali, Y.; Ehsan, A.; Islam, T. Sociodemographic determinants of use and misuse of antibiotics in commercial poultry farms in Bangladesh. *IJID Reg.* **2023**, *7*, 146–158. [CrossRef] [PubMed]

21. Uddin, B.; Alam, M.N.; Hasan, M.; Hossain, S.M.B.; Debnath, M.; Begum, R.; Samad, M.A.; Hoque, S.F.; Chowdhury, S.R.; Rahman, M.; et al. Molecular Detection of Colistin Resistance *mcr-1* Gene in Multidrug-Resistant *Escherichia coli* Isolated from Chicken. *Antibiotics* **2022**, *11*, 97. [CrossRef] [PubMed]

22. Mandal, A.K.; Talukder, S.; Hasan, M.; Tasmim, S.T.; Parvin, M.S.; Ali, Y.; Islam, T. Author response for 'Epidemiology and antimicrobial resistance of Escherichia coli in broiler chickens, farmworkers, and farm sewage in Bangladesh'. *Vet. Med. Sci.* **2022**, *8*, 187–199. [CrossRef] [PubMed]

23. Nandi, S.; Sultana, M.; Hossain, M. Prevalence and Characterization of Multidrug-Resistant Zoonotic Enterobacter spp. in Poultry of Bangladesh. *Foodborne Pathog. Dis.* **2013**, *10*, 420–427. [CrossRef]

24. Fazal, M.A.; Nath, C.; Islam, M.S.; Hasib, F.Y.; Reza, M.M.B.; Devnath, H.S.; Nahid-Ibn-Rahman, M.; Ahad, A. Isolation and identification of multidrug-resistant Escherichia coli from cattle, sheep, poultry and human in Cumilla, Bangladesh. *Malays. J. Microbiol.* **2022**, *18*, 227–234.

25. Rahman, A.; Khairalla, A. Susceptibility and Multidrug Resistance Patterns of Escherichia coli Isolated from Cloacal Swabs of Live Broiler Chickens in Bangladesh. *Pathogens* **2019**, *8*, 118.

26. Clinical and Laboratory Standards Institute (CLSI), USA. CLSI M100-ED32:2022 Performance Standards for Antimicrobial Susceptibility Testing, 32nd ed. 2022. Available online: https://clsi.org/media/wi0pmpke/m100ed32_sample.pdf (accessed on 15 May 2023).

27. Ibrahim, N.; Chantziaras, I.; Mohsin, M.A.S.; Boyen, F.; Fournié, G.; Islam, S.S.; Berge, A.C.; Caekebeke, N.; Joosten, P.; Dewulf, J. Quantitative and qualitative analysis of antimicrobial usage and biosecurity in broilers and Sonali poultry farms in Bangladesh. *Prev. Vet. Med.* **2023**, *217*, 105968.

28. Islam, M.S.; Hossain, M.J.; Sobur, M.A.; Punom, S.A.; Rahman, A.M.M.; Rahman, M.T. A Systematic Review on the Occurrence of Antimicrobial-Resistant Escherichia coli in Poultry and Poultry Environments in Bangladesh between 2010 and 2021. *BioMed Res. Int.* **2023**, *2023*, 2425564. [CrossRef]

29. Al Amin, A.; Hoque, M.N.; Siddiki, A.Z.; Saha, S.; Kamal, M. Antimicrobial resistance situation in animal health of Bangladesh. *Vet World* **2020**, *13*, 2713–2727. [CrossRef] [PubMed]

30. Boyen, F.; Vangroenweghe, F.; Butaye, P.; De Graef, E.; Castryck, F.; Heylen, P.; Vanrobaeys, M.; Haesebrouck, F. Disk prediffusion is a reliable method for testing colistin susceptibility in porcine E. coli strains. *Veter- Microbiol.* **2010**, *144*, 359–362. [CrossRef] [PubMed]

31. World Health Organization (WHO). *Integrated Surveillance of Antimicrobial Resistance in Foodborne Bacteria: Application of a One Health Approach: Guidance from the WHO Advisory Group on Integrated Surveillanec of Antimicrobial Resistance (AGISAR)*; World Health Organization: Geneva, Switzerland, 2017. Available online: https://www.who.int/publications/i/item/9789241515528 (accessed on 25 May 2023).

32. Ahmed, S.; Das, T.; Islam, Z.; Herrero-Fresno, A.; Biswas, P.K.; Olsen, J.E. High prevalence of mcr-1-encoded colistin resistance in commensal Escherichia coli from broiler chicken in Bangladesh. *Sci. Rep.* **2020**, *10*, 18637.

33. *Ban on All Forms of Combination Drug of Colistin and Ciprofloxacin*; Directorate General of Drug Administration: Dhaka, Bangladesh, 2019.

34. *Circular of All Forms Colistin Ban*; Directorate General of Drug Administration: Dhaka, Bangladesh, 2022.

35. Parvin, M.S.; Ali, Y.; Mandal, A.K.; Talukder, S.; Islam, T. Sink survey to investigate multidrug resistance pattern of common foodborne bacteria from wholesale chicken markets in Dhaka city of Bangladesh. *Sci. Rep.* **2022**, *12*, 10818. [PubMed]

36. Chantziaras, I.; Boyen, F.; Callens, B.; Dewulf, J. Correlation between veterinary antimicrobial use and antimicrobial resistance in food-producing animals: a report on seven countries. *J. Antimicrob. Chemother.* **2014**, *69*, 827–834. [CrossRef] [PubMed]

37. *Judicious Therapeutic Use of Antimicrobials*; Association American Veterinary Medical: Schaumburg, IL, USA, 2004.

38. More, S.J. European Perspectives on Efforts to Reduce Antimicrobial Usage in Food Animal Production. *Ir. Vet. J.* **2020**, *73*, 2. [CrossRef] [PubMed]

39. Al Azad, M.A.R.; Rahman, M.M.; Amin, R.; Begum, M.I.A.; Fries, R.; Husna, A.; Khairalla, A.S.; Badruzzaman, A.T.M.; El Zowalaty, M.E.; Lampang, K.N.; et al. Prevalence of antimicrobial resistance of Escherichia coli isolated from broiler at Rajshahi region, Bangladesh. *Pathogens* **2019**, *8*, 118. [CrossRef] [PubMed]

40. Schrauwen, E.J.A.; Huizinga, P.; Van Spreuwel, N.; Verhulst, C.; Bergh, M.F.Q.K.-V.D.; Kluytmans, J.A.J.W. High prevalence of the mcr-1 gene in retail chicken meat in the Netherlands in 2015. *Antimicrob Resist. Infect Control* **2017**, *6*, 83. [CrossRef] [PubMed]

41. Lazarus, B.; Paterson, D.L.; Mollinger, J.L.; Rogers, B.A. Do human extraintestinal Escherichia coli infections resistant to expanded-spectrum cephalosporins originate from food-producing animals? A systematic review. *Clin. Infect. Dis.* **2014**, *60*, 439–452. [CrossRef] [PubMed]

42. Collignon, P.J.; McEwen, S.A. One Health—Its Importance in Helping to Better Control Antimicrobial Resistance. *Trop. Med. Infect. Dis.* **2019**, *4*, 22. [CrossRef]

43. Tang, K.L.; Caffrey, N.P.; Nóbrega, D.B.; Cork, S.C.; E Ronksley, P.; Barkema, H.W.; Polachek, A.J.; Ganshorn, H.; Sharma, N.; Kellner, J.D.; et al. Restricting the use of antibiotics in food-producing animals and its associations with antibiotic resistance in food-producing animals and human beings: a systematic review and meta-analysis. *Lancet Planet. Health* **2017**, *1*, e316–e327. [CrossRef] [PubMed]

44. Levy, S.B. Balancing the drug-resistance equation. *Trends Microbiol.* **1994**, *2*, 341–342. [CrossRef] [PubMed]

45. Austin, D.J.; Kristinsson, K.G.; Anderson, R.M. The relationship between the volume of antimicrobial consumption in human communities and the frequency of resistance. *Proc. Natl. Acad. Sci. USA* **1999**, *96*, 1152–1156. [CrossRef] [PubMed]

46. Handel, A.; Regoes, R.R.; Antia, R. The Role of Compensatory Mutations in the Emergence of Drug Resistance. *PLoS Comput. Biol.* **2006**, *2*, e137. [CrossRef]

47. Nilsson, O.; Börjesson, S.; Landén, A.; Bengtsson, B. Vertical transmission of Escherichia coli carrying plasmid-mediated AmpC (pAmpC) through the broiler production pyramid. *J. Antimicrob. Chemother.* **2014**, *69*, 1497–1500. [CrossRef]

48. Davies, J.; Davies, D. Origins and evolution of antibiotic resistance. *Microbiol. Mol. Biol. Rev.* **2010**, *74*, 417–433. [CrossRef]

49. Checkley, W.; Buckley, G.; Gilman, R.H.; Assis, A.M.; Guerrant, R.L.; Morris, S.S.; Mølbak, K.; Valentiner-Branth, P.; Lanata, C.F.; E Black, R.; et al. Multi-country analysis of the effects of diarrhoea on childhood stunting. *Leuk. Res.* **2008**, *37*, 816–830. [CrossRef]

50. Vera, L.; Boyen, F.; Visscher, A.; Vandenbroucke, V.; Vanantwerpen, G.; Govaere, J. Limitations of a chromogenic agar plate for the identifying bacteria isolated from equine endometritis samples. *Equine Veter- J.* **2018**, *51*, 266–269. [CrossRef]

51. European Committee on Antimicrobial Susceptibility Testing (EUCAST). MIC Distributions. 2022. Available online: https://www.eucast.org/mic_distributions_and_ecoffs/ (accessed on 15 May 2023).

52. European Committee on Antimicrobial Susceptibility Testing (EUCAST). 2021. Available online: www.eucast.org (accessed on 15 May 2023).

53. Cornaglia, G.; Hryniewicz, W.; Jarlier, V.; Kahlmeter, G.; Mittermayer, H.; Stratchounski, L.; Baquero, F. European recommendations for antimicrobial resistance surveillance. *Clin. Microbiol. Infect.* **2004**, *10*, 349–383. [CrossRef]

54. Callens, B.; Persoons, D.; Maes, D.; Laanen, M.; Postma, M.; Boyen, F.; Haesebrouck, F.; Butaye, P.; Catry, B.; Dewulf, J. Prophylactic and metaphylactic antimicrobial use in Belgian fattening pig herds. *Prev. Veter- Med.* **2012**, *106*, 53–62. [CrossRef]

55.   European Food Safety Authority (EFSA). Antimicrobial Resistance Remains Commonly Detected in Bacteria in Humans, Animals and Food: EFSA-ECDC Report. 2014. Available online: https://www.efsa.europa.eu/en/press/news/140325 (accessed on 15 May 2023).

Article

The Impact of Low-Level Benzalkonium Chloride Exposure on *Staphylococcus* spp. Strains and Control by Photoinactivation

Erika C. R. Bonsaglia [1], Gustavo H. Calvo [1], Daniel O. Sordelli [1], Nathalia C. C. Silva [2], Vera L. M. Rall [3], Adriana Casas [4] and Fernanda Buzzola [1,*]

[1] Instituto de Investigaciones en Microbiología y Parasitología Médica (IMPaM), Facultad de Medicina, Universidad de Buenos Aires, Consejo Nacional de Investigaciones Científicas y Técnicas, Buenos Aires 1121, Argentina; ebonsaglia@fmed.uba.ar (E.C.R.B.); gustavohcalvo@gmail.com (G.H.C.); sordelli@fmed.uba.ar (D.O.S.)

[2] Department of Food Science, Faculty of Food Engineering (FEA), University of Campinas (UNICAMP), Campinas 13083-862, Brazil; ncirone@unicamp.br

[3] Department of Chemical and Biological Sciences, Institute of Biosciences, Sao Paulo State University, Botucatu 18618-691, Brazil; vera.rall@unesp.br

[4] Centro de Investigaciones sobre Porfirinas y Porfirias (CIPYP), Hospital de Clínicas José de San Martín, Universidad de Buenos Aires, Consejo Nacional de Investigaciones Científicas y Técnicas, Buenos Aires 1121, Argentina; adriana@qb.fcen.uba.ar

* Correspondence: ferbuz@fmed.uba.ar

Abstract: Exposure of bacteria to low concentrations of biocides can facilitate horizontal gene transfer, which may lead to bacterial adaptive responses and resistance to antimicrobial agents. The emergence of antibacterial resistance not only poses a significant concern to the dairy industry but also adds to the complexity and cost of mastitis treatment. This study was aimed to evaluate how selective stress induced by benzalkonium chloride (BC) promotes antibiotic non-susceptibility in *Staphylococcus* spp. In addition, we investigated the efficacy of photodynamic inactivation (PDI) in both resistant and susceptible strains. The study determined the minimum inhibitory concentration (MIC) of BC using the broth microdilution method for different *Staphylococcus* strains. The experiments involved pairing strains carrying the *qac*A/*qac*C resistance genes with susceptible strains and exposing them to subinhibitory concentrations of BC for 72 h. The recovered isolates were tested for MIC BC and subjected to disc diffusion tests to assess changes in susceptibility patterns. The results demonstrated that subinhibitory concentrations of BC could select strains with reduced susceptibility and antibiotic resistance, particularly in the presence of *S. pasteuri*. The results of PDI mediated by toluidine blue (100 μM) followed by 60 min irradiation (total light dose of 2.5 J/cm^2) were highly effective, showing complete inactivation for some bacterial strains and a reduction of up to 5 logs in others.

Keywords: *Staphylococcus* sp.; antimicrobial resistance; photodynamic inactivation

Citation: Bonsaglia, E.C.R.; Calvo, G.H.; Sordelli, D.O.; Silva, N.C.C.; Rall, V.L.M.; Casas, A.; Buzzola, F. The Impact of Low-Level Benzalkonium Chloride Exposure on *Staphylococcus* spp. Strains and Control by Photoinactivation. *Antibiotics* **2023**, *12*, 1244. https://doi.org/10.3390/antibiotics12081244

Academic Editor: Nahla O. Eltai

Received: 29 June 2023
Revised: 25 July 2023
Accepted: 26 July 2023
Published: 28 July 2023

1. Introduction

Intramammary infection in cows caused by *Staphylococcus* spp. is the most concerning disease that affects the dairy chain, causing vast economic losses worldwide. Bovine mastitis is conditioned by diverse factors and, for this reason, requires specific treatment according to the infecting agent, thus making disease control even more difficult, generating high costs for the producer [1]. *S. aureus* and *Staphylococcus non-aureus* showing resistance to antibiotics and biocides have been isolated in the dairy chain. It is known that the inappropriate use of antimicrobial agents in the dairy industry has contributed to the emergence of multi-resistant pathogens [2].

Antimicrobial agents and biocides have a wide range of applications worldwide, and are an important tool for management of unwanted bacterial growth. However, antibiotic and biocide effectiveness is decreasing due to the emergence of resistance through different mechanisms across different lineages and species. Indeed *Staphylococcus* spp. has a remarkable ability to acquire resistance to antimicrobial agents, either through mutation of chromosomal bacterial genes or through the incorporation of resistance genes through transfer from other microorganisms [3]. It has been proposed that there is a link between antibiotics and resistance to disinfectants such as quaternary ammonium compounds (QACs). In fact, QACs are frequently used in the food industry, including dairy environments, for disinfection of the environment and equipment due to their broad antimicrobial spectrum against bacteria, fungi, and viruses. Many QAC resistance genes (*qacA, qacB, smr, qacG, qacH, qacJ*) have already been identified in bacteria isolated from different sources. Due to their location on mobile genetic elements, the interaction between different species of *Staphylococcus* is facilitated [2]. QAC efflux pump determinants are usually found on multidrug resistance plasmids in *Staphylococcus spp.* strains. In this regard, previous studies have found the *qacA/B* genes located on the same plasmid that confers resistance to β-lactams in clinically- and in food-derived strains with the potential for uptake by plasmid-free *S. aureus*, indicating its ability to transfer under selective stress [4,5].

In the current scenario, the emergence of multi-resistant strains becomes a challenge for the pharmaceutical industry. The recurrent problem of lack of options for controlling these pathogens makes essential to seek new therapeutic alternatives and to review practices that may influence the emergence of antibiotic-resistant bacteria. Photodynamic inactivation (PDI), also known as antimicrobial photodynamic therapy, is an emerging therapy that involves the use of a photosensitizer agent with specific wavelength light to generate reactive oxygen species with the goal of destroying microbial agents. This procedure lacks genotoxic and mutagenic effects, preventing the development of bacterial resistance, and had exhibited excellent results in pathogens isolated from bovine mastitis [6].

The present study demonstrates that exposure to benzalkonium chloride (BC) subinhibitory concentrations select *S. aureus* strains with reduced susceptibility, and also confers an increase in antibiotic resistance. In addition, we evaluate and underline the effects of PDI on mutant strains that presented multi-resistance to antibiotics and BC.

2. Results

2.1. Exposure to Subinhibitory Concentration of Benzalkonium Chloride

Resistant strains of *S. xylosus* and *S. pasteuri* were, respectively, paired with the susceptible strains of *S. aureus* ATCC29213 and ATCC6538 (Figure 1). All tested strains were recovered after 72 h of passage in a 1/2 × MIC of BC and were renamed according to Figure 2. Control group strains were also recovered after 72 h.

2.2. Agar Diffusion and MIC of Benzalkonium Chloride

After BC sub-MIC assays, all strains were subjected to antimicrobial agar diffusion testing using cefoxitin, cefazolin, clindamycin, and penicillin discs. Oxacillin MIC values were determined by E-test. Table 1 shows the antimicrobial susceptibility obtained before and after exposure to sub-MIC BC. Phenotypic alteration occurred in only one strain (Table 1). Regarding the *S. aureus* strains that were grown together with *S. xylosus*, only the strain named 8 (M6538.x) showed a two-fold increase (7.8 µg/mL) in MIC of BC value compared to that of its parental ATCC6538 strain (3.9 µg/mL). Moreover, when the *S. aureus* ATCC6538 strain was confronted with *S. pasteuri*, the recovered mutant exhibited a four-fold increase in resistance (15.7 µg/mL) compared to the parental MIC of BC value (3.9 µg/mL). In addition, mutant number 10 of the *S. aureus* ATCC29213 strain (named M29213.p) showed a two-fold increase (15.7 µg/mL) of the BC MIC compared to value of its parental strain (7.8 µg/mL) (Table 1). None of the recovered strains of the control group showed alterations in MIC or antibiotic susceptibility and, for this reason, they were not included in Table 1.

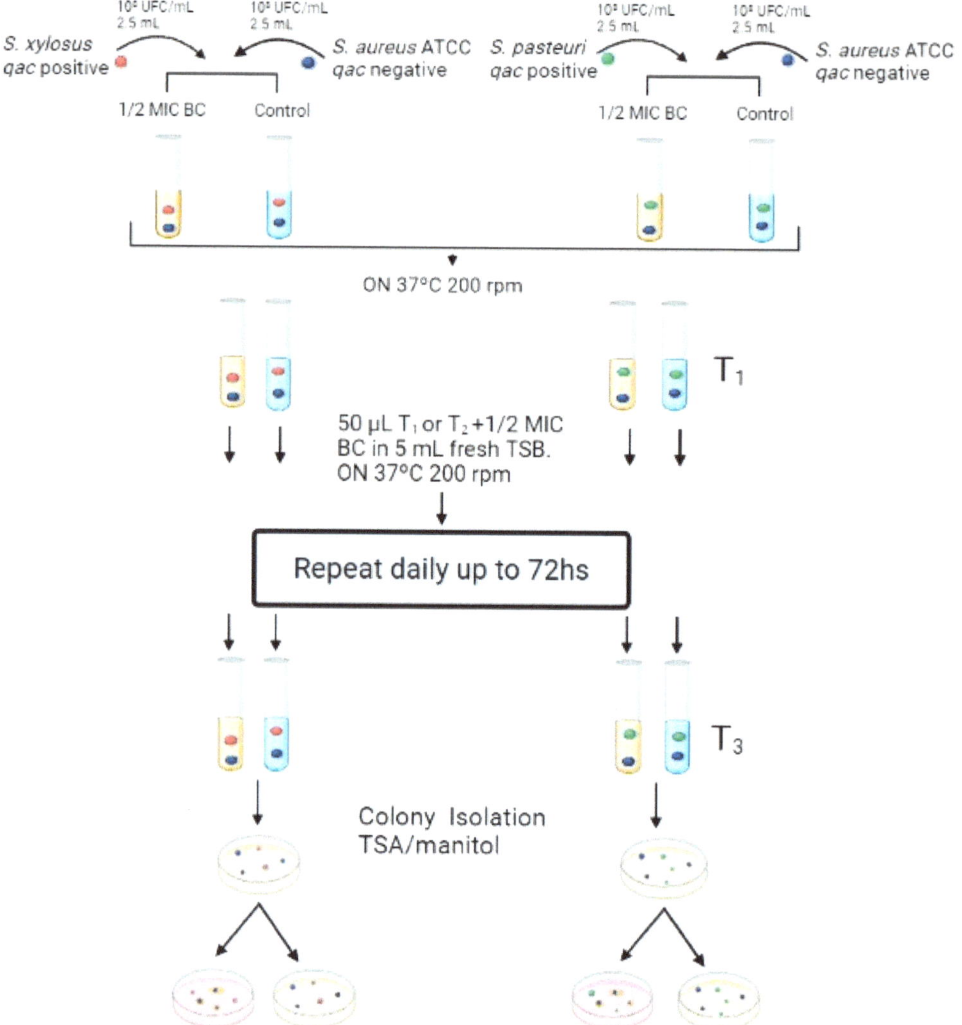

Figure 1. Schematic representation of benzalkonium chloride exposure assay. *S. xylosus* carries *qacA/B* and *mecA* genes, and *S. pasteuri* carries *qacC* gene. *S. aureus* ATCC6538 and ATCC29213 strains are susceptible to BC and antibiotics. T1: 24 h post co-culture. T2: 24 h post subculture of T1. T3: 24 h post subculture of T2. Created in BioRender.com.

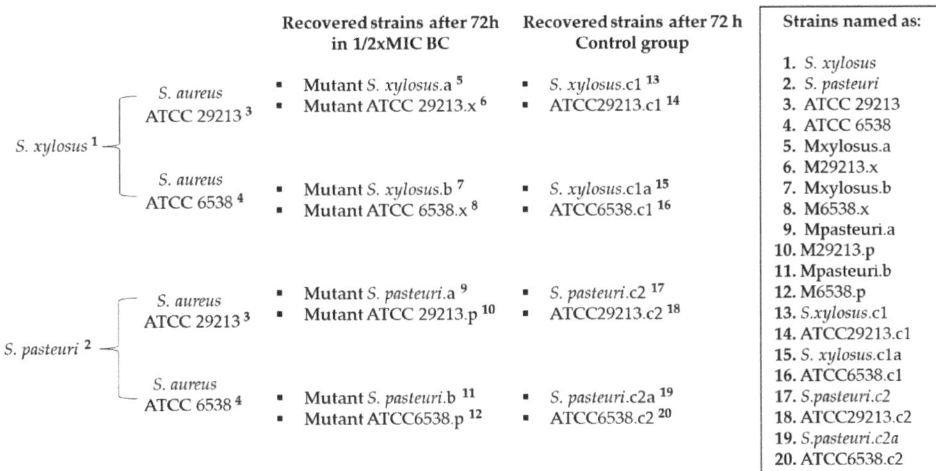

Figure 2. Nomenclature used for the strains and control group recovered after 72 h of exposure to subinhibitory concentration of benzalkonium chloride.

Table 1. Susceptibility to antimicrobial agents and benzalkonium chloride before (grey) and after subinhibitory concentration exposition.

Strains	MIC BC µg/mL	MIC Oxacilin µg/mL	Clindamicin	Cefalotin	Cefazolin	Penicillin
1. *S. xylosus*	7.8	6	R	S	S	S
5. Mxylosus.a	7.8	6	R	S	S	S
7. Mxylosus.b	7.8	6	R	S	S	S
2. *S. pasteuri*	7.8	6	R	S	S	S
9. Mpasteuri.a	15.7	6	R	S	S	S
11. Mpasteuri.b	15.7	6	R	S	S	S
3. ATCC 29213	7.8	0.125	S	S	S	S
6. M29213.x	7.8	0.125	S	S	S	S
10. M29213.p	15.7	0.125	S	S	S	S
4. ATCC 6538	3.9	0.094	S	S	S	S
8. M6538.x	7.8	0.094	S	S	S	S
12. M6538.p	15.7	0.094	R	R	R	R

BC: Benzalkonium chloride; R: resistant; S: susceptible.

2.3. Photodynamic Inactivation (PDI) of Planktonic Cultures

All bacterial strains exposed to PDI mediated by TB (TB-PDI) showed a very high reduction in their populations. There were differences between the effect of the treatment among strains, but even in the strain that showed the least reduction, the number of bacteria decreased from 1.25×10^7 to 3×10^4 CFU/mL total bacteria, which represents a 99.76% mortality in the bacterial population (Figure 3, strain number 4). The *S. xylosus* and *S. pasteuri* strains named as 1, 7, 2 and 11 (Figure 2) were more susceptible to the TB-PDI treatment. Moreover, strains 7, 2 and 11 showed virtually total bacterial death after TB-PDI. *S. aureus* strains named 3, 10, 4 and 12 (Figure 2) appeared to be less susceptible to the treatment, although the bacterial mortality observed was around 99%. The total CFU/mL of the strains before and after treatment, and the percentage of bacterial death, are shown in Table 2.

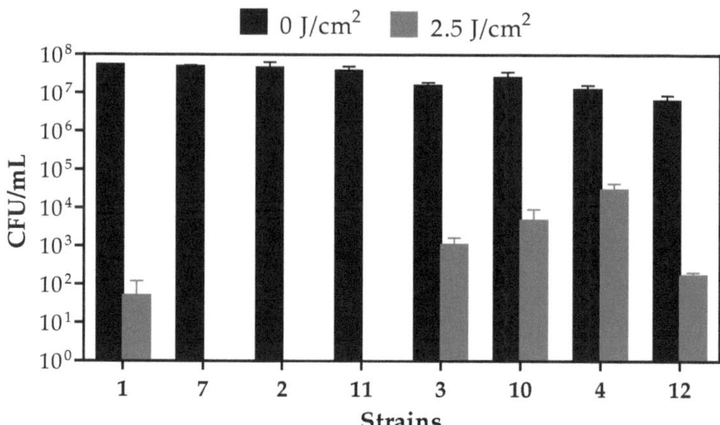

Figure 3. Total viable bacteria (CFU/mL) exposed to a light dose of 0 or 2.5 J/cm^2. Strains were named as 1: *S. xylosus*; 7: *Mxylosus.b*; 2: *S. pasteuri*; 11: *Mpasteuri.b*; 3: *S. aureus* ATCC29213; 10: M29213.p; 4: *S. aureus* ATCC6538; 12: M6538.p. Control assays irradiated without TB and non-treated bacteria showed values of CFU/mL ranging from 1×10^7 to 5×10^7 CFU/mL. Thus, no toxic per se effect of TB was observed. Each bar represents the mean ± SD values of 2 independent experiments performed in triplicate. Statistical significance of TB-PDI (grey bars) vs. control TB non-irradiated (black bars): $p = 0.0026$, two-way ANOVA and D'Agostino–Pearson normality test $p > 0.05$.

Table 2. Effect of TB-PDI on bacterial survival of the different strains studied.

Strain	CFU/mL		% Death
	No Light	Light	
1	5.5×10^7	5×10^1	99.99
7	5×10^7	0	100
2	4.55×10^7	0	100
11	3.8×10^7	0	100
3	1.55×10^7	1.1×10^3	99.99
10	2.5×10^7	4.8×10^3	99.98
4	1.25×10^7	3×10^4	99.76
12	6.25×10^6	1.75×10^2	99.99

It is worth mentioning that the three strains (named 11, 10 and 12) with high MIC of BC values were shown to be susceptible to TB-PDI. Therefore, this therapy may be useful to combat strains bearing the *qac* genes. None of the strains studied showed intrinsic photosensitivity in the absence of the photosensitizer TB.

3. Discussion

The growing restrictions on antimicrobial use and the expansion of multi-resistant bacteria highlight the necessity to explore alternative or complementary methods to control mastitis in dairy herds [7,8]. One of the major challenges is that *Staphylococcus* spp., besides possessing a wide array of virulence factors that hinder its elimination, also exhibits diverse responses to the presence of antimicrobial agents. Acquired resistance is one of these mechanisms, which can arise through mutations in normal genes due to external and internal factors, or by acquiring genetic information from other microorganisms, enabling the bacteria to survive in harsh environments [9]. Environmentally relevant concentrations of biocides are likely below those required to inhibit microbial growth. In our present study, we observed that susceptible strains to BC, when exposed to subinhibitory concentrations of this agent in the presence of resistant strains, exhibited changes in susceptibility not only to BC but also to several antibiotics. These findings support the evidence that prolonged use of BC may facilitate the emergence of antibiotic-resistant strains and compromise the efficacy of other antimicrobial agents, thereby diminishing the ability to control infections [10]. Quaternary ammonium-based sanitizers are commonly used to clean dairy equipment and environments. However, as our study reveals, the prolonged and/or improper use of these sanitizers can foster the development of more resilient bacterial clones, particularly in *S. aureus*. Hence, it is crucial to implement appropriate cleaning and disinfection strategies and to regularly assess the efficacy of these products to minimize the risk of selecting and disseminating resistant strains. A previous study by Weber et al. [11] has shown that the improper use or dilution of sanitizers can exert selective pressure on microorganisms, thus effectively contributing to the rise in bacterial resistance. Furthermore, the presence of these resistance genes in mobile elements, such as plasmids, enables rapid and facilitated transfer between different *Staphylococcus* species. In our study, we detected changes in susceptibility patterns of susceptible strains exposed to BC within a period as short as 72 h. Interestingly, the control group strains did not exhibit any changes, indicating that these events may occur more readily under stress conditions. However, several factors can influence the success of this transfer. Our results revealed that among the strains confronted with *S. xylosus*, only one displayed a change in susceptibility, while both *S. aureus* ATCC strains demonstrated resistance acquisition when exposed to *S. pasteuri*. This suggests that both *S. aureus* ATCC strains have good receptivity, and that *S. pasteuri* may be a more efficient donor compared to *S. xylosus*. However, to substantiate this claim, further studies involving different types of plasmids would be necessary, as the observed variability could be influenced by factors such as the specific plasmid type, the donor and recipient strains involved, as well as their interactions [12]. Based upon our findings, it can be inferred that adaptive mutations initially select strains with increased tolerance to chemical stress, subsequently leading to a selective pressure that favors the development of a cross-resistance phenotype to various other antimicrobials. The definition of tolerant strains varies among researchers. According to Gerba [13], tolerant strains are those exhibiting any increase in minimum inhibitory concentration (MIC) compared to their control. However, other authors define tolerant strains as those capable of surviving treatment without an increase in MIC, with tolerance preceding the development of resistance [14]. After conducting the strain confrontations, we observed both scenarios mentioned earlier in the mutants, with variations observed between the species involved. For instance, strain *S. aureus* ATCC29213 did not exhibit an increase in the MIC of BC when confronted with *S. xylosus*, but was able to survive for 72 h in subinhibitory concentration. However, when confronted with *S. pasteuri*, its MIC increased twice compared to its initial value. On the other hand, *S. aureus* ATCC6538 displayed an increase in MIC in both situations, with a four-fold higher increase observed in the presence of *S. pasteuri*. Notably, *S. aureus* ATCC6538 demonstrated cross-resistance potential, as it exhibited changes in resistance phenotypes to all four tested antibiotics. Therefore, we understand that tolerance may be associated with metabolic changes that enable survival in environments. For tolerant strains that do not exhibit altered antibiotic phenotypes, it is only a matter of time before they acquire cross-resistance to other agents. While all strains require an adaptation period, there are cases where the emergence of resistant strains may occur more rapidly within a population. The excessive use of antimicrobials in milk production can lead to the selection of drug-resistant bacterial strains and contribute to increased selective pressure within the bacterial population present in the environment. Consequently, there is a growing concern regarding the spread of these resistant bacteria, which can compromise milk quality and the health of animals and humans involved in the production chain. Therefore, the identification of an effective mastitis treatment that does not promote the selection of drug-resistant populations is extremely necessary. Photodynamic therapy has demonstrated remarkable efficacy in the treatment of microbial diseases in humans [15,16].

However, limited research has explored its application in bovine mastitis pathogens and drug-resistant bacteria. Silva et al. [17] evaluated the effect of photoinactivation on strains of clinical and subclinical mastitis isolated from sheep, and concluded that this treatment is promissory for bovine mastitis. Furthermore, Sellera et al. [6] demonstrated the efficacy of PDI on different antibiotic-resistant species isolated from bovine mastitis. These studies provide evidence that photoinactivation therapy holds great promise as an alternative for treating mastitis in animals, as positive outcomes have been consistently achieved irrespective of the parameters and photosensitizers employed. In our study, we have shown that TB-PDI can effectively reduce bacterial populations that were resistant to BC under experimental conditions. Strains that acquired resistance to biocides were more sensitive to TB-PDI treatment than the same strains used as controls. This might be due to the accumulation of diverse negative stimuli upon the bacterial population. Nevertheless, all the strains studied showed very high sensitivity towards PDI treatment, which positions PDI as a very promising multipurpose alternative: on one hand, to kill resistant strains of bacteria that cause mastitis in herds and that function as a resistant gene reservoir (which can be passed into non-resistant strains in certain conditions); on the other hand, to kill bacteria that acquire resistance genes from the environment. In any event, the final result will be the reduction of the total number of bacteria from animals and from equipment utilized in the dairy industry.

4. Materials and Methods

4.1. Bacterial Strains

Experiments were conducted using two reference strains *Staphylococcus aureus* ATCC 6538 and ATCC 29213. *S. aureus* ATCC6538 is conventionally used to evaluate the efficacy of biocidal agents [18], and *S. aureus* ATCC29213 represents a methicillin-sensitive strain that serves as a standard quality-control strain in laboratory testing [19]. Additionally, two wild *Staphylococcus* spp. strains isolated from dairies in previous studies [20] were used to perform the experiments: (a) *S. xylosus* which carry the *qacA/B* and *mecA* genes; and (b) *S. pasteuri* which carry the *qacC* and *lsaB* genes.

4.2. Determination of BC Minimum Inhibitory Concentration

The microdilution broth method was used to determine the minimum inhibitory concentration (MIC) of BC, performed according to the Clinical and Laboratory Standards Institute (CLSI) guidelines for MIC testing of *Staphylococcal* species [21]. All strains were inoculated in Mueller–Hinton broth and incubated at 37 °C for 24 h. Afterwards, cultures were standardized using the McFarland 0.5 scale (1×10^8 CFU/mL) as the starting inoculate, and an aliquot of 100 µL of standardized bacteria was added. Then, the 96-well plates were coated with 100 µL serial dilutions of the sanitizer tested. The dilution ranges applied for BC spanned from 1.95 µg/mL to 250 µg/mL. Media-only aliquots were added as sterility checks and as a positive control of antibiotic-mediated killing. All test sample volumes were 200 µL/mL per well with duplicates. The plates were incubated for 24 h at 37 °C under static conditions. Following incubation, the turbidity was measured using a microplate reader (Multiskan EX, Thermo Electron Corporation, Waltham, MA, USA) at 595 nm optical density. The lowest concentrations of BC that resulted in a greater reduction in turbidity compared to the respective positive-growth controls were defined as the MIC. All MIC tests were performed in three biological replicates.

4.3. Benzalkonium Chloride Exposure Assays

The subinhibitory concentration (subMIC) of BC was defined based on the MIC values determined previously. The experiments were performed by confronting strains carrying the *qacA/qacC* resistance genes (*S. xylosus* and *S. pasteuri*) with sensitive strains (*S. aureus* ATCC 29213 and ATCC 6538) according to Karatzas et al. [22], with modifications. Overnight cultures were standardized to 8.5×10^8 CFU/mL in 5 mL of TSB. Following this, a mixture containing 2.5 mL of each strain to be confronted (one resistant and one sensitive strain) was added to two separate tubes. One tube served as a control, while the other tube had $1/2\times$ MIC of BC added. The tubes were incubated in agitation (200 rpm) at 37 °C for 24 h. The next day, from all tubes that showed bacterial growth, an aliquot of 50 μL was transferred to a new tube with fresh medium using the same $1/2 \times$ MIC concentration of BC and incubated under the same conditions. These passages were performed for up to 72 h. After this period, an aliquot was plated on trypticase soy agar (TSA) and mannitol agar, incubated for 24 h, and colonies with different morphologies in TSA and mannitol were separated onto a new plate and frozen for subsequent coagulase and phenotypic testing (Figure 1). The isolates recovered after 72 h were subjected to determination of the MIC of BC to verify resistance transfer. The disk diffusion test was also performed to evaluate changes in susceptibility patterns. All bacteria were stored in TSB medium with 20% glycerol at -20 °C until use.

4.4. Antimicrobial Susceptibility Testing by Agar Disk Diffusion

Agar disk diffusion was performed in all strains before and after of subMIC BC exposure assays using clindamycin (2 μg), cefalotin (30 μg), penicillin (10 U) and cefazolin (30 μg) according to CLSI standards [21]. In addition, oxacillin MIC were determined using E-test strips (Liofilchem, MTS™, Roseto degli Abruzzi, Italy) with a 0.5 McFarland standard inoculum on Mueller–Hinton agar plates (Britania, Buenos Aires, Argentina) according to the manufacturer's manual.

4.5. Light Source

An array of three white, fluorescent lamps (Osram, Buenos Aires, Argentina) was employed as a non-coherent light source, with an emission spectrum ranging from 400 to 700 nm. Light power was measured with a Yellow Springs Kettering 65 radiometer (Yellow Springs, OH, USA). The light dose of 2.5 J/cm^2 was obtained by applying 60 min of light exposure.

4.6. Photosensitizer

Toluidine blue (TB) (Sigma Chem. Co., St. Louis, MO, USA) was employed as a photosensitizer at a concentration of 100 μM for treatment of bacterial suspensions.

4.7. Photoinactivation of Planktonic Cultures

To evaluate the efficacy of TB-PDI, we chose the strains that showed the greatest changes in susceptibility patterns to BC and antibiotics (Table 1). From an overnight culture, a tube containing 10 mL of TSB with a bacterial concentration of O.D. 0.05 was prepared and incubated at 37 °C under constant agitation of 200 rpm until reaching an O.D. 0.2. An aliquot of 10 μL was taken for control dilutions of inoculum. Then, the bacterial suspensions were centrifuged at 10,000 rpm for 10 min at 4 °C, washed and suspended in 10 mL of sterile PBS. A volume of 1 mL of each bacterial suspension was added to a tube and incubated with TB in darkness for 30 min at room temperature. After this period, aliquots of 200 μL of each culture were plated by duplicate in a 96-well microplate. Then, the plate placed on a glass slide was irradiated for 60 min from below at 16 cm from the light source, reaching a maximum light dose of 2.5 J/cm^2. The viable bacteria cell number was determined by quantitative plating on TSA, the experimental condition consisted of bacterial suspension treated with TB and exposed to light. The following conditions were used as controls: (1) bacterial suspension treated with TB without receiving light; (2) non-TB treated bacterial suspension exposed to light; and (3) untreated bacterial suspension.

4.8. Statistical Analysis

For TB-PDI statistical analysis, a D'Agostino–Pearson normality test and a two-way ANOVA were performed using the GraphPad Prism 8 software, and $p < 0.05$ values were considered statistically significant.

5. Conclusions

Based on our study, the use of subinhibitory concentrations of BC in dairy environments can lead to an increased dissemination of resistance, promoting cross-resistance transfer and further exacerbating the issue of antibiotic resistance. However, we observed that the photodynamic therapy (TB-PDI) can be a promising alternative to control the spread of strains with decreased sensitivity to biocides and antibiotics, as well as a treatment option for mastitis, irrespective of the phenotype and species involved.

Author Contributions: Conceptualization, E.C.R.B. and F.B.; methodology, E.C.R.B.; N.C.C.S., V.L.M.R.; investigation, E.C.R.B.; G.H.C.; resources, F.B.; writing—original draft preparation, E.C.R.B.; writing—review and editing, N.C.C.S., V.L.M.R., A.C. and F.B.; supervision, D.O.S.; project administration, F.B.; funding acquisition, F.B. All authors have read and agreed to the published version of the manuscript.

Funding: This research was funded by ANPCyT grant number PICT 2019-2883, CONICET grant number PIP2021-11220200102843CO and PUE085.

Institutional Review Board Statement: Not applicable.

Informed Consent Statement: Not applicable.

Data Availability Statement: Data will be made available upon request.

Acknowledgments: E.C.R.B is fellow of Consejo Nacional de Investigaciones Científicas y Técnicas (CONICET). We also thank Lorena Medina for her valuable technical assistance.

Conflicts of Interest: The authors declare no conflict of interest.

References

1. De Buck, J.; Ha, V.; Naushad, S.; Nobrega, D.B.; Luby, C.; Middleton, J.R.; De Vliegher, S.; Barkema, H.W. Non-Aureus Staphylococci and Bovine Udder Health: Current Understanding and Knowledge Gaps. *Front. Vet. Sci.* **2021**, *8*, 658031. [CrossRef] [PubMed]
2. Capita, R.; Buzón-Durán, L.; Riesco-Peláez, F.; Alonso-Calleja, C. Effect of Sub-Lethal Concentrations of Biocides on the Structural Parameters and Viability of the Biofilms Formed by *Salmonella* Typhimurium. *Foodborne Pathog. Dis.* **2017**, *14*, 350–356. [CrossRef]
3. Klibi, A.; Maaroufi, A.; Torres, C.; Jouini, A. Detection and Characterization of Methicillin-Resistant and Susceptible Coagulase-Negative Staphylococci in Milk from Cows with Clinical Mastitis in Tunisia. *Int. J. Antimicrob. Agents* **2018**, *52*, 930–935. [CrossRef]
4. Sidhu, M.S.; Heir, E.; Sørum, H.; Holck, A. Genetic Linkage Between Resistance to Quaternary Ammonium Compounds and β-Lactam Antibiotics in Food-Related *Staphylococcus* spp. *Microb. Drug Resist.* **2001**, *7*, 363–371. [CrossRef]
5. Paul, D.; Chakraborty, R.; Mandal, S.M. Biocides and Health-Care Agents Are More than Just Antibiotics: Inducing Cross to Co-Resistance in Microbes. *Ecotoxicol. Environ. Saf.* **2019**, *174*, 601–610. [CrossRef]
6. Sellera, F.P.; Sabino, C.P.; Ribeiro, M.S.; Gargano, R.G.; Benites, N.R.; Melville, P.A.; Pogliani, F.C. In Vitro Photoinactivation of Bovine Mastitis Related Pathogens. *Photodiagn. Photodyn. Ther.* **2016**, *13*, 276–281. [CrossRef] [PubMed]
7. Ajose, D.J.; Oluwarinde, B.O.; Abolarinwa, T.O.; Fri, J.; Montso, K.P.; Fayemi, O.E.; Aremu, A.O.; Ateba, C.N. Combating Bovine Mastitis in the Dairy Sector in an Era of Antimicrobial Resistance: Ethno-Veterinary Medicinal Option as a Viable Alternative Approach. *Front. Vet. Sci.* **2022**, *9*, 800322. [CrossRef]
8. Rainard, P.; Foucras, G. A Critical Appraisal of Probiotics for Mastitis Control. *Front. Vet. Sci.* **2018**, *5*, 251. [CrossRef]
9. Chan, M.K.L.; Koo, S.H.; Quek, Q.; Pang, W.S.; Jiang, B.; Ng, L.S.Y.; Tan, S.H.; Tan, T.Y. Development of a Real-Time Assay to Determine the Frequency of Qac Genes in Methicillin Resistant Staphylococcus Aureus. *J. Microbiol. Methods* **2018**, *153*, 133–138. [CrossRef]
10. Hegstad, K.; Langsrud, S.; Lunestad, B.T.; Scheie, A.A.; Sunde, M.; Yazdankhah, S.P. Does the Wide Use of Quaternary Ammonium Compounds Enhance the Selection and Spread of Antimicrobial Resistance and Thus Threaten Our Health? *Microb. Drug Resist.* **2010**, *16*, 91–104. [CrossRef] [PubMed]
11. Weber, D.J.; Rutala, W.A.; Sickbert-Bennett, E.E. Outbreaks Associated with Contaminated Antiseptics and Disinfectants. *Antimicrob. Agents Chemother.* **2007**, *51*, 4217–4224. [CrossRef]
12. Dionisio, F.; Matic, I.; Radman, M.; Rodrigues, O.R.; Taddei, F. Plasmids Spread Very Fast in Heterogeneous Bacterial Communities. *Genetics* **2002**, *162*, 1525–1532. [CrossRef]
13. Gerba, C.P. Quaternary Ammonium Biocides: Efficacy in Application. *Appl. Environ. Microbiol.* **2015**, *81*, 464–469. [CrossRef] [PubMed]
14. Levin-Reisman, I.; Ronin, I.; Gefen, O.; Braniss, I.; Shoresh, N.; Balaban, N.Q. Antibiotic Tolerance Facilitates the Evolution of Resistance. *Science* **2017**, *355*, 826–830. [CrossRef]

15. Ning, X.; He, G.; Zeng, W.; Xia, Y. The Photosensitizer-Based Therapies Enhance the Repairing of Skin Wounds. *Front. Med.* **2022**, *9*, 915548. [CrossRef] [PubMed]
16. Hoorijani, M.N.; Rostami, H.; Pourhajibagher, M.; Chiniforush, N.; Heidari, M.; Pourakbari, B.; Kazemian, H.; Davari, K.; Amini, V.; Raoofian, R.; et al. The Effect of Antimicrobial Photodynamic Therapy on the Expression of Novel Methicillin Resistance Markers Determined Using CDNA-AFLP Approach in Staphylococcus Aureus. *Photodiagn. Photodyn. Ther.* **2017**, *19*, 249–255. [CrossRef]
17. Silva, L.O.; Da Silva Souza, K.L.; De Jesus Beloti, L.; Neto, W.M.R.; Núñez, S.C.; Frias, D.F.R. Use of Photodynamic Therapy and Photobiomodulation as Alternatives for Microbial Control on Clinical and Subclinical Mastitis in Sheep. *Lasers Med. Sci.* **2022**, *37*, 2305–2310. [CrossRef] [PubMed]
18. Schug, A.R.; Bartel, A.; Scholtzek, A.D.; Meurer, M.; Brombach, J.; Hensel, V.; Fanning, S.; Schwarz, S.; Feßler, A.T. Biocide Susceptibility Testing of Bacteria: Development of a Broth Microdilution Method. *Vet. Microbiol.* **2020**, *248*, 108791. [CrossRef] [PubMed]
19. Soni, I.; Chakrapani, H.; Chopra, S. Draft Genome Sequence of Methicillin-Sensitive Staphylococcus Aureus ATCC 29213. *Genome Announc.* **2015**, *3*, e01095-15. [CrossRef]
20. Abreu, A.C.D.S.; Crippa, B.L.; Souza, V.V.M.A.D.; Nuñez, K.V.M.; Almeida, J.M.D.; Rodrigues, M.X.; Silva, N.C.C. Assessment of Sanitiser Efficacy against Staphylococcus Spp. Isolated from Minas Frescal Cheese Producers in São Paulo, Brazil. *Int. Dairy J.* **2021**, *123*, 105171. [CrossRef]
21. CLSI. *2019 Performance Standards for Antimicrobial Disk and Dilution Susceptibility Tests for Bacteria Isolated from Animals*, 5th ed.; CLSI Standard VET01; Clinical and Laboratory Standards Institute: Wayne, PA, USA, 2019.
22. Karatzas, K.A.G.; Webber, M.A.; Jorgensen, F.; Woodward, M.J.; Piddock, L.J.V.; Humphrey, T.J. Prolonged Treatment of Salmonella Enterica Serovar Typhimurium with Commercial Disinfectants Selects for Multiple Antibiotic Resistance, Increased Efflux and Reduced Invasiveness. *J. Antimicrob. Chemother.* **2007**, *60*, 947–955. [CrossRef] [PubMed]

Article

Epidemiology of Antimicrobial Resistance Genes in *Staphyloccocus aureus* Isolates from a Public Database in a One Health Perspective—Sample Characteristics and Isolates' Sources

Francesca Zaghen [1,2], Valerio Massimo Sora [1,2,*], Gabriele Meroni [1], Giulia Laterza [1,2], Piera Anna Martino [1], Alessio Soggiu [1], Luigi Bonizzi [1] and Alfonso Zecconi [1]

[1] Department of Biomedical, Surgical and Dental Sciences-One Health Unit, School of Medicine, University of Milan, Via Pascal 36, 20133 Milan, Italy
[2] Department of Clinical and Community Sciences, School of Medicine, University of Milan, Via Celoria 22, 20133 Milan, Italy
* Correspondence: valerio.sora@unimi.it; Tel.: +39-3358120996

Abstract: *Staphylococcus aureus* is considered one of the most widespread bacterial pathogens for both animals and humans, being the causative agent of various diseases like food poisoning, respiratory tract infections, nosocomial bacteremia, and surgical site and cardiovascular infections in humans, as well as clinical and subclinical mastitis, dermatitis, and suppurative infections in animals. Thanks to their genetic flexibility, several virulent and drug-resistant strains have evolved mainly due to horizontal gene transfer and insurgence of point mutations. Infections caused by the colonization of such strains are particularly problematic due to frequently occurring antibiotic resistance, particulary methicillin-resistant *S. aureus* (MRSA), and are characterized by increased mortality, morbidity, and hospitalization rates compared to those caused by methicillin-sensitive *S. aureus* (MSSA). *S. aureus* infections in humans and animals are a prime example of a disease that may be managed by a One Health strategy. In fact, *S. aureus* is a significant target for control efforts due to its zoonotic potential, the frequency of its illnesses in both humans and animals, and the threat posed by *S. aureus* antibiotic resistance globally. The results of an epidemiological analysis on a worldwide public database (NCBI Pathogen Detection Isolate Browser; NPDIB) of 35,026 *S. aureus* isolates were described. We considered the diffusion of antibiotic resistance genes (ARGs), in both human and animal setting, and the results may be considered alarming. The result of this study allowed us to identify the presence of clusters with specific ARG patterns, and that these clusters are associated with different sources of isolation (e.g., human, non-human).

Keywords: *S. aureus*; One Health; antimicrobial resistance; molecular epidemiology

Citation: Zaghen, F.; Sora, V.M.; Meroni, G.; Laterza, G.; Martino, P.A.; Soggiu, A.; Bonizzi, L.; Zecconi, A. Epidemiology of Antimicrobial Resistance Genes in *Staphyloccocus aureus* Isolates from a Public Database in a One Health Perspective—Sample Characteristics and Isolates' Sources. *Antibiotics* **2023**, *12*, 1225. https://doi.org/10.3390/antibiotics12071225

Academic Editor: Nahla O. Eltai

Received: 30 June 2023
Revised: 20 July 2023
Accepted: 23 July 2023
Published: 24 July 2023

1. Introduction

Staphylococcus aureus is considered one of the most widespread bacterial pathogens for both animals and humans, being a causative agent of various diseases like food poisoning, respiratory tract infections, nosocomial bacteremia, and surgical site and cardiovascular infections in humans, as well as clinical and subclinical mastitis, dermatitis, and suppurative infections in animals [1,2]. *S. aureus* has the ability to produce a variety of virulence factors that cause tissue injury, immune evasion, colonization, cell–cell interactions, and adhesion [3–5]. Due to their genetic flexibility, several virulent and drug-resistant strains have evolved mainly through horizontal gene transfer and insurgence of point mutations [6]. In particular, methicillin-resistant *S. aureus* (MRSA) is characterized by increased mortality, morbidity, and hospitalization rates compared to those caused by methicillin-sensitive *S. aureus* (MSSA) [7,8]. As stated in the latest ECDC (European Centre for Disease Prevention and Control) annual epidemiological report, MRSA strains have a population-weighted

EU/EEA (Europena Union/European Economic Area) mean prevalence of 15.8% [9]. Over time, *S. aureus* evolved resistance to various antibiotics, including conventional betalactam antibiotics (e.g., penicillin and its derivatives) [10] as well as to the most recent ones such as vancomicyin [11]. In addition to specific antibiotic resistance, biofilm also contributes to nonspecific antibiotic resistance, which is a common feature in many biofilm-associated *S. aureus* infections [12].

S. aureus infections in humans and animals are a good example of a disease that may be managed by a One Health strategy. In fact, *S. aureus* is a significant target for control efforts due to its zoonotic potential, the frequency of illnesses caused in both humans and animals, and the threat posed by *S. aureus* antibiotic resistance globally. These programs should take into account the genetic and phenotypic traits of the bacteria, the epidemiology of the illness, and a strategy that takes into account isolates from both humans and animals as well as the possible risk related to bacteria and ARGs spreading via the environment. To the best of our knowledge, the literature lacks studies investigating the possible relation between molecular antibiotic resistance patterns and sources of bacterial isolation (species, clinical status, organ...). With the aim of contibuting to fill this gap, we present the results of an epidemiological analysis of a worldwide public database (NCBI Pathogen Detection Isolate Browser; NPDIB) including 35,026 *S. aureus* isolates to better characterize the origin of antibiotic-resistant isolates.

2. Results

2.1. Data Description

The public database at the date 30 April 2022 included 35,026 isolates; among them, 16,787 (47.9%) were classified as clinical and human-associated (HUA); 2091 (6.0%) were isolated from animals, farms, or environmental sources, and classified as non-human-associated (NHA); 15,355 (43.8%) had an unknown origin (UNK); and 793 (2.3%) were from other human sources. Due to the relative low frequency of isolates classified as other human sources and their heterogeneity, these latter isolates were not furthermore considered in the epidemiological analyses.

The total number of isolates after data polishing was brought to 34,233, of which 2091 (6.1%) isolates belonged to the NHA class, 16,787 (49.0%) belonged to the HUA class, and 15,355 (44.9%) belonged to the UNK class.

Table 1 and Figure 1 report the distribution of the sources of the isolates classified by the three categories considered (clinical, animal/farms/environment and unknown).

Table 1. Sources of isolates' distribution.

Category	Source	Isolates (N)	Relative Frequency among Category (%)
Non-Human-Associated (NHA)	Animal	725	34.7
	Environment	256	12.2
	Farm	201	9.6
	Food	909	43.5
Human-Associated (HUA)	Blood	4298	25.6
	Respiratory sources	7366	43.9
	Skin	1338	8.0
	Wound	704	4.2
	Abscess	312	1.8
	Other districts	1846	11.0
	Other sources	923	5.5
Unknown (UNK)	Unknown sources	14,247	92.8
	Other districts	432	2.8
	Other sources	676	4.4

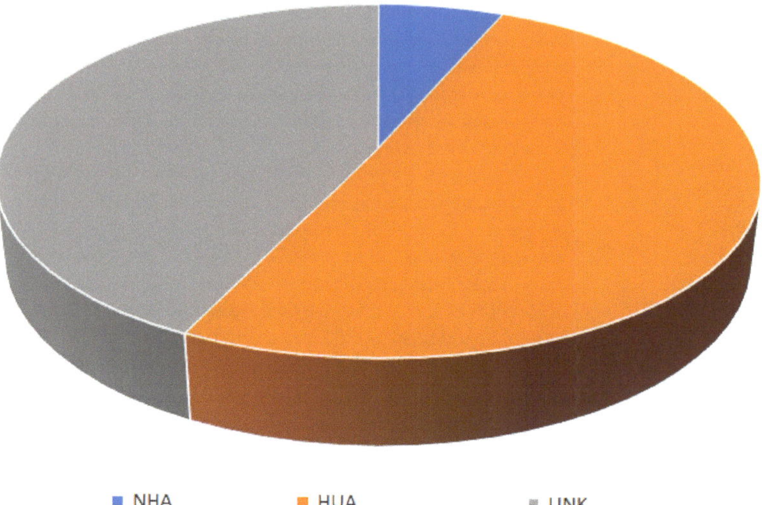

Figure 1. Distribution of *S. aureus* isolates by source (NHA = non-human-associated, HUA = human associated, UNK = unknown origin).

2.2. Resistance Gene Distribution

S. aureus is a well-known pathogen that carries a wide variety of ARGs. The database reported 67 different ARGs, and in this study, we considered 39 ARGs, excluding the ones having a prevalence <2%. The frequencies ranged from 2% of *dfrS1* to 99.9% of *mepA* (Table 2). We provided a detailed description of the antibiotic classes' resistance and related mechanism of all the ARGs defined in the database in Supplementary Table S1.

Table 2. Antibiotic resistance genes' frequencies (%) of the NPDIB isolates (threshold for prevalence >2%) [1].

Gene	TOTAL	NHA [2]	HUA	UNK
mepA	99.9	100.0 [a3]	100.0 [b]	99.9 [a]
tet (38)	99.0	99.4 [a]	98.4 [b]	99.6 [a]
blaI	82.1	72.8 [a]	80.5 [b]	85.0 [c]
blaR1	72.4	66.4 [a]	75.4 [b]	70.0 [c]
mecA	67.1	49.9 [a]	73.5 [b]	62.4 [c]
blaZ	66.7	66.4 [a]	64.5 [a]	69.2 [b]
fos(B)	62.1	45.2 [a]	71.4 [b]	54.3 [c]
mecR1	56.2	29.9 [a]	66.7 [b]	48.2 [c]
parC	51.0	24.2 [a]	59.0 [b]	45.8 [c]
gyrA	48.5	19.2 [a]	57.1 [b]	43.1 [c]
murA	44.0	54.8 [a]	43.0 [b]	43.5 [b]
Abc-f	40.4	34.3 [a]	46.6 [b]	34.4 [a]
glpT	27.3	42.2 [a]	20.4 [b]	32.9 [c]
ant(9)-la	26.8	6.4 [a]	36.0 [b]	19.5 [c]
erm(A)	26.6	4.7 [a]	35.9 [b]	19.4 [c]
ant(6)-la	21.4	24.8 [a]	28.8 [b]	12.7 [c]
mecI	20.7	4.0 [a]	30.2 [b]	12.5 [c]
aadD1	20.5	12.1 [a]	28.6 [b]	12.9 [a]
aph(3′)-lla	20.4	18.1 [a]	27.9 [b]	12.5 [c]

Table 2. *Cont.*

Gene	TOTAL	NHA [2]	HUA	UNK
aac(6′)-le/aph(2″)-la	17.8	12.7 [a]	18.1 [b]	18.2 [b]
sat4	17.5	8.9 [a]	25.0 [b]	10.5 [a]
erm(C)	16.6	16.8 [a]	15.4 [b]	17.9 [a]
blaPC1	16.5	9.9 [a]	16.3 [b]	17.6 [c]
bleO	15.2	3.2 [a]	21.5 [b]	9.9 [c]
Tet(K)	14.2	19.5 [a]	10.9 [b]	17.1 [c]
msr(A)	13.4	9.1 [a]	19.7 [b]	7.0 [c]
mph(C)	12.9	9.0 [a]	19.2 [b]	6.5 [c]
dfrG	10.7	14.3 [a]	11.4 [b]	9.4 [c]
tet(M)	10.6	11.4 [a]	8.2 [b]	13.1 [a]
parE	6.2	1.2 [a]	6.3 [b]	6.8 [b]
rpoB	5.3	3.9 [a]	6.6 [b]	4.1 [a]
mup(A)	4.2	0.5 [a]	6.0 [b]	2.6 [c]
tet(L)	3.2	8.8 [a]	2.4 [b]	3.2 [c]
erm(B)	3.0	12.7 [a]	2.2 [b]	2.5 [b]
ileS	2.9	0.4 [a]	2.8 [b]	3.4 [c]
fex(A)	2.4	9.8 [a]	2.2 [b]	1.5 [c]
fusC	2.4	0.2 [a]	2.1 [b]	2.9 [c]
catA	2.3	4.4 [a]	2.9 [b]	1.5 [c]
dfrS1	2.0	1.0 [a]	2.3 [b]	1.8 [c]

[1] <2% aac(6′)-le; aph(2″)-l; apmA; cat-TC; cfr; dfrB; dfrF; dfr(K); erm(T); fosD; fosY; fus(A); fusB; lnu(A); mecC; mprF; rpIC_G152D; spd; tet(C); vanA; vanH-A; vanR-A; vanS-A; vanX-A; vanY-A; vanZ-A; vga(A); walK. [2] NHA = non-human-associated, HUA = human-associated, UNK = unknown origin. [3] values with different superscript among lines statistically differ at χ^2 test or Fisher's exact test ($\alpha = 0.05$).

The resistance genes shown in Table 2 belong to 13 different classes of antibiotics: tetracyclines, penams, phosphonic acid, fluoroquinolones, aminoglycosides, nucleosides, macrolides, glycopeptides, diaminopyrimidines, rifamycines, mupirocines, phenicols, and fusidanes. *BlaI, blaR1, mecR1,* and *mecI* genes were excluded from the study since they are regulatory genes of *blaZ* and *mecA,* respectively.

Regarding ARG diffusion, we identified the most widespread ARG for each antibiotic class: among the ARGs for amynoglycosides, the gene with the highest frequency is *ant(9)-la,* with a total of 9179 (17.8%) positive isolates; *dfrG* for diaminopyrimidines, with 3657 (10.7%) positive isolates; *parC* for the fluoroquinolones class, with 17,446 (51%) positive isolates; *mup(A)* for the mupirocines class, with 1421 (4.2%) positive isolates; *mecA* for the penams class (67.1%). The relative frequencies for *fex(A)* and *catA,* belonging to the phenicols class, are 806 (2.4%) and 795 (2.3%) positive isolates, respectively; in the phosphonic acid class, *fos(B)* is the most prevalent gene, with 21,257 (62.1%) positive isolates; *erm(C)* is the most frequent gene identified in the macrolides class, with 5685 (16.6%) cases of positivity; and finally, *mepA* is the most represented gene of the tetracyclines class, with 34,205 (99.9%) positive isolates.

2.3. Cluster Analysis

Cluster analysis was performed to identify possible AMR patterns, which can be considered the expression of the specific ARG asset of different isolates, thus allowing for the analysis of a potential relationship with other factors such as source, geographical origin, and others. The analysis identified seven different clusters based on the presence of the ARGs described in Table 2. Table 3 describes the composition of the clusters based on ARG frequencies and Table 4 contains information on isolation category and the total number of isolates per cluster.

Table 3. ARG frequencies according to the antibiotic class and cluster; *n* = number of positive isolates. The genes with the highest frequency for each antibiotic class are represented in bold types.

Antibiotic Class	Gene	Cluster						
	ARG (*n*)	1	2	3	4	5	6	7
aminoglycosides	***ant(9)-la* (9179)**	0.3%	0.3%	0.1%	95.7%	20%	94.7%	9.4%
	ant(6)-la (7310)	10.6%	0.1%	0%	0.3%	94.3%	99.2%	8.7%
	aadD1 (7033)	17.7%	1.1%	3.3%	70.5%	19.9%	23.6%	5.3%
	aph(3′)-lla (6984)	0.7%	0.5%	0.2%	2.9%	96.6%	99.3%	7.9%
	aac(6′)-le/aph(2″)-la (6102)	18.3%	1.7%	4%	23.3%	10.6%	92.1%	14.3%
antibiotics targeting protein synthesis	***Abc-f* (13,816)**	49.7%	34.2%	52.1%	59.3%	53.2%	24.3%	13.3%
diaminopyrimidines	***dfrG* (3657)**	16.4%	2.2%	6.2%	0.8%	6.9%	64.2%	9%
	dfrS1 (681)	2.7%	0.6%	1.4%	4.9%	0.7%	1.1%	1.3%
fluoroquinolones	***parC* (17,446)**	97.7%	13.3%	1.4%	92.2%	64.6%	96.0%	18.7%
	gyrA (16,601)	95.0%	10.3%	0.2%	91.4%	61.1%	95.8%	14.6%
	parE (2126)	15.2%	0.8%	0.3%	13.5%	3.7%	14.3%	0.3%
fusidanes	***fusC* (807)**	2.7%	3.1%	7.6%	0.9%	1.2%	0.7%	0.7%
glycopeptides	***bleO* (5197)**	5.9%	0.1%	0.6%	70.1%	16.2%	1%	0.6%
mupirocines	***mup(A)* (1421)**	5%	1.1%	1.5%	4.0%	7.0%	23.1%	0.4%
	ileS (1000)	1.6%	0.4%	0.1%	11.9%	1.5%	4.9%	0.4%
nucleosides	***sat4* (5986)**	0.7%	0.3%	0%	1.8%	87.2%	98.7%	1.4%
penams	***mecA* (22,968)**	94.4%	18.2%	32.0%	93.3%	91.3%	98.2%	53.5%
	blaZ (22,850)	96.1%	0.4%	99.4%	56.5%	97.6%	27.1%	60.1%
	blaPC1 (5641)	7.9%	0.1%	0.2%	10.7%	1.7%	66.2%	39.3%
phenicols	***fex(A)* (806)**	10%	0.4%	0.8%	0%	0%	0%	2.6%
	catA (795)	1.5%	0.4%	0.9%	1.6%	2.3%	4.6%	4.6%
phosphonic acid	***fos(B)* (21,257)**	33.1%	51%	64.2%	96.9%	93.3%	99.5%	34.1%
	murA (15,048)	10.3%	48.4%	5.5%	16.1%	57.6%	73.3%	93.8%
	glpT (9354)	0.3%	34.3%	1.7%	8%	2%	0.3%	92.6%
rifamycines	***rpoB* (1819)**	3.1%	1.8%	0.9%	13.5%	2.5%	25.1%	1.6%
macrolides	*mph(C)* (4401)	4.2%	0.6%	0.6%	6.2%	87.7%	1.9%	0.3%
	msr(A) (4571)	4.4%	0.9%	1.5%	6.4%	87.8%	2.1%	1.3%
	erm(A) (112)	0.3%	0.2%	0.1%	95.6%	19.9%	95.2%	8.5%
	***erm(C)* (5685)**	50.7%	4.8%	10.5%	7.8%	11.4%	3.3%	15.7%
	erm(B) (1015)	2.5%	0.9%	1%	0%	0.2%	0%	9.8%
tetracyclines	*tet (38)* (33,893)	99.6%	99.9%	100%	99.6%	93.9%	98.5%	100%
	tet(k) (4869)	8.5%	3.4%	9.1%	3.5%	8.4%	40.2%	30.7%
	tet (L) (1087)	9.9%	0.3%	0.7%	0.3%	2.6%	0.1%	4.7%
	tet(m) (3621)	1.1%	1.6%	1.6%	11.9%	1%	43.9%	22.3%
	***mepA* (34,205)**	99.9%	100%	99.9%	100%	100%	100%	99.8%

Table 4. Distribution of the isolates among clusters based on source (human, animal, and unknown).

Source of the Isolates	Cluster						
	1	2	3	4	5	6	7
NHA [1]	189 (3.5%) [a2]	528 (13.8%) [a]	417 (8.5%) [a]	66 (1.1%) [a]	192 (4.6%) [a]	4 (0.2%) [a]	695 (8.7%) [a]
HUA	2361 (43.8%) [b]	1643 (43%) [b]	1993 (40.6%) [b]	3735 (63.8%) [b]	2916 (69.1%) [b]	1497 (72.5%) [b]	2642 (33.1%) [b]
UNK	2845 (52.7%) [c]	1653 (43.2%) [c]	2494 (50.9%) [c]	2050 (35%) [c]	1110 (26.3%) [c]	564 (27.3%) [c]	4639 (58.2%) [c]
TOT	5395 (100%)	3824 (100%)	4904 (100%)	5851 (100%)	4218 (100%)	2065 (100%)	7976 (100%)

[1] NHA = non-human-associated, HUA = human-associated, UNK = unknown origin, TOT = total. [2] values with different superscript among columns statistically differ (α = 0.05).

In Table 3, we can see that the distribution of the frequencies of the most represented ARG in each cluster and for each antibiotic class, apart for *mepA*, is not homogeneous: among all clusters, cluster 1 has higher frequencies for *dfrG*, *parC*, *fex(A)*, and *erm(C)*; cluster 2 does not have high frequencies for any of the considered ARGs; cluster 3 shows an abundance of *fusC*; cluster 4 is characterized by high frequencies of the genes *ant(9)-la*, *abc-f*, *bleO*, *sat4*, *mecA*, *fos(B)*, and *rpoB*; cluster 5 has higher frequencies of *abc-f*, *bleO*, *sat4*, *mecA*, and *fos(B)*; in cluster 6, the more abundant ARGs are *ant(9)-la*, *dfrG*, *parC*, *mup(A)*, *mecA*, and *fosB*; and finally, cluster 7 is characterized by an higher frequency of the gene *fex(A)*. When specific genes were considered, *ant(9)-la* is concentrated mainly in cluster 4 and 6; *abc-f* in cluster 4 and 5; *dfrG* and *parC* are more prevalent in cluster 1 and 6; *fusC* is mostly predominant in cluster 3; *bleO* and *sat4* have higher frequencies in cluster 4 and 5; *mup(A)* has the highest prevalence in cluster 6; *mecA* has the highest prevalence in cluster 4, 5, and 6; *fex(A)* is nearly absent in all clusters apart from cluster 1 and 7; *fos(B)* has high frequencies throughout all clusters, particularly in cluster 4, 5, and 6; *rpoB* is concentrated in cluster 4 and 6; and finally, *erm(C)* has an high prevalence in cluster 1.

In Figure 2, we tried to graphically represent the ARG rates divided for relative antibiotic class and clusters to visually describe the ARG distribution among all seven clusters.

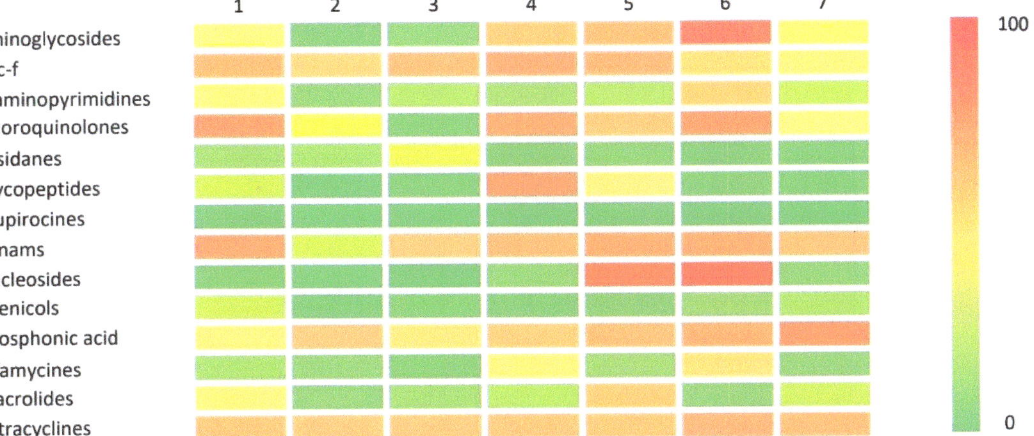

Figure 2. Graphical representation (heat map) of the antibiotic classes' related ARG frequencies in each cluster.

In Table 4 and Figure 3, we reported the distribution of the isolates among the clusters based on the isolation source. Clusters 4, 5, and 6 showed the highest frequency of HUA isolates with values >60%; clusters 2, 3, and 7 showed the highest frequencies of NHA isolates, even though this category is still not the predominant one in these clusters. The UNK category, including all the isolates that are not characterized regarding the source of isolation, comprises roughly 50% of all clusters apart from clusters 4, 5, and 6. The characteristics of the different clusters in relation to the isolation source classes of the isolates shown as clusters 2 and 7 have the highest proportion of NHA isolates (25.3% and 33.2% of total NHA isolates, respectively), while clusters 4 and 5 have the highest proportion of HUA isolates (22.2% and 17.4% of total HUA isolates, respectively). UNK isolates have the highest proportion in cluster 7 (30.2% of total UNK isolates). For all the clusters, a significant statistical difference was observed among the frequency of isolates classified by source.

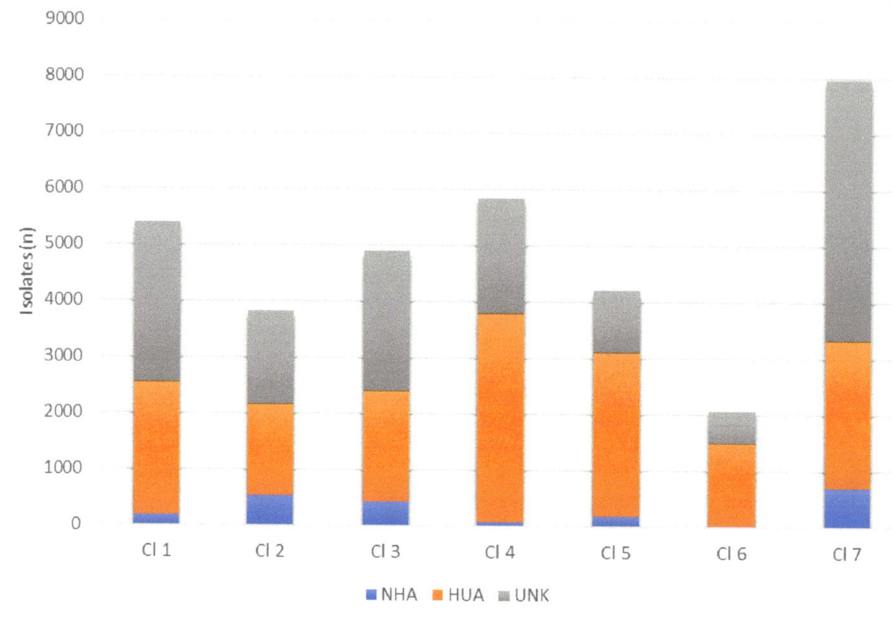

Figure 3. Distributions of isolates by source (NHA = non-human-associated, HUA = human-associated, UNK = unknown origin) among the seven clusters identified.

3. Discussion

3.1. Relevance of the Dataset

The NPDIB database can be considered an important source of data, since it is an open access database that gathers information like date and source of isolation, geographical localization, and AMR genes' presence of different bacterial isolates from all over the world. Thanks to its characteristics, it can be used as a reliable source of data to monitor AMR spread and perform epidemiological analyses with a statistical power that cannot be achieved through conventional methods. This information source does not precisely correspond to epidemiological guidelines (i.e., random sampling) because is based on voluntary upload of data. The remarkably high number of *S. aureus* isolates in the database, however, may be regarded as representative of the community of these bacteria linked to human or animal pathologies.

3.2. Antimicrobial Resistance Genes

Overall, 35 ARGs, divided into thirteen antibiotic classes, have been considered for the analyzes conducted in this study. The summary characteristics of these ARGs may be described as follows.

ant(9)-la is the gene responsible for the expression of aminoglycoside O-nucleotidyltransferase which catalyzes the adenylation of an AMP group from a substrate to the aminoglycoside molecule to make it inactive and convey specific resistance to spectinomycin [13,14].

abc-f is responsible for the expression of antibiotic resistance ABC-F proteins responsible for mediating resistance to a wide variety of antibiotics targeting the 50S ribosomal subunit by dissociating bound antibiotic molecules from the ribosome [15].

dfrG encodes the synthesis of dihydrofolate reductase and is mainly associated with trimethoprim resistance [16].

S. aureus resistance to fluoroquinolones comes from mutations in various genes, such as *parC*, resulting in the synthesis of altered proteins, in the quinolone resistance-determining region, which are less susceptible or insensible to this class of antibiotics [10,16].

fusC is responsible for the expression of FusC protein, which is a protein that actively protects the elongation factor-G from fusic acid molecules, enabling the pathogen to exert resistance to the fusidanes class of antibiotics [17].

Bleomycin and the related antibiotics phleomycin and tallysomycin function as DNA-breaking molecules capable of killing both procaryotic and eucaryotic cells at low concentrations. Plasmid-mediated resistance to bleomycin is widely spread among both clinically relevant Gram-negative and Gram-positive bacteria resistant to aminoglycosides [18]. The mechanism of resistance consists in the binding of the acidic bleomycin resistance proteins (BRPs), encoded by the *ble* genes, via electrostatic interactions to the bleomycin molecule, characterized by a basic pH, to prevent DNA cleavage [19].

The *mupA* gene, which is another name for the *ileS2* gene, is responsible for producing an isoleucyl-t-RNA synthetase that is resistant to mupirocin, an antibiotic that reversibly binds to the active site of bacterial and archaeal isoleucyl-t-RNA synthetase and competes with isoleucine and ATP or Ile-AMP [16,20].

sat4 is a gene that confers resistance to streptothricin, an antibiotic that causes mRNA mistranslation and protein synthesis inhibition by interacting with the ribosome and encoding for the streptothricin acetyltransferase Sat4 [21].

The penicillin-binding proteins (PBPs) involved in peptidoglycan production are closely related to the mechanism of methicillin resistance. Additionally, MRSA strains produce PBP2A, which replaces PBP2's transpeptidase activity and takes over the activity of other, inactivated PBPs. The PBP2 transpeptidase domain is made inactive by penams, but the PBP2 transglycosylase domain is still active, and in the case of MRSA strains, works in conjunction with the PBP2A transpeptidase to enable the synthesis of the cell wall [22].

The gene *fexA* encodes a protein of 475 amino acids with fourteen transmembrane domains, which represents an efflux protein of the major facilitator superfamily, FexA, able to actively remove chloramphenicol from the bacterial cell [23].

Fosfomycin is an inhibitor of peptidoglycan synthesis, and the cause of phosphomycin resistance in *S. aureus* is the synthesis of the metalloenzyme FosB, encoded by the *fosB* gene, which catalyzes the $Mg_2{}^+$-dependent attachment of L-cysteine to the phosphomycin ring [24].

Rifampicin inhibits transcription by interfering with the beta subunit of RNA polymerase. Resistance to rifampicin in *S. aureus* is determined by mutations in the *rpoB* gene encoding the B subunit of RNA polymerase. The most common are mutations that cause amino acid sequence changes in the RpoB protein, leading to a reduced affinity of the enzyme for the antibiotic [25].

The resistance mechanisms to macrolides are various in *S. aureus*. The most common one involves antibiotic's target site modification, and it is carried out by the enzyme adenylyl-N-methyltransferase Erm (erythromycin ribosome methylation). The gene encoding Erm methylase synthetase may be expressed in a constitutive manner, in which case strains show resistance to all macrolides, or in an inducible manner, in which case resistance occurs only to antibiotics that are inducers of methylase synthesis. Resistance to the other macrolide-class antibiotics requires the presence of an inducer, which may be erythromycin or another macrolide. Inducible resistance to macrolides in *S. aureus* is most often determined by the *ermA* or *ermC* genes [26,27].

Tetracyclines inhibit protein synthesis by interfering with the 30S subunit of the ribosome, and the mechanism of resistance to tetracyclines in *S. aureus* usually involves active removal of the antibiotic from the bacterial cell and ribosomal protection. The MepA efflux pump, encoded by the *mepA* gene, is a part of the multidrug and toxic extrusion (MATE) family, and the decreased susceptibility to antibiotics (mainly fluoroquinolones, tetracycline), biocides, and dyes may indirectly be associated with overexpression of these pumps [16,28].

Despite concerns about the risk of an increasing frequency of vancomycin resistance, we observed frequencies below 2% (precisely 0.1% for *vanA*, 0.1% for *vanH-A*, 0.1% for *vanR-A*, 0.1% for *vanS-A*, 0.1% for *vanX-A*, 0.1% for *vanY-A*, and 0.1% for *vanZ-A*) for this antimicrobial molecule. The presence of vancomycin (VAN) resistance genes' lowers the ability of the pathogen to spread because it restricts other biological functions of the bacterial cells, like being able to efficiently replicate and spread from one host to another; this biological concept can be referred to as "fitness cost" [29]. Indeed, several vancomycin-resistant (VRSA) and intermediate resistance to vancomycin (VISA) strains have appeared but have not spread throughout the population, making VAN still the antibiotic of last resort for MRSA infections [11,30]. These results partially support a recent work published by Wu and colleagues [31] stating that vancomycin resistance has increased globally in the past years, but overall frequency of resistant isolates can still be considered rather low.

3.3. Antimicrobial Resistance Genes' Pattern

The different isolates may carry different distributions of AMR genes; therefore, the pattern of resistance may represent a more appropriate description of the features of the isolates in relation to human and animal diseases. The resistance patterns of the clusters shown as ARGs related to abc-f proteins, phosphonic acid, and tetracyclines are evenly spread across all clusters, which is in accordance with several studies investigating rates of resistance to these antibiotics [32–34]. Cluster 2 can be identified as the cluster with lowest presence of ARGs, while clusters 4, 5, and 6 have high rates of presence of ARGs related to the nine different antibiotic classes. Indeed, clusters 4, 5, and 6 have high rates of positivity for ARGs related to aminoglycosides, abc-f proteins, fluoroquinolones, penams, nucleosides, phosphonic acid, and tetracyclines, while high rates of ARGs related to glycopeptides and rifamycines characterize only cluster 4, high rates of ARGs related to glycopeptides and macrolides characterize only cluster 5, and high rates of ARGs related to diaminopyrimidines and rifamycines only characterize cluster 6. ARGs related to diaminopyrimidines, fusidanes, mupirocines, and phenicols have a low level of prevalence, with the only exception being diaminopyrimidines in cluster 6. The results related to mupirocines and phenicols are in accordance with other published studies [35,36], while the results that we obtained for diaminopyrimidines and fusidanes are discordant from what has been reported in the literature [37–39]. Penam-related ARGs are frequently observed in every cluster, with frequencies similar to the ones stated in the Global Antimicrobial Resistance and Use Surveillance System (GLASS) Report of 2022 [40], with the only exception of cluster 2, where the observed frequency is exceptionally low. Mupirocines and phenicols are the only two classes of antibiotics where the frequencies of the related ARGs can be considered nonrelevant in all clusters. Comparing the clusters' ARG frequencies and the source of isolation composition, we observed that the clusters composed mainly by HUA isolates (cluster 4, 5 and 6) had higher frequencies of ARGs related to several different antibiotic classes, while clusters 2 and 7, characterized by a higher number of NHA isolates, show low-to-intermediate frequencies of ARG positivity. This information could be indicative of a different distribution of antibiotic resistance genetic elements between isolates from animals, food, the environment, and humans.

This database enables us to work with a superb amount of global data on ARG epidemiology. However, a potential weakness of this study is the fact that NHA isolates are far fewer than HUA and UNK isolates, so the imbalance among these categories could be a source of bias in the analysis. From a One Health perspective, the relative low frequency of isolates from animal sources should not be attributed to a low prevalence of illnesses in animals, but to a low frequency of upload of animal-related isolates, supporting the need to implement this database with more information on animal-derived isolates. Moreover, *S. aureus* has been known as an extremely important pathogen both in human and animal health [41,42], and the ability to study its AMR epidemiology with a One Health approach is of paramount importance to improve surveillance programs.

Finally, it should be also taken into account when interpreting these data that there are increasing evidences of genotype–phenotype discrepancies so that genomic AMR data should always be paired with phenotypic data, especially in clinical settings [43,44].

4. Materials and Methods

4.1. NCBI Pathogen Detection Isolate Browser and Antibacterial Data (NPDIB)

Approximately one million isolates from 53 different bacteria are currently available in the NCBI pathogen detection isolate browser (NPDIB). The parameters selected to perform an epidemiological study on *S. aureus* strains uploaded to this database were retrieved from a previous study [45]. Briefly, the data were exported into Microsoft Excel and the identification data were organized into columns in a matrix. Each AMR gene was associated with a column, which was filled with 1 if the gene was discovered in the sample and 0 if it was not. The information in the other columns were changed to align the formats and switch out text entries for numbers.

4.2. Statistical Analysis

All the data were analyzed on SPSS 28.0.1.1 (IBM Corp., Armonk, NY, USA, 2022). We applied $\chi2$ test with Bonferroni adjustment to analyze the frequency distribution. Fisher's exact test was applied instead of $\chi2$ test when cell numerosity was below 6.

To classify isolates based on the different combination of AMR genes, cluster analysis was applied with the following parameters: squared Euclidean distance, Ward's agglomeration method, and truncation at 20% of total distance [46]. Cluster analysis is a multivariate technique allowing one to group isolates based on the characteristics they possess (e.g., AMR genes).

5. Conclusions

The analysis that we performed on the NPDIB database of globally collected *S. aureus* isolates could represent a useful tool to constantly monitor the evolution of ARG spread throughout different countries and environments, perfectly following the One Health paradigm. The results of this study allowed us to identify the presence of clusters with specific ARGs pattern, and that these clusters may be associated to different sources of isolation (e.g., human, non-human). Indeed, cluster analysis allowed to identify the clusters with isolates with higher frequency of AMR genes, and to associate them with their source. The presence of a significant higher frequency of HUA isolates among the clusters with higher AMR pattern, suggests that these isolates have higher risks for human health, and the specific AMR pattern should be considered in presence of clinical outcome. Moreover, the large differences in the source of isolates among the different clusters suggest that the development of surveillance and/or preventive programs should consider these differences to increase the efficacy of these programs.

This latter result supports the importance of characterizing the isolates not only for the presence of gene of importance, but also for their source of isolation (species, organ...). Using a voluntary-based database obviously has its drawbacks, such as it does not follow strict epidemiological guidelines for collecting isolates and the incompleteness of the majority of the data, but even taking into account these critical points, it cannot be overlooked that useful information can be gathered by such a large amount of data. We strongly believe that the attention of both public health and veterinary authorities should focus on implementing the use of this database to further increase the quantity and quality of the uploaded data, making it a useful tool to better adjust surveillance plans and contrast the ever-growing threat of AMR worldwide.

Supplementary Materials: The following supporting information can be downloaded at: https://www.mdpi.com/article/10.3390/antibiotics12071225/s1. Table S1: drug class resistance, resistance mechanism induced by antibiotic resistance genes and AMR gene family of the genes found in *S. aureus* isolates of NPDIB.

Author Contributions: Conceptualization, G.M. and V.M.S.; methodology, A.Z. and A.S.; formal analysis, A.Z.; resources, F.Z.; data curation, G.L. and A.S.; writing—original draft preparation, A.Z., V.M.S. and F.Z.; writing—review and editing, G.M., P.A.M. and A.S.; funding acquisition, A.Z. and L.B. All authors have read and agreed to the published version of the manuscript.

Funding: This work supported by FEASR—Programma di Sviluppo Rurale 2014–2020 Misura 16.1 project MOOH, PRIN 2017MZ5KWM_001 project "SAFE MILK: OMICS SCIENCE FOR MILK SAFETY AND QUALITY" and University of Milan PSR2021 line 6 One Health Action Hub.

Institutional Review Board Statement: Not applicable.

Informed Consent Statement: Not applicable.

Data Availability Statement: Publicly available datasets were analyzed in this study. This data can be found here: https://www.ncbi.nlm.nih.gov/pathogens/isolates/#taxgroup_name:%22Staphyloc occus%20aureus%22 (accessed on 30 April 2022).

Conflicts of Interest: The authors declare no conflict of interest.

References

1. Haaber, J.; Penadés, J.R.; Ingmer, H. Transfer of Antibiotic Resistance in *Staphylococcus aureus*. *Trends Microbiol.* **2017**, *25*, 893–905. [CrossRef] [PubMed]
2. Cheung, G.Y.C.; Bae, J.S.; Otto, M. Pathogenicity and virulence of *Staphylococcus aureus*. *Virulence* **2021**, *12*, 547–569. [CrossRef] [PubMed]
3. Holden, M.T.G.; Feil, E.J.; Lindsay, J.A.; Peacock, S.J.; Day, N.P.J.; Enright, M.C.; Foster, T.J.; Moore, C.E.; Hurst, L.; Atkin, R.; et al. Complete genomes of two clinical *Staphylococcus aureus* strains: Evidence for the evolution of virulence and drug resistance. *Proc. Natl. Acad. Sci. USA* **2004**, *101*, 9786–9791. [CrossRef]
4. Zecconi, A.; Cesaris, L.; Liandris, E.; Daprà, V.; Piccinini, R. Role of several *Staphylococcus aureus* virulence factors on the inflammatory response in bovine mammary gland. *Microb. Pathog.* **2006**, *40*, 177–183. [CrossRef]
5. Zecconi, A.; Scali, F. *Staphylococcus aureus* virulence factors in evasion from innate immune defenses in human and animal diseases. *Immunol. Lett.* **2013**, *150*, 12–22. [CrossRef]
6. Foster, T.J. Antibiotic resistance in *Staphylococcus aureus*. Current status and future prospects. *FEMS Microbiol. Rev.* **2017**, *41*, 430–449. [CrossRef]
7. Turner, N.A.; Sharma-Kuinkel, B.K.; Maskarinec, S.A.; Eichenberger, E.M.; Shah, P.P.; Carugati, M.; Holland, T.L.; Fowler, V.G. Methicillin-resistant *Staphylococcus aureus*: An overview of basic and clinical research. *Nat. Rev. Microbiol.* **2019**, *17*, 203–218. [CrossRef] [PubMed]
8. Ippolito, G.; Leone, S.; Lauria, F.N.; Nicastri, E.; Wenzel, R.P. Methicillin-resistant *Staphylococcus aureus*: The superbug. *Int. J. Infect. Dis.* **2010**, *14*, 7–11. [CrossRef] [PubMed]
9. European Centre for Disease Prevention and Control. *Antimicrobial Resistance in the EU/EEA (EARS-Net)—Annual Epidemiological Report 2020*; ECDC: Solna, Sweden, 2022.
10. Lowy, F.D. Antimicrobial resistance: The example of *Staphylococcus aureus*. *J. Clin. Investig.* **2003**, *111*, 1265–1273. [CrossRef] [PubMed]
11. McGuinness, W.A.; Malachowa, N.; DeLeo, F.R. Vancomycin Resistance in *Staphylococcus aureus*. *Yale J. Biol. Med.* **2017**, *90*, 269–281. [PubMed]
12. Otto, M. Staphylococcal biofilms. *Microbiol. Spectr.* **2018**, *6*, 139–148. [CrossRef] [PubMed]
13. Sheng, X.; Lu, W.; Li, A.; Lu, J.; Song, C.; Xu, J.; Dong, Y.; Fu, C.; Lin, X.; Zhu, M. ANT(9)-Ic, a Novel Chromosomally Encoded Aminoglycoside. *Microbiol. Spectr.* **2023**, *11*, 4–10. [CrossRef] [PubMed]
14. Ramirez, M.S.; Tolmasky, M.E. Aminoglycoside Modifying Enzymes. *Drug Resist. Updat.* **2010**, *13*, 151–171. [CrossRef] [PubMed]
15. Sharkey, L.K.R.; Edwards, T.A.; O'Neill, A.J. ABC-F proteins mediate antibiotic resistance through ribosomal protection. *mBio* **2016**, *7*, e01975-15. [CrossRef] [PubMed]
16. Mlynarczyk-Bonikowska, B.; Kowalewski, C.; Krolak-Ulinska, A.; Marusza, W. Molecular Mechanisms of Drug Resistance in *Staphylococcus aureus*. *Int. J. Mol. Sci.* **2022**, *23*, 8088. [CrossRef]
17. Lannergård, J.; Norström, T.; Hughes, D. Genetic determinants of resistance to fusidic acid among clinical bacteremia isolates of *Staphylococcus aureus*. *Antimicrob. Agents Chemother.* **2009**, *53*, 2059–2065. [CrossRef]
18. Gatignol, A.; Durand, H.; Tiraby, G. Bleomycin resistance conferred by a drug-binding protein. *FEBS Lett.* **1988**, *230*, 171–175. [CrossRef]
19. Mori, T.; Mizuta, S.; Suenaga, H.; Miyazaki, K. Metagenomic screening for bleomycin resistance genes. *Appl. Environ. Microbiol.* **2008**, *74*, 6803–6805. [CrossRef]

20. Chung, S.; Kim, S.; Ryu, S.H.; Hwang, K.Y.; Cho, Y. Structural basis for the antibiotic resistance of eukaryotic isoleucyl-trna synthetase. *Mol. Cells* **2020**, *43*, 350–359. [CrossRef]

21. Burckhardt, R.M.; Escalante-Semerena, J.C. Insights into the function of the Nacetyltransferase SatA that detoxifies streptothricin in *Bacillus subtilis* and *Bacillus anthraci*. *Appl. Environ. Microbiol.* **2019**, *85*, 1–20. [CrossRef]

22. Moreillon, P. New and emerging treatment of *Staphylococcus aureus* infections in the hospital setting. *Clin. Microbiol. Infect.* **2008**, *14*, 32–41. [CrossRef]

23. Kehrenberg, C.; Schwarz, S. Distribution of florfenicol resistance genes fexA and cfr among chloramphenicol-resistant Staphylococcus isolates. *Antimicrob. Agents Chemother.* **2006**, *50*, 1156–1163. [CrossRef]

24. Wright, G.D. Bacterial resistance to antibiotics: Enzymatic degradation and modification. *Adv. Drug Deliv. Rev.* **2005**, *57*, 1451–1470. [CrossRef]

25. Aubry-Damon, H.; Soussy, C.J.; Courvalin, P. Characterization of mutations in the *rpoB* gene that confer rifampin resistance in *Staphylococcus aureus*. *Antimicrob. Agents Chemother.* **1998**, *42*, 2590–2594. [CrossRef]

26. Mlynarczyk, B.; Mlynarczyk, A.; Kmera-Muszynska, M.; Majewski, S.; Mlynarczyk, G. Mechanisms of resistance to antimicrobial drugs in pathogenic Gram-positive cocci. *Mini Rev. Med. Chem.* **2010**, *10*, 928–937. [CrossRef]

27. Weisblum, B. Macrolide resistance. *Drug Resist. Updat.* **1998**, *1*, 29–41. [CrossRef] [PubMed]

28. Hassanzadeh, S.; Ganjloo, S.; Pourmand, M.R.; Mashhadi, R.; Ghazvini, K. Epidemiology of efflux pumps genes mediating resistance among *Staphylococcus aureus*; A systematic review. *Microb. Pathog.* **2020**, *139*, 103850. [CrossRef]

29. Rajer, F.; Sandegren, L. The Role of Antibiotic Resistance Genes in the Fitness Cost of Multiresistance Plasmids. *mBio* **2022**, *13*, e0355221. [CrossRef] [PubMed]

30. Hiramatsu, K.; Kayayama, Y.; Matsuo, M.; Aiba, Y.; Saito, M.; Hishinuma, T.; Iwamoto, A. Vancomycin-intermediate resistance in *Staphylococcus aureus*. *J. Glob. Antimicrob. Resist.* **2014**, *2*, 213–224. [CrossRef] [PubMed]

31. Wu, Q.; Sabokroo, N.; Wang, Y.; Hashemian, M.; Karamollahi, S.; Kouhsari, E. Systematic review and meta-analysis of the epidemiology of vancomycin-resistance *Staphylococcus aureus* isolates. *Antimicrob. Resist. Infect. Control* **2021**, *10*, 101. [CrossRef]

32. Emaneini, M.; Bigverdi, R.; Kalantar, D.; Soroush, S.; Jabalameli, F.; Noorazar Khoshgnab, B.; Asadollahi, P.; Taherikalani, M. Distribution of genes encoding tetracycline resistance and aminoglycoside modifying enzymes in *Staphylococcus aureus* strains isolated from a burn center. *Ann. Burns Fire Disasters* **2013**, *26*, 76–80.

33. Miklasińska-Majdanik, M. Mechanisms of resistance to macrolide antibiotics among *Staphylococcus aureus*. *Antibiotics* **2021**, *10*, 1406. [CrossRef] [PubMed]

34. Falagas, M.E.; Athanasaki, F.; Voulgaris, G.L.; Triarides, N.A.; Vardakas, K.Z. Resistance to fosfomycin: Mechanisms, Frequency and Clinical Consequences. *Int. J. Antimicrob. Agents* **2019**, *53*, 22–28. [CrossRef] [PubMed]

35. Chaturvedi, P.; Singh, A.K.; Shukla, S.; Agarwal, L. Prevalence of mupirocin resistant *Staphylococcus aureus* isolates among patients admitted to a tertiary care hospital. *N. Am. J. Med. Sci.* **2014**, *6*, 403–407. [CrossRef] [PubMed]

36. Fayyaz, M.; Mirza, I.A.; Ahmed, Z.; Abbasi, S.A.; Hussain, A.; Ali, S. In Vitro Susceptibility of Chloramphenicol Against Methicillin-Resistant *Staphylococcus aureus*. *J. Coll. Physicians Surg. Pak.* **2013**, *23*, 637–640.

37. Budri, P.E.; Shore, A.C.; Coleman, D.C.; Kinnevey, P.M.; Humpreys, H.; Fitzgerald-Hughes, D. Observational cross-sectional study of nasal staphylococcal species of medical students of diverse geographical origin, prior to healthcare exposure: Prevalence of SCCmec, fusC, fusB and the arginine catabolite mobile element (ACME) in the absence of sele. *BMJ Open* **2018**, *8*, e020391. [CrossRef]

38. Nurjadi, D.; Olalekan, A.O.; Layer, F.; Shittu, A.O.; Alabi, A.; Ghebremedhin, B.; Schaumburg, F.; Hofmann-Eifler, J.; Van Genderen, P.J.J.; Caumes, E.; et al. Emergence of trimethoprim resistance gene dfrG in *Staphylococcus aureus* causing human infection and colonization in sub-Saharan Africa and its import to Europe. *J. Antimicrob. Chemother.* **2014**, *69*, 2361–2368. [CrossRef]

39. Nurjadi, D.; Schäfer, J.; Friedrich-Jänicke, B.; Mueller, A.; Neumayr, A.; Calvo-Cano, A.; Goorhuis, A.; Molhoek, N.; Lagler, H.; Kantele, A.; et al. Predominance of *dfrG* as determinant of trimethoprim resistance in imported *Staphylococcus aureus*. *Clin. Microbiol. Infect.* **2015**, *21*, 1095.e5–1095.e9. [CrossRef]

40. World Health Organization. *Global Antimicrobial Resistance and Use Surveillance System (GLASS) Report 2022*; World Health Organization: Geneva, Switzerland, 2022.

41. Zetola, N.; Francis, J.S.; Nuermberger, E.L.; Bishai, W.R. Community-acquired meticillin-resistant *Staphylococcus aureus*: An emerging threat. *Lancet Infect. Dis.* **2005**, *5*, 275–286. [CrossRef]

42. Haag, A.F.; Ross Fitzgerald, J.; Penadés, J.R. *Staphylococcus aureus* in animals. *Gram-Positive Pathog.* **2019**, *7*, 731–746. [CrossRef]

43. Yee, R.; Bard, J.D.; Simner, P.J. The genotype-to-phenotype dilemma: How should laboratories approach discordant susceptibility results? *J. Clin. Microbiol.* **2021**, *59*, e00138-20. [CrossRef] [PubMed]

44. Feldgarden, M.; Brover, V.; Haft, D.H.; Prasad, A.B.; Slotta, D.J.; Tolstoy, I.; Tyson, G.H.; Zhao, S.; Hsu, C.H.; McDermott, P.F.; et al. Validating the AMRFINder tool and resistance gene database by using antimicrobial resistance genotype-phenotype correlations in a collection of isolates. *Antimicrob. Agents Chemother.* **2019**, *63*, e00483-19. [CrossRef] [PubMed]

45. Meroni, G.; Sora, V.M.; Martino, P.A.; Sbernini, A.; Laterza, G.; Zaghen, F.; Soggiu, A.; Zecconi, A. Epidemiology of Antimicrobial Resistance Genes in *Streptococcus agalactiae* Sequences from a Public Database in a One Health Perspective. *Antibiotics* **2022**, *11*, 1236. [CrossRef] [PubMed]
46. Everitt, B.S.; Landau, S.; Leese, M.; Stahl, D. *Cluster Analysis*, 5th ed.; John Wiley & Sons: Hoboken, NJ, USA, 2011; ISBN 978-0-470-74991-3.

Article

Serotyping and Antimicrobial Resistance Profiling of Multidrug-Resistant Non-Typhoidal *Salmonella* from Farm Animals in Hunan, China

Zhuohui Zhang [1,2,†], Jiyun Li [1,2,†] , Rushun Zhou [3], Qianqian Xu [1,2], Shiyin Qu [1,2], Hongguang Lin [1,2], Yan Wang [1,2] , Pishun Li [1,2] and Xiaofeng Zheng [1,2,*]

1   College of Veterinary Medicine, Hunan Agricultural University, Changsha 410128, China; zh_zhang@stu.hunau.edu.cn (Z.Z.); lijiyuny@foxmail.com (J.L.)

2   Hunan Engineering Technology Research Center of Veterinary Drugs, Hunan Agricultural University, Changsha 410128, China

3   Hunan Provincial Institution of Veterinary Drug and Feed Control, Changsha 410006, China

*   Correspondence: zheng.x@hunau.edu.cn

†   These authors equally contributed to this work.

Citation: Zhang, Z.; Li, J.; Zhou, R.; Xu, Q.; Qu, S.; Lin, H.; Wang, Y.; Li, P.; Zheng, X. Serotyping and Antimicrobial Resistance Profiling of Multidrug-Resistant Non-Typhoidal *Salmonella* from Farm Animals in Hunan, China. *Antibiotics* **2023**, *12*, 1178. https://doi.org/10.3390/antibiotics12071178

Academic Editor: Nahla O. Eltai

Received: 6 June 2023
Revised: 3 July 2023
Accepted: 10 July 2023
Published: 12 July 2023

Abstract: Non-typhoidal *Salmonella* (NTS) is a foodborne pathogen and a prevalent causative agent for disease outbreaks globally. The *Salmonella* enterica serovar 4,[5],12:i:- (S.4,[5],12:i:-) belongs to the monophasic variant of *Salmonella typhimurium*, which is of current global concern. In this study, the epidemiology and genomic characterization of S. 4,[5],12:i:- isolates from 17 livestock farms in Hunan Province between 2019 and 2020, as well as their susceptibility to 14 antimicrobial agents, were profiled. Twelve *Salmonella* serotypes were identified using the White–Kauffmann–Le Minor scheme, and whole-genome sequencing analyses were conducted based on these isolates. Overall, 107 *Salmonella* strains were isolated, of which 73% (78/107) were multidrug resistant. Resistance to tetracycline (85.05%) was found to be the most prevalent, followed by the *oqxAB* and *aac(6′)-Ib-cr* genes. *S. typhimurium* (monophasic) 4,[5],12:i:- was the most common serotype, followed by *S. typhimurium* and *S. derby*. Most antimicrobial-resistant strains were isolated from pigs, indicating that they could be important reservoirs of resistant non-typhoidal *Salmonella* strains. The presence of similar genetic environments in S. 4,[5],12:i:- indicates both vertical and horizontal transmission of resistance plasmids, which may promote the spread of drug resistance genes. Appropriate measures should be taken to curb the prevalence of S. 4,[5],12:i:-.

Keywords: chicken farm; foodborne pathogen; horizontal transmission; multidrug resistance; pig farm; *Salmonella typhimurium* monophasic variant; vertical transmission

1. Introduction

Salmonella is a Gram-negative, rod-shaped bacterium that is facultatively anaerobic and belongs to the Enterobacteriaceae family. These bacteria are generally mobile, without a capsule, non-spore-forming, and are able to colonize the digestive tracts of many vertebrates. *Salmonella* is one of the most important zoonotic pathogens and a causative agent for foodborne gastroenteritis in humans, domestic animals, and wildlife worldwide [1]. In the USA, *Salmonella* was estimated to cause illness in 46,623 patients annually across 53 states in 2016 [2]. Salmonellosis is the second most commonly reported gastrointestinal infection in the European Union (EU), with 91,662 confirmed human salmonellosis cases in all member states in 2017 [3]. *Salmonella* is also reportedly responsible for approximately 70–80% of foodborne pathogenic outbreaks in China [4].

Salmonella can be classified as typhoidal or non-typhoidal (NTS) based on its ability to cause specific pathologies in humans [5]. It is notable that non-typhoidal *Salmonella* is the main pathogen causing diarrhea and responsible for approximately 153 million cases of gastroenteritis as well as 57,000 deaths globally per annum [6]. Recently, the incidence of non-typhoidal salmonellosis was reported to be 626.5 cases per 100,000 persons in China [7].

Generally, animal farms are cultivable environments for the replication and persistence of *Salmonella*, and livestock are considered to be natural reservoirs for this bacterium [8]. Because fluoroquinolones, cephalosporins, azithromycin, and carbapenems are critically important antibiotics for the treatment of salmonellosis, emerging resistances to these drug classes are of paramount concern [9]. As a result, the World Health Organization (WHO) has deemed fluoroquinolone-resistant *Salmonella* to be priority pathogens, which urgently calls for new antimicrobials [10].

More than 2600 distinct serovars of *Salmonella* have been identified. In Salmonella, the resistance profiles vary in different serovars [11]. However, in recent decades, the monophasic variant of *S. typhimurium*, has emerged as a new multidrug-resistant serovar. This variant lacks the second-phase flagellar antigen (encoded by *fljB*) and produces a unique antigenic formula 4,[5],12:i:-. It has been frequently isolated from husbandry animals, food products, and humans in many countries and territories worldwide [12] and has become one of the major serotypes responsible for human diarrhea. In 2017, *S. ntyphimurium* (monophasic) was the third most commonly identified serovar among human cases of salmonellosis reported in the European Union after *S. enteritidis* and *S. typhimurium* [3]. A recent study pointed out that the prevalence of S. 4,[5],12:i:- has increased and become the second most frequently identified serotype in outpatients in the Henan Province of China [13].

To provide a further understanding of multidrug resistance and the distribution of drug resistance genes in *S. typhimurium* (monophasic), we collected 107 *Salmonella* isolates from 17 livestock farms (9 chicken farms and 8 pig farms) in Hunan Province, China. Serological typing and molecular epidemiological analyses were used to comprehensively profile the diversity of *Salmonella* isolates in their serotypes, drug resistances, and genotypic characteristics. These data will aid in the development of scientific strategies for the future prevention and control of *Salmonella*.

2. Results

2.1. Salmonella Strains Isolation

A total of 107 non-typhoidal *Salmonella* isolates were collected, with an isolation rate of 5.1% (52/1017) in samples from chickens and 4.5% (55/1223) in samples from pigs, in 2019 and 2020; the isolation rate in 2019 (7.8%) was higher than that in 2020 (2.8%) (Table 1). The 107 collected isolates were sampled in geographically different sites, as follows: Changsha (n = 15); Liuyang (n = 20); Ningxiang (n = 22); Changde (n = 30); Zhuzhou (n = 10); Chenzhou (n = 6); Leiyang (n = 3); and Xiangtan (n = 1) (Supplementary Materials).

Table 1. Number of strains with resistance phenotypes.

Antibiotic Agent	2019 (n = 79)	2020 (n = 28)
Penicillin: Ampicillin	56 (70.9%)	20 (71.4%)
β-lactams combination: Amoxicillin/Clavulanic	3 (3.9%)	2 (7.1%)
Aminoglycosides: Gentamicin	36 (45.6%)	2 (7.1%)
Spectinomycin	23 (29.1%)	13 (46.4%)
Tetracyclines: Tetracycline	27 (34.2%)	21 (75.0%)
Florfenicol	59 (74.7%)	13 (46.4%)
Sulfafurazole	52 (65.8%)	16 (57.1%)
Trimethoprim/sulfamethoxazole	47 (59.5%)	13 (46.4%)
Ceftiofur	9 (11.4%)	-
Ceftazidime	2 (2.5%)	-
Enrofloxacin	43 (54.4%)	4 (14.3%)
Ofloxacin	17 (21.5%)	-
Meropenem	2 (2.5%)	-
Apramycin	-	-
Polymyxins: Colistin	5 (6.3%)	-
Mequindox	-	-

2.2. Prevalence of Salmonella Serovar

The 12 serovars were identified in 107 isolates (Table 2), including *S. typhimurium* (monophasic) (*n* = 34), *S. typhimurium* (*n* = 21), and S. derby (*n* = 17). Others serovars included *S. rissen* (*n* = 13), *S. enteritidis* (*n* = 11), *S. kentucky* (*n* = 2), *S. Iindiana* (*n* = 2), *S. london* (*n* = 1), *S. thompson* (*n* = 1), *S. meleagridis* (*n* = 1), and *S. readings* (*n* = 1). The dominant serotypes in chickens were *S. typhimurium* (28.9%), and *S. enteritidis* (21.2%), while the dominant serotypes in pigs were *S. typhimurium* (monophasic) (49.1%) and *S. derby* (21.8%) (Supplementary Materials).

Table 2. Number of *Salmonella* serovars.

Serotypes	In Total	Pig	Poultry
Potential monophasic variant of Typhimurium	34	27	7
Typhimurium	20	6	15
Derby	17	12	5
Rissen	13	9	4
Enteritidis	11	-	11
Apeyeme	3	-	3
Kentucky	2	-	2
Indiana	2	-	2
London	1	1	-
Meleagridis	1	-	1
Thompson	1	-	1
Reading	1	-	1
In total	107	55	52

2.3. Antibiotic Susceptibility Testing

Fourteen antibiotics were selected for the susceptibility test, including tetracycline, ampicillin, amoxicillin/clavulanic acid, ceftazidime, ceftiofur, gentamicin, florfenicol, enrofloxacin, ofloxacin, spectinomycin, gentamicin, meropenem, trimethoprim/sulfamethoxazole, sulfafurazole, and colistin (Table 3). The results showed that 73% (78/107) of the *Salmonella* isolates were resistant to three or more antimicrobial agents. The *Salmonella* strains were mostly found to be resistant to tetracycline (84.1%), followed by ampicillin (70.9%), florfenicol (66.4%), sulfafurazole (63.6%), spectinomycin (62.6%), trimethoprim/sulfamethoxazole (56.1%), enrofloxacin (43.9%), gentamicin (34.6%), ofloxacin (15.9%), and colistin (10.3%). However, all isolates were less frequently resistant to ceftiofur (8.4%), ceftazidime (5.6%), amoxicillin (4.7%), and meropenem (1.9%) (Table 2). Interestingly, resistance to ceftazidime (*n* = 2), meropenem (*n* = 2), and colistin (*n* = 11) was observed only in *Salmonella* isolated from chickens. Overall, the multidrug-resistant profile of the pig-origin *Salmonella* (*n* = 51) was higher than that of chicken-origin *Salmonella* (*n* = 27). In addition, the resistance rates of the pig-derived strains to six drugs (tetracycline, florfenicol, sulfafurazole, trimethoprim/sulfamethoxazole, enrofloxacin, and colistin) were all higher when compared with the chicken-derived strains (Table 4).

Table 3. Antimicrobial resistance phenotype of 107 *Salmonella* isolates.

Antibiotic Agent	Abbreviation	Antibiotic Concentration Range (µg/mL)	Breakpoint Interpretive Criteria (µg/mL)			Results in Percentage (%)		
			S	I	R	S	I	R
Penicillin: Ampicillin	AMP	0–512	≤8	16	≥32	32 (29.9%)	0	75 (70%)
β-lactams combination: Amoxicillin/Clavulanic	AMC	0.5/0.25–256/128	≤8/4	16/8	≥32/16	76 (71%)	26 (24.3%)	5 (4.7%)
Aminoglycosides: Gentamicin	GEN	0.25–128	≤4	8	≥16	51 (47.7%)	19 (17.8%)	37 (34.6%)
Spectinomycin	STP	0–512	≤32	64	≥128	14 (13%)	26 (24%)	67 (62.6%)
Tetracyclines: Tetracycline	TET	0–512	≤4	8	≥16	17 (15.9%)	0	90 (84.1%)
Florfenicol	FFC	0–256	≤4	8	≥16	36 (33.6%)	0	71 (66.4%)
Sulfafurazole	SOX	0–512	≤256	-	≥512	39 (36.5%)	-	68 (63.6%)
Co-trimoxazole	SXT	0–32/608	≤2/38	-	≥4/76	47 (43.9%)	-	60 (56%)
Ceftiofur	EFT	0.12–256	≤2	4	≥8	96 (89.7%)	1 (0.9%)	10 (9.4%)
Ceftazidime	CAZ	0.12–256	≤4	8	≥16	97 (90.7%)	4 (3.7%)	6 (5.6%)
Enrofloxacin	ENR	0.01–32	≤0.25	0.5–1	≥2	18 (16.8%)	42 (39.3%)	47 (43.9%)
Ofloxacin	OFX	0.03–64	≤2	4	≥8	57 (53.3%)	33 (30.8%)	17 (15.9%)
Meropenem	MEM	0.03–10	≤1	2	≥4	104 (97.2%)	0	3 (2.8%)
Apramycin	APR	0–64	-	-	-	-	-	-
Polymyxins: Colistin	CL	0.12–256	≤2	-	≥4	95 (88.8%)	-	12 (11.2%)
Mequindox	NA	1–512	-	-	-	-	-	-

Table 4. Different resistance phenotypes between chickens and pigs.

Antimicrobial Agents	Pig (*n* = 55)	Chicken (*n* = 52)	*p*-Value
Ampicillin	40 (72.7%)	35 (67.3%)	0.6903
Amoxicillin/Clavulanic	2 (3.6%)	3 (5.8%)	0.6013
Gentamicin	17 (30.9%)	21 (40.4%)	0.3059
Spectinomycin	50 (90.9%)	18 (34.6%)	1.47
Tetracycline	54 (98.2%)	37 (71.2%)	<0.0005
Florfenicol	51 (92.7%)	21 (40.4%)	<0.0005
Sulfafurazole	50 (90.9%)	19 (36.5%)	<0.0005
Trimethoprim/sulfamethoxazole	46 (83.6%)	14 (26.9%)	<0.0005
Ceftiofur	1 (1.8%)	8 (15.4%)	0.0115
Ceftazidime	0 (0%)	2 (3.9%)	0.1420
Enrofloxacin	33 (60.0%)	14 (26.9%)	0.0005
Ofloxacin	7 (12.7%)	10 (19.2%)	0.3576
Meropenem	0 (0%)	2 (3.9%)	0.1420
Apramycin	-	-	
Colistin	0 (0%)	11 (21.2%)	0.0003
Mequindox	-	-	

2.4. Antibiotic Resistance Gene and Plasmid Profiles

A total of 46 antimicrobial resistance genes (ARGs) were detected in the *Salmonella* isolates (Figure 1). Most ARGs (e.g., *aac(6′)-Iaa*, *oqxAB*, *aac(6′)-Ib-cr*, *qnrS1*, and *qnrS2*) are associated with resistance to aminoglycosides and quinolones. The aminoglycoside gene *aac(6′)-Iaa* was detected in all isolates. In addition, three plasmid-mediated quinolone resistance (PMQR) genes [*oqxA* (*n* = 50), *oqxB* (*n* = 50), *aac(6′)Ib-cr* (*n* = 45)] and two variants of the qnrS gene family *qnrS2* (*n* = 40) and *qnrS1* (*n* = 20) were detected. Among the genes encoding β-lactamases, the majority of CTX-M-type genes were *blaCTX-M-27*, *blaCTX-M-55* and *blaCTX-M-65*. The frequency of occurrence of the majority of CTX-M-type genes was lower than 4.7%, which is consistent with the finding of less resistance to β-lactam antibiotics in all isolates (Figures 1 and 2). The *Salmonella* isolates that were positive for the tetracycline resistance gene *tet(A)* accounted for 73.8% and carried the sulfonamide resistance gene *sul2* (42%) (Figure 2). Additionally, the sulfonamide resistance gene sul3 was detected in more than half of the isolates.

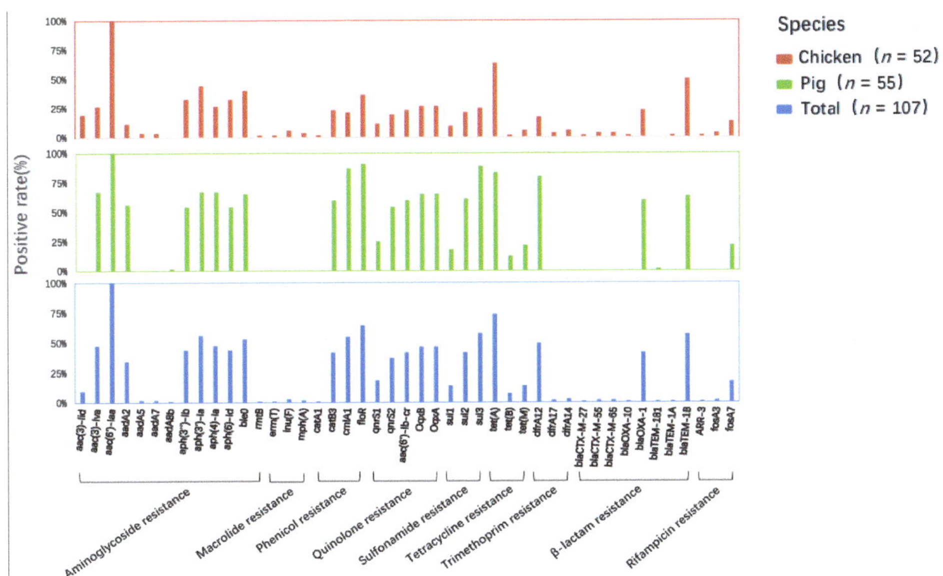

Figure 1. Different positive rates of resistance genes between chickens and pigs.

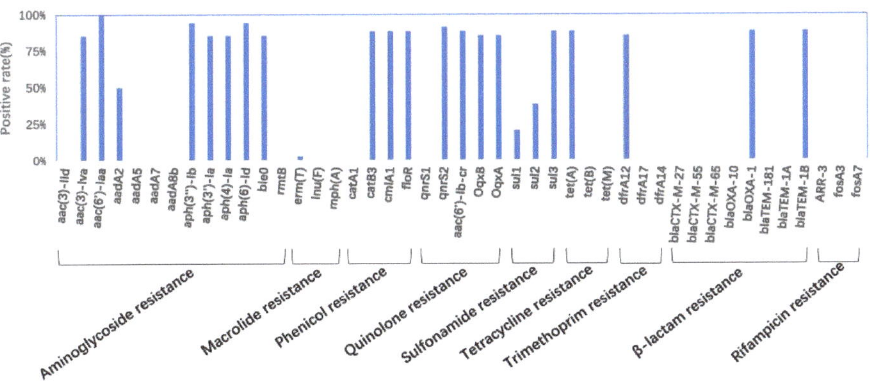

Figure 2. Positive rates of diverse resistance genes among different S. 4,[5],12:i:- isolates.

The results of the plasmid replicons in 107 *Salmonella* isolates are presented in the Supplementary Materials. The results show that the most abundant plasmid replicons were IncHI2 (33.6%, 36/107), IncHI2A (33.6%, 36/107), IncX1 (15.9%, 17/107), IncFII (S) (15.0%;16/107), and IncFIB(S) (15.0%, 16/107).

2.5. Characteristic of S. 4,[5],12:i:-

Among all isolates, 34 S. 4,[5],12:i:- isolates (31.8%, 34/107) were identified according to serotyping. Isolates with similar sequence types were grouped based on their phylogenetic relationship, and all S. 4,[5],12:i:- isolates belonged exclusively to ST34 (Supplementary Materials). It was the most common serovar in our sampling (Table 4), particularly from pigs. An antimicrobial susceptibility test of these 34 S. 4,[5],12:i:- isolates against 14 antimicrobial compounds showed resistance phenotypes in all isolates. The most common resistance was to tetracycline (100%), followed by spectinomycin (91.2%), florfenicol (91.2%), ampicillin (91.2%), trimethoprim/sulfamethoxazole (88.2%), enrofloxacin (76.5%), sulfafurazole (70.6%), gentamicin (47.1%), and ofloxacin (20.6%). The S. 4,[5],12:i:- isolates were less frequently resistant to ceftiofur (2.9%) and colistin (2.9%), and none of them were resistant to amoxicillin/clavulanic acid, meropenem, or ceftazidime (Table 5).

Table 5. Different resistance phenotypes among S. 4,[5],12:i:- isolates.

Numbers	4,[5],12:i:- (*n* = 34)	
Antibiotic agent	Drug resistant	Multidrug resistant (*n* = 31)
Tetracyclines: Tetracycline	34 (100%)	31
β-lactam: Ampicillin	31 (91.2%)	31
Chloramphenicols: Florfenicol	31 (91.2%)	31
Aminocyclitols: Spectinomycin	31 (91.2%)	31
Sulfonamides: trimethoprim/sulfamethoxazole	30 (88.2%)	30
Fluoroquinolones: Enrofloxacin	26 (76.5%)	26
Sulfonamides: Sulfafurazole	24 (70.6%)	22
Aminoglycosides: Gentamicin	16 (47.1%)	16
Fluoroquinolones: Ofloxacin	7 (20.6%)	7
Cephalosporins: Ceftiofur	1 (2.9%)	1
Polymyxins: Colistin	1 (2.9%)	1
β-lactams combination: Amoxicillin/Clavulanic	-	-
Carbapenems: Meropenem	-	-
Cephalosporins: Ceftazidime	-	-

2.6. Phylogenetic Analysis

To investigate the genomic relationships among the isolates, a phylogenetic tree was established based on core single-nucleotide polymorphism (SNP) analysis (Figure 3). Although there were differences in the hosts, timing, and sites of the collection, these strains still exhibited relatively close genetic relationships. The close genetic relationship between the isolates at different time points on the same farm or city also proved the existence of clonal transmission. For example, the high similarity in genomic data in strains S90, S91, S156, S157, and S191–S193 conceivably indicated that certain hosts had driven the horizontal transmission. It is concerning that clonal transmission probably happened alongside horizontal transmission in this study.

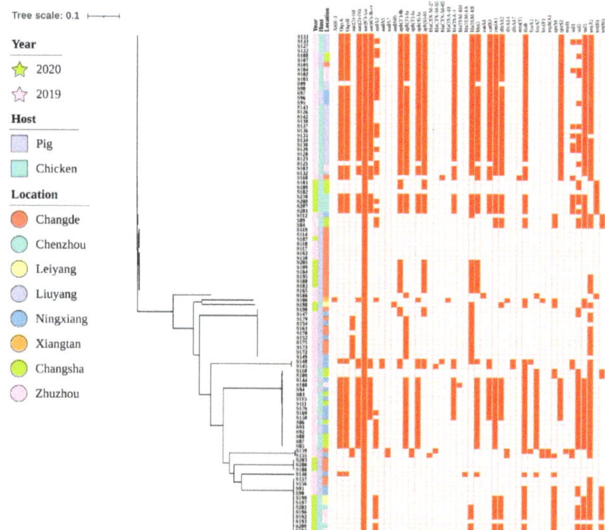

Figure 3. Bayesian phylogenetic analysis of 107 isolates (**left**) and the correlation with the distribution of resistance genes (**right**). Location and time are represented in the middle. To further evaluate possible transmission events, we coded the farms in the same city with one color (defined as intra-farm association events).

3. Discussion

Poultry and livestock farms are considered to be favorable biotopes for the accumulation of pathogens like *Salmonella*, which cause huge economic losses in many countries, including China [14]. Hunan Province is one of the most concentrated areas for farm animals in China [15]. Disease control and prevention during breeding mainly depend on the use of antibiotics. The irrational use of antibiotics, however, has contributed to the emergence of multidrug-resistant bacteria under selective antimicrobial pressure.

In this study, resistance phenotypes varied in the identified isolates, as 73% were resistant to at least three classes of antimicrobials (considered multidrug-resistant), which is remarkably higher than that observed in previous reports conducted in China [7] and lower than the resistance rates reported in Argentina and Australia [16,17]. Most of the resistant strains were isolated from pigs, which indicates that pigs could be important reservoirs of resistant non-typhoidal *Salmonella* strains. Among the identified drug resistances, resistance to tetracycline (85.1%) and ampicillin (70.1%) was predominant. These results are in agreement with previous studies on *Salmonella* isolates obtained from food animal farms in Xinjiang, China [18]. Resistance to quinolones and β-lactams was also recognized in many *Salmonella* isolates in this study, which is in agreement with previous reports [19,20]. Quinolone resistance, and that to ciprofloxacin in particular, has become a common issue in China and other countries, particularly resistance to ciprofloxacin [21]. The high prevalence of such genes is regarded as a significant threat to public health since these antimicrobials are currently used for frontline therapy against salmonellosis in humans [22].

PMQR genes are very common in farms [23], and this study showed that *oqxAB* and *aac(6′)-Ib-cr* were the main PMQR genes. All *oqxA*-positive isolates were screened for *oqxB. oqxAB*, and *aac(6′)-Ib-cr* commonly coexisted in the same strain, and 36 strains were found to carry them simultaneously. Carrying two or more PMQR genes in the same strain normally leads to resistance to nalidixic acid and decreases the susceptibility to fluoroquinolones like ciprofloxacin [24]. The presence of different ARGs based on whole-genome sequencing analysis demonstrated that the *aac(6′)-Iaa* gene, which mediates the resistance to aminoglycosides, was detected in all of the studied isolates. This was consistent with previous Chinese and South Korean studies [25]. The high levels of resistance to quinolones detected in this study may be due to the acquisition of PMQR genes through horizontally transferable elements, as well as mutagenesis in genes affecting the DNA gyrase and DNA topoisomerase IV genes [26].

Determination of serovars and multilocus sequence typing (MLST) patterns showed the dominance of S. 4,[5],12:i:- among the collected *Salmonella* isolates, especially those from pigs. This is consistent with the results of previous studies, in most of which S. 4,[5],12:i:- strains originated from pigs and pork products [27]. Intriguingly, the isolates from other sources such as chicken or the environment were less resistant to antibiotics of clinical importance, suggesting that pigs might be important reservoirs of resistant S.4,[5],12:i:- strains. The multidrug resistance of S. 4,[5],12:i:- is primarily associated with antimicrobials from seven classes. The detection of the *aac(6′)-Ib-cr, oqxA, oqxB, qnrS1*, and *qnrS2* genes has been reported in different serovars, including isolates of S. 4,[5],12:i:-, S. London, S. Indiana, S. Thompson, S. Kentucky, and S. Enteritidis. The presence of these genes enhances the adaptability of S. 4,[5],12:i:- to drugs, promoting the broader dissemination of such resistance genes [23]. In this study, 13 different plasmid replicons were identified among 107 *Salmonella* isolates. The most abundant plasmids were IncHI2A, IncHI2, and IncX1 (Supplementary Materials). IncHI2A and IncHI2 were predominant in S.4,[5],12:i:-. Interestingly, these plasmids were found to be associated with resistance to different antimicrobial classes, including β-lactams, aminoglycosides, sulfafurazole, tetracyclines, and polymyxins [28]. Consequently, these plasmids may increase the risk related to the horizontal transmission of these antimicrobial resistance genes in animal food chains, leading to severe disease in humans [29].

Phenotypic and genotypic resistance of most tested antibiotics showed high coherence, but tetracycline and quinolone resistance showed moderate coherence. The obtained results are consistent with previous reports on *Salmonella* isolates from dead poultry that revealed that drug resistance gene expression patterns and drug resistance spectra were remarkably similar among strains in Shandong [14]. Similarly, a large-scale study reported high levels of coherence between phenotypic and genotypic resistance for all tested antibiotics [30]. Hence, performing phenotypic verification on the collected isolates was necessary to avoid potential bias caused by genomic analysis.

Phylogenetic analysis showed that isolates of the same serovar with similar sequence types were closely clustered. Notably, the exact inter-farm transmission event occurred among ST34, which may be due to farms' lack of strict hygiene standards for handling. Hence, it is essential to improve hygiene and sanitizing procedures. Additional documentation of the traceability of inputs and outputs that may carry disease sources on each farm can help reduce the persistence and spread of *Salmonella* between poultry farms. However, we found that the detection rates of isolates and drug resistances were lower in 2020 than in 2019; one potential explanation is that the Ministry of Agriculture of the People's Republic of China formulated plans to ban or reduce the use of specific antimicrobials in 2019 [7]. This indicates that strengthening veterinary medicine management could help to effectively prevent the development of antibiotic-resistant bacteria.

In conclusion, in this study, the high prevalence of multidrug-resistant non-typhoidal *Salmonella* in the studied samples and its severe risk to human health were reported. The results indicate that, in the future, we must continue monitoring *Salmonella* serovars and conduct strategic control plans based on whole-genome sequencing. The application of an antimicrobial management plan for the rational use of essential antimicrobials in animal farms will also be vital to help control the spread and prevalence of drug resistance genes and to provide reliable human health protection measures.

4. Materials and Methods

4.1. Sample Collection and Isolation of Salmonella Strains

In 2019 and 2020, 2240 fecal samples were collected from 17 chicken and pig farms located in nine cities of Hunan Province: Changsha, Leiyang, Hengyang, Xiangtan, Ningxiang, Changde, Zhuzhou, Chenzhou, and Liuyang. *Salmonella* was isolated as described previously [19,20]. Briefly, cotton swab samples were subjected to pre-enrichment in buffered peptone water and then enriched in a modified semisolid Rappaport Vassiliadis plate, and colonies were isolated on xylose lysine deoxy-cholate agar. Subsequently, the isolated strains were confirmed via the amplification of the invA gene using the following primers according to a previously described protocol [31]: F:5′-ACAGTGCTCGTTTACGACCTGAAT-3′ and R:5′-AGACGACTGGTACT-GATCGATAAT-3′. Specifically, 25 PCR cycles with an annealing temperature of 56 °C were performed with Taq Polymerase (Tsingke Biotechnology Co., Ltd., Beijing, China) on a thermal cycler. Colonies confirmed as *Salmonella* were inoculated into Luria–Bertani broth for minimal inhibitory concentration (MIC) determination and genomic DNA preparation.

4.2. Detection of the Salmonella Strain Serotypes

The confirmed isolates were serotyped via slide agglutination using commercially available antisera kits (Tianrun BioPharmaceutical Co. Ltd., Ningbo, China) with O and H antigen-specific sera. Serovar results were interpreted according to the Kauffmann–White–Le Minor scheme [32].

4.3. Antibiotic Susceptibility Testing

The MIC of 14 antibiotics (nine classes) was determined using the microdilution broth method according to the criteria recommended by the Clinical and Laboratory Standards Institute [33]. *Escherichia coli* ATCC25922 was used as a quality control strain.

4.4. Whole-Genome Sequencing, De Novo Assembly, and Annotation

Genomic DNA was extracted from the *Salmonella* isolates and purified using the TIANamp Bacteria DNA Kit (Tiangen Biotech Co., Beijing, China) according to the manufacturer's instructions. Protein quality was assessed via gel electrophoresis and quantified using a Qubit Fluorometer 2.0 (Invitrogen, Waltham, MA, USA; Life Technologies, Carlsbad, CA, USA). Whole genome sequencing was performed using Annoroad Gene Technology (Beijing, China) on a NovaSeq 6000 S4 sequencing platform with the NovaSeq 6000 S4 Reagent kit V1.5. Bacterial genome assembly was performed using the SPAdes software (version 3.11) [34].

4.5. Antibiotic Resistance Genes and Phylogenetic Analysis

The ARGs for *Salmonella* strains were analyzed using the Center for Genomic Epidemiology (ResFinder tools). The relationship with non-typhoidal *Salmonella* isolates was evaluated using core-genome alignments and phylogenetic trees were constructed using Parsnp (neighbor-joining method)and visualized using the online tool (iTOL 6.5.7) [35].

Supplementary Materials: The following supporting information can be downloaded at: https://www.mdpi.com/article/10.3390/antibiotics12071178/s1, Table S1. MIC, serotypes and plasmid of 107 non-typhoidal *Salmonella* isolates.

Author Contributions: Conceptualization and methodology, X.Z. and J.L.; investigation, Z.Z., J.L. and R.Z.; data curation and formal analysis, H.L., Q.X., Y.W. and S.Q.; writing—original draft, Z.Z.; writing—review and editing, X.Z., Z.Z., J.L., Y.W. and P.L. All authors have read and agreed to the published version of the manuscript.

Funding: This research was funded by grants from the Department of Science and Technology of Hunan Province to Xiaofeng Zheng (2019RS1050 and 2021JJ30354) and a grant from the Department of Science and Technology of Hunan Province to Pishun Li (2023JJ30291).

Institutional Review Board Statement: Not applicable.

Informed Consent Statement: Not applicable.

Data Availability Statement: All genome assemblies of all strains were deposited in GenBank under the BioProject accession number PRJNA993584.

Conflicts of Interest: The authors declare no conflict of interest.

References

1. Hendriksen, R.S.; Vieira, A.R.; Karlsmose, S.; Lo Fo Wong, D.M.; Jensen, A.B.; Wegener, H.C.; Aarestrup, F.M. Global monitoring of Salmonella serovar distribution from the World Health Organization Global Foodborne Infections Network Country Data Bank: Results of quality assured laboratories from 2001 to 2007. *Foodborne Pathog. Dis.* **2011**, *8*, 887–900. [CrossRef] [PubMed]
2. CDC. National Enteric Disease Surveillance: Salmonella Annual Report. 2016. Available online: https://www.census.gov/geo/pdfs/maps-data/maps/ (accessed on 24 April 2021).
3. EFSA; ECDC. European Food Safety Authority (EFSA); European Centre for Disease Prevention and Control (ECDC)The European Union summary report on trends and sources of zoonoses, zoonotic agents and food-borne outbreaks in 2017. *EFSA J.* **2018**, *16*, e05500. [CrossRef]
4. Zhou, K.; Wu, B.; Pan, H.; Paudyal, N.; Jiang, J.; Zhang, L.; Li, Y.; Yue, M. ONE Health Approach to Address Zoonotic Brucellosis: A Spatiotemporal Associations Study Between Animals and Humans. *Front. Vet. Sci.* **2020**, *7*, 521. [CrossRef]
5. Arnold, M.E.; Papadopoulou, C.; Davies, R.H.; Carrique-Mas, J.J.; Evans, S.J.; Hoinville, L.J. Estimation of *Salmonella* prevalence in UK egg-laying holdings. *Prev. Vet. Med.* **2010**, *94*, 306–309. [CrossRef] [PubMed]
6. Kirk, M.D.; Pires, S.M.; Black, R.E.; Caipo, M.; Crump, J.A.; Devleesschauwer, B.; Dopfer, D.; Fazil, A.; Fischer-Walker, C.L.; Hald, T.; et al. World Health Organization Estimates of the Global and Regional Disease Burden of 22 Foodborne Bacterial, Protozoal, and Viral Diseases, 2010: A Data Synthesis. *PLoS Med.* **2015**, *12*, e1001921. [CrossRef]
7. Xu, Y.; Zhou, X.; Jiang, Z.; Qi, Y.; Ed-Dra, A.; Yue, M. Epidemiological Investigation and Antimicrobial Resistance Profiles of *Salmonella* Isolated From Breeder Chicken Hatcheries in Henan, China. *Front. Cell. Infect. Microbiol.* **2020**, *10*, 497. [CrossRef]
8. Bonardi, S. *Salmonella* in the pork production chain and its impact on human health in the European Union. *Epidemiol. Infect.* **2017**, *145*, 1513–1526. [CrossRef]
9. WHO Guidelines Approved by the Guidelines Review Committee. In *WHO Guidelines on Use of Medically Important Antimicrobials in Food-Producing Animals*; World Health Organization: Geneva, Switzerland, 2017.

10. Cuypers, W.L.; Jacobs, J.; Wong, V.; Klemm, E.J.; Deborggraeve, S.; Van Puyvelde, S. Fluoroquinolone resistance in *Salmonella*: Insights by whole-genome sequencing. *Microb. Genom.* **2018**, *4*, e000195. [CrossRef]

11. Ito, J.; Sugimoto, R.; Nakaoka, H.; Yamada, S.; Kimura, T.; Hayano, T.; Inoue, I. Systematic identification and characterization of regulatory elements derived from human endogenous retroviruses. *PLoS Genet.* **2017**, *13*, e1006883. [CrossRef]

12. Campos, J.; Mourao, J.; Peixe, L.; Antunes, P. Non-typhoidal *Salmonella* in the Pig Production Chain: A Comprehensive Analysis of Its Impact on Human Health. *Pathogens* **2019**, *8*, 19. [CrossRef]

13. Sun, H.; Wan, Y.; Du, P.; Bai, L. The Epidemiology of Monophasic *Salmonella* Typhimurium. *Foodborne Pathog. Dis.* **2020**, *17*, 87–97. [CrossRef] [PubMed]

14. Li, Y.; Kang, X.; Ed-Dra, A.; Zhou, X.; Jia, C.; Müller, A.; Liu, Y.; Kehrenberg, C.; Yue, M. Genome-Based Assessment of Antimicrobial Resistance and Virulence Potential of Isolates of Non-Pullorum/Gallinarum *Salmonella* Serovars Recovered from Dead Poultry in China. *Microbiol. Spectr.* **2022**, *10*, e0096522. [CrossRef] [PubMed]

15. *China Agriculture Statistical Report*; Ministry of Agriculture and Rural Affairs: Beijing, China, 2021.

16. Vico, J.P.; Lorenzutti, A.M.; Zogbi, A.P.; Aleu, G.; Sanchez, I.C.; Caffer, M.I.; Rosmini, M.R.; Mainar-Jaime, R.C. Prevalence, associated risk factors, and antimicrobial resistance profiles of non-typhoidal *Salmonella* in large scale swine production in Cordoba, Argentina. *Res. Vet. Sci.* **2020**, *130*, 161–169. [CrossRef]

17. Sia, C.M.; Baines, S.L.; Valcanis, M.; Lee, D.Y.J.; Gonçalves da Silva, A.; Ballard, S.A.; Easton, M.; Seemann, T.; Howden, B.P.; Ingle, D.J.; et al. Genomic diversity of antimicrobial resistance in non-typhoidal *Salmonella* in Victoria, Australia. *Microb. Genom.* **2021**, *7*, 000725. [CrossRef]

18. Liu, Y.; Jiang, J.; Ed-Dra, A.; Li, X.; Peng, X.; Xia, L.; Guo, Q.; Yao, G.; Yue, M. Prevalence and genomic investigation of *Salmonella* isolates recovered from animal food-chain in Xinjiang, China. *Food. Res. Int.* **2021**, *142*, 110198. [CrossRef] [PubMed]

19. Liu, Q.; Chen, W.; Elbediwi, M.; Pan, H.; Wang, L.; Zhou, C.; Zhao, B.; Xu, X.; Li, D.; Yan, X.; et al. Characterization of *Salmonella* Resistome and Plasmidome in Pork Production System in Jiangsu, China. *Front. Vet. Sci.* **2020**, *7*, 617. [CrossRef]

20. Jiang, Z.; Paudyal, N.; Xu, Y.; Deng, T.; Li, F.; Pan, H.; Peng, X.; He, Q.; Yue, M. Antibiotic Resistance Profiles of *Salmonella* Recovered From Finishing Pigs and Slaughter Facilities in Henan, China. *Front. Microbiol.* **2019**, *10*, 1513. [CrossRef] [PubMed]

21. Chen, H.; Song, J.; Zeng, X.; Chen, D.; Chen, R.; Qiu, C.; Zhou, K. National Prevalence of *Salmonella* enterica Serotype Kentucky ST198 with High-Level Resistance to Ciprofloxacin and Extended-Spectrum Cephalosporins in China, 2013 to 2017. *mSystems* **2021**, *6*, e00935-20. [CrossRef]

22. Elbediwi, M.; Li, Y.; Paudyal, N.; Pan, H.; Li, X.; Xie, S.; Rajkovic, A.; Feng, Y.; Fang, W.; Rankin, S.C.; et al. Global Burden of Colistin-Resistant Bacteria: Mobilized Colistin Resistance Genes Study (1980–2018). *Microorganisms* **2019**, *7*, 461. [CrossRef]

23. He, J.; Sun, F.; Sun, D.; Wang, Z.; Jin, S.; Pan, Z.; Xu, Z.; Chen, X.; Jiao, X. Multidrug resistance and prevalence of quinolone resistance genes of *Salmonella* enterica serotypes 4,[5],12:i:- in China. *Int. J. Food Microbiol.* **2020**, *330*, 108692. [CrossRef]

24. Zhang, J.; Yang, X.; Kuang, D.; Shi, X.; Xiao, W.; Zhang, J.; Gu, Z.; Xu, X.; Meng, J. Prevalence of antimicrobial resistance of non-typhoidal *Salmonella* serovars in retail aquaculture products. *Int. J. Food Microbiol.* **2015**, *210*, 47–52. [CrossRef]

25. Hyeon, J.Y.; Li, S.; Mann, D.A.; Zhang, S.; Kim, K.J.; Lee, D.H.; Deng, X.; Song, C.S. Whole-Genome Sequencing Analysis of *Salmonella* Enterica Serotype Enteritidis Isolated from Poultry Sources in South Korea, 2010–2017. *Pathogens* **2021**, *10*, 45. [CrossRef]

26. Vidovic, S.; An, R.; Rendahl, A. Molecular and Physiological Characterization of Fluoroquinolone-Highly Resistant *Salmonella* Enteritidis Strains. *Front. Microbiol.* **2019**, *10*, 729. [CrossRef]

27. Hauser, E.; Tietze, E.; Helmuth, R.; Junker, E.; Blank, K.; Prager, R.; Rabsch, W.; Appel, B.; Fruth, A.; Malorny, B. Pork contaminated with *Salmonella* enterica serovar 4,[5],12:i:-, an emerging health risk for humans. *Appl. Environ. Microbiol.* **2010**, *76*, 4601–4610. [CrossRef] [PubMed]

28. Elbediwi, M.; Pan, H.; Biswas, S.; Li, Y.; Yue, M. Emerging colistin resistance in *Salmonella* enterica serovar Newport isolates from human infections. *Emerg. Microbes. Infect.* **2020**, *9*, 535–538. [CrossRef] [PubMed]

29. Diaconu, E.L.; Alba, P.; Feltrin, F.; Di Matteo, P.; Iurescia, M.; Chelli, E.; Donati, V.; Marani, I.; Giacomi, A.; Franco, A.; et al. Emergence of IncHI2 Plasmids With Mobilized Colistin Resistance (mcr)-9 Gene in ESBL-Producing, Multidrug-Resistant *Salmonella* Typhimurium and Its Monophasic Variant ST34 From Food-Producing Animals in Italy. *Front. Microbiol.* **2021**, *12*, 705230. [CrossRef]

30. Neuert, S.; Nair, S.; Day, M.R.; Doumith, M.; Ashton, P.M.; Mellor, K.C.; Jenkins, C.; Hopkins, K.L.; Woodford, N.; de Pinna, E.; et al. Prediction of Phenotypic Antimicrobial Resistance Profiles From Whole Genome Sequences of Non-typhoidal *Salmonella* enterica. *Front. Microbiol.* **2018**, *9*, 592. [CrossRef] [PubMed]

31. Zhu, C.; Yue, M.; Rankin, S.; Weill, F.X.; Frey, J.; Schifferli, D.M. One-Step Identification of Five Prominent Chicken *Salmonella* Serovars and Biotypes. *J. Clin. Microbiol.* **2015**, *53*, 3881–3883. [CrossRef] [PubMed]

32. Guibourdenche, M.; Roggentin, P.; Mikoleit, M.; Fields, P.I.; Bockemuhl, J.; Grimont, P.A.; Weill, F.X. Supplement 2003–2007 (No. 47) to the White-Kauffmann-Le Minor scheme. *Res. Microbiol.* **2010**, *161*, 26–29. [CrossRef] [PubMed]

33. CLSI. *Performance Standards for Antimicrobial Susceptibility Testing, 30th Edn.CLSI Supplement M100*; Clinical and Laboratory Standards Institute: Wayne, PA, USA, 2020.

34. Bankevich, A.; Nurk, S.; Antipov, D.; Gurevich, A.A.; Dvorkin, M.; Kulikov, A.S.; Lesin, V.M.; Nikolenko, S.I.; Pham, S.; Prjibelski, A.D.; et al. SPAdes: A new genome assembly algorithm and its applications to single-cell sequencing. *J. Comput. Biol.* **2012**, *19*, 455–477. [CrossRef]
35. Hu, J.; Yang, J.; Chen, W.; Liu, Z.; Zhao, Q.; Yang, H.; Sun, Z.; Chen, X.; Li, J. Prevalence and Characteristics of mcr-1-Producing Escherichia coli in Three Kinds of Poultry in Changsha, China. *Front. Microbiol.* **2022**, *13*, 840520. [CrossRef] [PubMed]

Brief Report

Storage of Extended Boar Semen at 5 °C Inhibits Growth of Multi-Drug Resistant *Serratia marcescens* and *Klebsiella oxytoca* while Maintaining High Sperm Quality

Isabel Katharina Maaßen [1], Anne-Marie Luther [1], Jutta Verspohl [2] and Dagmar Waberski [1,*]

[1] Unit for Reproductive Medicine, Clinic for Pigs and Small Ruminants, University of Veterinary Medicine Hannover, Foundation, Bünteweg 15, D-30559 Hannover, Germany; isabel.katharina.maassen@tiho-hannover.de (I.K.M.); anne-marie.luther@tiho-hannover.de (A.-M.L.)

[2] Institute for Microbiology, University of Veterinary Medicine Hannover, Foundation, Bischofsholer Damm 15, D-30173 Hannover, Germany; jutta.verspohl@tiho-hannover.de

* Correspondence: dagmar.waberski@tiho-hannover.de

Abstract: Multi-drug antibiotic resistance of *Serratia (S.) marcescens* and *Klebsiella (K.) oxytoca* in boar semen is an emerging threat to pig reproduction and the environment. The aim of this study is to examine the efficiency of a novel hypothermic preservation method to inhibit the growth of these bacterial species in extended boar semen and to maintain the sperm quality. The semen samples extended in an antibiotic-free Androstar Premium extender were spiked with ~10^2 CFU/mL of *S. marcescens* or *K.oxytoca*. Storage at 5 °C for 144 h inhibited the growth of both bacterial species and maintained the sperm quality, whereas bacterial counts increased to more than 10^{10} CFU/mL in the 17 °C samples used as positive controls. This was accompanied by an increase in the sperm agglutination and the loss of motility and membrane integrity. We conclude that hypothermic storage is a promising tool to combat resistant bacteria in boar semen and to contribute to the One Health approach.

Keywords: *Serratia marcescens*; *Klebsiella oxytoca*; boar semen; antibiotic resistance; semen preservation

check for **updates**

Citation: Maaßen, I.K.; Luther, A.-M.; Verspohl, J.; Waberski, D. Storage of Extended Boar Semen at 5 °C Inhibits Growth of Multi-Drug Resistant *Serratia marcescens* and *Klebsiella oxytoca* while Maintaining High Sperm Quality. *Antibiotics* **2023**, *12*, 857. https://doi.org/10.3390/antibiotics12050857

Academic Editor: Nahla O. Eltai

Received: 6 April 2023
Revised: 26 April 2023
Accepted: 3 May 2023
Published: 5 May 2023

1. Introduction

Artificial insemination (AI) with liquid-preserved semen is the most used biotechnology in pig reproduction worldwide [1]. Due to the high chilling sensitivity of boar spermatozoa [2], porcine semen is commonly stored between 16 and 18 °C. The relatively high storage temperature poses the risk of bacterial growth, thus enforcing the use of antibiotics in semen extenders. The continuous use of antibiotics, together with disinfectants in semen collection centers, has favored the generation and spread of multi-drug resistances [3]. Most of the bacteria usually occurring in the raw semen are commensal [4], belonging to the *Enterobacter*, and neither impact the sow's health nor the sperm quality if their amount is below 10^7 CFU/mL [5–8]. However, the contamination of extended semen with the opportunistic pathogens *Serratia marcescens* and *Klebsiella oxytoca* were identified as bacteria of high concern due to their fast growth in extended semen, together with a high spermicidal effect and multi-drug resistance [9–12]. These Gram-negative bacterial species, like many other bacterial species identified in preserved boar semen, may either originate from the animal or may enter the semen from the environment in the stable or laboratory [5,13,14]. *Serratia marcescens* and *K. oxytoca* are well known as nosocomial bacteria, typically acquiring resistances in locations with high exposure to antibiotics and disinfectants [15,16], such as in hospitals and laboratories of AI centers. The threat of bacterial contamination has increased the overuse of antibiotics in semen extenders, resulting in the loss of efficient antimicrobial control. Notably, the high rate of semen backflow from the sow's reproductive tract after insemination [17] poses the risk of antibiotics and resistant bacteria entering the environment and, from there, the human food chain. These situations have stimulated intense research for alternatives to antibiotics in boar semen extenders [3,18], of which, to date, none have become established in the AI industry, either due to the lack of broad-spectrum efficiency, sperm toxicity, or high time and cost intensity. Recently, hypothermic storage of boar semen was proposed as an innovative preservation concept to keep bacteria below spermicidal levels [19]. Antibiotic-free semen storage at 5 °C proved to be efficient for inhibiting the growth of commensal bacteria naturally occurring in boar semen, and to maintain the high sperm quality and fertility in vivo [19–21]. Whether the hypothermic semen storage is also effective against multi-drug resistant bacteria remains to be shown. Beyond this background, this study aims to examine whether semen storage at 5 °C efficiently inhibits the growth of multi-drug resistant *S. marcescens* and *K. oxytoca* to levels that do not affect the sperm quality.

2. Results

The results of Experiment 1 with *S. marcescens* are shown in Figure 1. The spiked semen samples stored at 17 °C showed exponential growth of *S. marcescens* in pure cultures to >10^{11} CFU/mL, whereas storage at 5 °C inhibited the bacterial count at ~10^3 CFU/mL (Figure 1A). In the un-spiked samples stored at 5 °C, the bacterial load decreased below <10^1 CFU/mL within 48 h ($p < 0.05$). At 24 h and 48 h, the sperm motility and membrane integrity were higher in the 17 °C samples compared to the 5 °C samples. At 72 h, the 17 °C samples showed an increase in the sperm agglutination (Figure 1B) and a decrease in the sperm motility (Figure 1C) and membrane integrity (Figure 1D). After 72 h storage, the sperm motility and membrane integrity were higher in the 5 °C samples compared to the 17 °C samples. At 144 h, the samples stored at 17 °C were not analyzed for motility and membrane integrity due to the high sperm agglutination. At all the time points, the sperm quality parameters did not differ between the 5 °C samples with and without added *S. marcescens*.

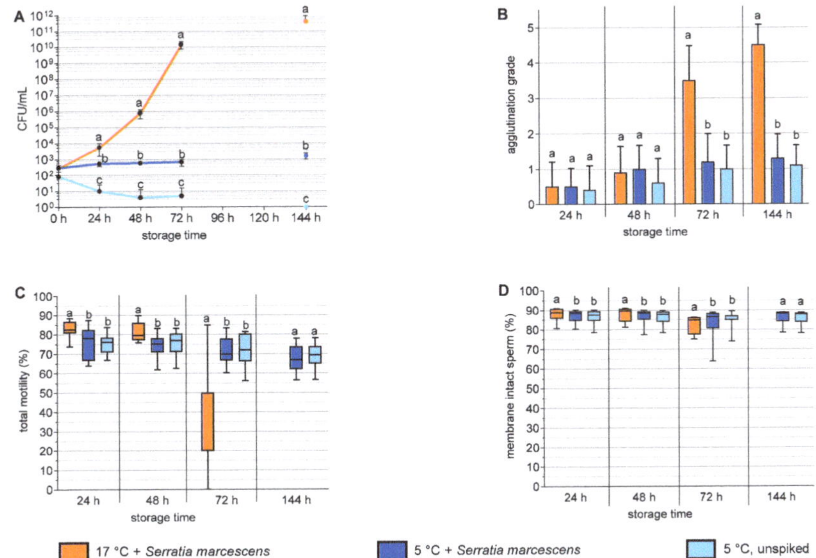

Figure 1. Bacterial growth (**A**), sperm agglutination (**B**), sperm motility (**C**), and sperm membrane integrity (**D**) in boar semen spiked with *Serratia marcescens* and stored at 5 °C or 17 °C in antibiotic-free Androstar Premium extender. At 144 h storage at 17 °C, motility and membrane integrity were not analyzed due to the high sperm agglutination. Experiment 1, $n = 10$ semen samples from 8 boars. a–c: Different lowercase letters indicate differences at a given storage time point ($p < 0.05$).

The results of Experiment 2 with *K. oxytoca* are shown in Figure 2. The spiked semen samples stored at 17 °C showed almost linear growth of *K. oxytoca* in pure cultures to ~10^{11} CFU/mL at 144 h, whereas storage at 5 °C inhibited the bacterial count to the initial spiking dose (Figure 2A). In the un-spiked samples stored at 5 °C, the bacterial load decreased below the detection limit (i.e., <10^1 CFU/mL) within 48 h ($p < 0.05$). At 24 h, 48 h, and 72 h, the sperm motility and membrane integrity were higher in the 17 °C samples compared to the 5 °C samples. At 72 h and 144 h, the spiked 17 °C samples showed higher sperm agglutination compared to both types of 5 °C samples (Figure 2B). At 144 h, the sperm motility (Figure 2C) was higher in the 5 °C samples compared to the 17 °C samples, whereas the percentage of membrane-intact spermatozoa (Figure 2D) was at the same high level (greater than 85%) in all three sample types. At all the time points, the sperm quality parameters did not differ between the 5 °C samples with and without added *K. oxytoca*.

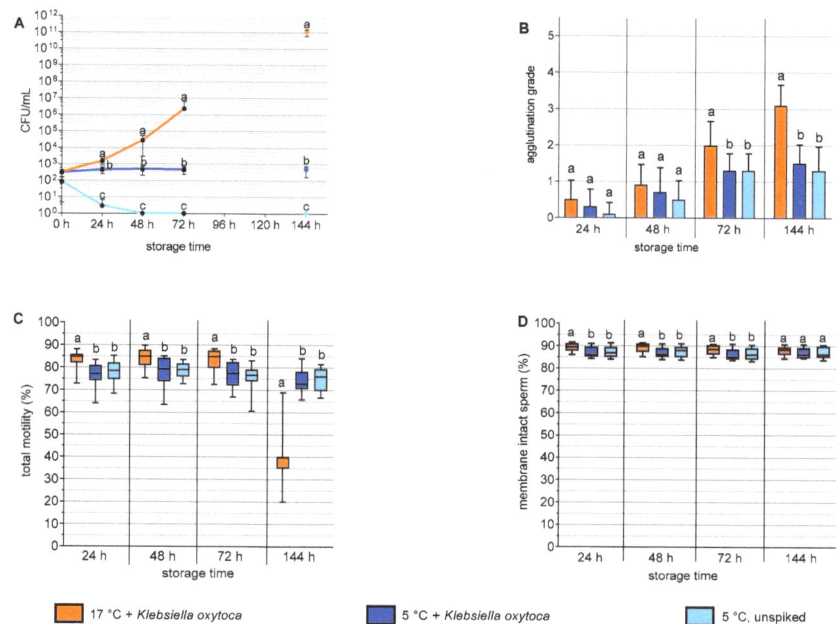

Figure 2. Bacterial growth (**A**), sperm agglutination (**B**), sperm motility (**C**), and sperm membrane integrity (**D**) in boar semen spiked with *Klebsiella oxytoca* and stored at 5 °C or 17 °C in antibiotic-free Androstar Premium extender. Experiment 2, *n* = 10 semen samples from 9 boars. a–c: Different lowercase letters indicate differences at a given storage time point ($p < 0.05$).

3. Discussion

The present study demonstrates that the preservation of extended boar semen at 5 °C efficiently inhibits the growth of multi-drug resistant *S. marcescens* and *K. oxytoca* during long-term storage. These data add to our previous observation that the growth of the two multi-drug bacterial species was inhibited at 5 °C compared to 17 °C in the semen-free extender medium [11]. In the present study, the inhibitory effect was shown using semen portions in the same way they would be used for insemination, i.e., in the presence of spermatozoa and seminal plasma. The latter is rich in proteins, lipids, and electrolytes [22], which provide nutrients for bacteria and thus could reduce the inhibitory effect of hypothermic semen storage. Here, by mimicking the contamination of semen portions with relatively high counts of two multi-drug resistant bacterial species isolated from boar semen, we showed that semen storage at 5 °C effectively kept the bacterial load far below spermicidal levels. By comparison of the bacterial growth at 5 °C storage in semen-free BTS (manufactured as a short-term extender) and the Androstar Premium (manufactured as a long-term extender), we previously demonstrated that the antimicrobial effect against *S. marcescens* and *K. oxytoca* is primarily caused by the low storage temperature and, to a lesser extent, the extender medium [11]. It is to note that the BTS extender does not protect the sperm against chilling injury [19], and therefore, is not applicable for 5 °C semen storage.

The thresholds for sperm damage were reported at $>10^7$/mL for *S. marcescens* and $>10^8$/mL for *K. oxytoca* [11]. In the present study, these levels were reached after 72 h of storage in the spiked 17 °C samples but not in the 5 °C samples. Similar to the observations with *Escherichia coli* [5] and *Enterobacter cloacae* [23], the sperm damage induced by *S. marcescens* and *K. oxytoca* was expressed at a bacteria/sperm ratio higher than 1:1.

Based on the studies with *E. coli* and *K. pneumonia* in human semen, bacterial adhesion to the sperm surface is regarded as a key event leading to increased sperm agglutination [24,25]. Similar to these observations, the sperm adherence of *S. marcescens* could be mediated by mannose-binding adhesion molecules detected on the pili and fimbria of this bacterial species [26,27] among further mechanisms, e.g., extracellular bacterial secretions or molecular aggregates as reported for *E. coli* [28]. The present study indicates that regardless of the cause for increased agglutination and the loss of sperm motility and membrane integrity, the sperm damaging effect is clearly related to the bacterial concentration, but not to the bacterial exposure time. This becomes evident in the spiked 5 °C samples, showing that the long-term exposition (144 h) to moderate counts of both *S. marcescens* and *K. oxytoca* did not induce a decline in the sperm quality. Confirming previous observations in the un-spiked 5 °C samples [19–21], the growth of commensal bacteria was well controlled below the detectable limits, but the loss of the sperm quality due to the chilling sensitivity of boar spermatozoa [2] is a challenge. Here, we used a recently established preservation protocol [19,29], considering that the type of extender medium and the slow controlled cooling are essential for high in vitro performance and fertility in vivo. Similar to the previous reports [19–21], there was a small decrease (less than 10 percent points) in the motility and membrane integrity in the 5 °C samples compared to the 17 °C samples with low bacterial counts in the first 48 h of storage. Notably, despite this initial chilling-associated loss of the sperm quality, the viable (i.e., membrane-intact) sperm population kept its functional integrity including its capacitation ability and mitochondrial activity during long-term storage [21]. This suggests that an impact on fertility is unlikely as long as sufficient numbers of viable spermatozoa are present, which has been previously confirmed by high fertility results in insemination trials under field conditions [19,21].

In conclusion, this study shows that the antibiotic-free storage of boar semen at 5 °C is effective not only against commensal bacteria naturally occurring in the raw semen, but also toward two multi-drug resistant bacterial species of highest concern in pig insemination. In the case of antibiotic ineffectiveness, currently, this seems to be the only option for an instant application in AI practice until the source of contamination is eliminated. When used routinely, antibiotic-free hypothermic semen storage presents a pathway toward sustainable pig reproduction in the sense of the One Health approach.

4. Materials and Methods

4.1. Semen Processing and Bacterial Inoculation

Semen was collected routinely once a week by trained personnel from nine mature, healthy boars (1 to 5 years of age) housed at the Unit for Reproductive Medicine, University of Veterinary Medicine Hannover, Hannover, Germany and treated in accordance with the European Commission Directive for Pig Welfare. After discarding the bulbourethral secretion, semen was extended to 20×10^6 spermatozoa/mL in a final volume of 100 mL with the commercial antibiotic-free semen extender Androstar Premium (Minitüb GmbH, Tiefenbach, Germany). Extended semen portions were spiked with ~10^2 CFU/mL *S. marcescens* (Experiment 1, $n = 10$ semen samples from 8 boars), or *K. oxytoca* (Experiment 2, $n = 10$ semen samples from 9 boars) isolated from commercial semen portions received from AI centers on the occasion of an annual semen quality control program conducted by our reference laboratory. Bacteria were cultured on Columbia agar with sheep blood (Oxoid Deutschland GmbH, Wesel, Germany) for 24 h at 37 °C before inoculation. Bacterial colonies were added to 2 mL extender medium and bacterial concentrations were adjusted after density photometry (SDM5, Minitüb, Tiefenbach, Germany). Immediately after inoculation (0 h), the bacterial count in the extended semen was determined. The bacterial species were identified with MALDI-TOF MS (microFlex LT, Bruker Daltonic, Bremen, Germany) and software Biotyper (Bruker Daltonic) before inoculation and at the end of semen storage. The Minimal Inhibitory Concentrations (MIC) for two common antibiotics in semen extenders, i.e., Gentamicin and Ampicillin, were evaluated by the microdilution method in the Institute for Microbiology at the University of Veterinary Medicine Hannover. For *S. marcescens*, MIC values for Gentamicin were $\geq$16 µg/mL, and for Ampicillin, 8 µg/mL. For *K. oxytoca*, MIC values for Gentamicin were $\geq$32 µg/mL, and for Ampicillin, $\geq$32 µg/mL. Both bacterial species were considered as multi-drug resistant based on the susceptibility test with 16 different antibiotics [11]. Spiked semen portions were either stored at 17 °C (positive control) or slowly cooled to 5 °C following a previously established cooling protocol for the hypothermic storage of boar semen [29]. Additional semen portions remained un-spiked and were stored at 5 °C after slow cooling (negative control). All samples were stored for 144 h in the dark.

4.2. Bacterial Count

Bacterial counts were determined from 10-fold serial dilutions in PBS ranging from 10^{-1} to 10^{-10} and were plated in volumes of 100 µL on Columbia agar with sheep blood. After incubation for 24 h at 37 °C under aerobic conditions, bacterial colonies were counted, and the total bacterial numbers were calculated. Bacterial counts were expressed as colony-forming units per milliliter (CFU/mL).

4.3. Spermatology

Semen was examined at 24 h, 48 h, 72 h, and 144 h of storage. Sperm agglutinations were assessed with phase contrast microscopy (Carl Zeiss Microscopy GmbH, Jena, Germany) at 200× magnification. At least three different fields were examined, and the degree of agglutination was scored between 0 and 5 according to the estimated percentage of agglutinated spermatozoa as follows: 0 = 0 to 5%, 1 = less than 20%, 2 = 20 to 40%, 3 = 40 to 60%, 4 = 60 to 80%, 5 = 80 to 100%.

Sperm motility was measured as the total number of motile spermatozoa with the computer-assisted semen analysis (CASA) system AndroVision® (Version 1.2, Minitüb GmbH, Tiefenbach, Germany). Subsamples were prewarmed at 38 °C for 30 min in a water bath under air and then filled in a 20 µL Leja chamber (Leja Products B.V., Nieuw Vennep, The Netherlands). At least 500 sperm were recorded with a frame rate of 30 pictures per 0.5 s. Motile spermatozoa were identified when their curved-line velocity was >24 µm/s and their amplitude of lateral head displacement was >1 µm. Sperm membrane integrity was assessed by flow cytometry using the Cyto Flex flow cytometer (Beckman Coulter GmbH, Krefeld, Germany) and the Cyt Expert 2.4. software (Beckman Coulter GmbH). Semen samples were stained in final concentrations with 1.3 µmol/L Hoechst 33342, 1.5 µmol/L propidium iodide (PI), and 2 µmol/L fluorescein conjugated peanut agglutinin (FITC-PNA). Fluorescence signals were detected in 10,000 events on the detectors FL-1 (450/45 nm BP), FL-2 (525/40 nm BP), and FL-3 (610/20 nm BP). Spermatozoa with intact plasma membranes and acrosomes were identified by a positive Hoechst stain and negative stainings for PI and FITC-PNA.

4.4. Statistical Analysis

Data analysis was performed with IBM SPSS Statistics Professional (SPSS Inc., IBM, Armonk, NY, USA). The normal distribution of data was checked with the Shapiro–Wilk Test. Data were then analyzed with the Friedman Test (XLSX). Pairwise comparisons were performed with the Wilcoxon Test and corrected with Holm Bonferroni. Values were considered as statistically significantly different when $p < 0.05$. Data are presented as mean and standard deviation (SD).

Author Contributions: Conceptualization, A.-M.L. and D.W.; methodology, A.-M.L. and J.V.; formal analysis, I.K.M. and A.-M.L.; investigation, I.K.M.; data curation, I.K.M.; writing—original draft preparation, I.K.M. and D.W.; writing—review and editing, D.W., A.-M.L., and J.V.; supervision, A.-M.L., D.W., and J.V.; project administration, D.W.; funding acquisition, D.W. All authors have read and agreed to the published version of the manuscript.

Funding: The project was supported by funds of the Federal Ministry of Food and Agriculture (BMEL) based on a decision of the Parliament of the Federal Republic of Germany via the Federal Office for Agriculture and Food (BLE) under the innovation support programme (281DT01B21). This Open Access publication was funded by the Deutsche Forschungsgemeinschaft (DFG, German Research Foundation)—491094227 "Open Access Publication Funding" and the University of Veterinary Medicine Hannover, Foundation.

Institutional Review Board Statement: All procedures involving animals were performed in accordance with the European Commission Directive for Pig Welfare and were approved by the Institutional Animal Welfare Committee of the University of Veterinary Medicine, Foundation, Hannover.

Informed Consent Statement: Not applicable.

Data Availability Statement: The data presented in this study are available upon request from the corresponding author.

Conflicts of Interest: The authors declare no conflict of interest.

References

1. Waberski, D.; Riesenbeck, M.A.; Schulze, K.F. Weitze, and L. Johnson. Application of Preserved Boar Semen for ArtificialInsemination: Past, Present and Future Challenges. *Theriogenology* **2019**, *137*, 2–7. [CrossRef]
2. White, I.G. Lipids and calcium uptake of sperm in relation to cold shock and preservation: A review. *Reprod. Fertil. Dev.* **1993**, *5*, 639–658. [CrossRef]
3. Schulze, M.; Nitsche-Melkus, E.; Hensel, B.; Jung, M.; Jakop, U. Antibiotics and their alternatives in Artificial Breeding in livestock. *Anim. Reprod. Sci.* **2020**, *220*, 106284. [CrossRef] [PubMed]
4. Ciornei, Ş.; Drugociu, D.; Ciornei, L.M.; Mareş, M.; Roşca, P. Total Aseptization of Boar Semen, to Increase the Biosecurity of Reproduction in Swine. *Molecules* **2021**, *26*, 6183. [CrossRef] [PubMed]
5. Althouse, G.C.; Kuster, C.E.; Clark, S.G.; Weisiger, R.M. Field Investigations of Bacterial Contaminants and Their Effects on Extended Porcine Semen. *Theriogenology* **2000**, *53*, 1167–1176. [CrossRef]
6. Pinart, E.; Domènech, E.; Bussalleu, E.; Yeste, M.; Bonet, S. A comparative study of the effects of Escherichia coli and Clostridium perfringens upon boar semen preserved in liquid storage. *Anim. Reprod. Sci.* **2017**, *177*, 65–78. [CrossRef]
7. Gączarzewicz, D.; Udała, J.; Piasecka, M.; Błaszczyk, B.; Stankiewicz, T. Bacterial Contamination of Boar Semen and its Relationship to Sperm Quality Preserved in Commercial Extender Containing Gentamicin Sulfate. *Pol. J. Vet. Sci.* **2016**, *19*, 451–459. [CrossRef]
8. Bussalleu, E.; Yeste, M.; Sepúlveda, L.; Torner, E.; Pinart, E.; Bonet, S. Effects of Different Concentrations of Enterotoxigenic and Verotoxigenic E. Coli on Boar Sperm Quality. *Anim. Reprod. Sci.* **2011**, *127*, 176–182. [CrossRef]
9. Althouse, G.C.; Pierdon, M.S.; Lu, K.G. Thermotemporal Dynamics of Contaminant Bacteria and Antimicrobials in Extended Porcine Semen. *Theriogenology* **2008**, *70*, 1317–1323. [CrossRef]
10. Costinar, L.; Herman, V.; Pitoiu, E.; Iancu, I.; Degi, J.; Hulea, A.; Pascu, C. Boar Semen Contamination: Identification of Gram-Negative Bacteria and Antimicrobial Resistance Profile. *Animals* **2021**, *12*, 43. [CrossRef]
11. Luther, A.-M.; Beckermann, C.; Nguyen, T.Q.; Verspohl, J.; Waberski, D. Growth Dynamic and Threshold Values for Spermicidal Effects of Multidrug-Resistant Bacteria in Extended Boar Semen. *Microorganisms* **2023**, *11*, 788. [CrossRef]
12. Althouse, G.C. Biological and chemical contaminants in extended porcine semen: Outcomes and diagnosis. *Anim. Reprod. Sci.* **2022**, *247*, 107086. [CrossRef]
13. Nitsche-Melkus, E.; Bortfeldt, R.; Jung, M.; Schulze, M. Impact of Hygiene on Bacterial Contamination in Extended Boar Semen: An Eight-Year Retrospective Study of 28 European AI Centers. *Theriogenology* **2019**, *146*, 133–139. [CrossRef] [PubMed]
14. Althouse, G.C. Sanitary Procedures for the Production of Extended Semen. *Reprod. Domest. Anim.* **2008**, *43* (Suppl. 2), 374–378. [CrossRef]
15. Tavares-Carreon, F.; De Anda-Mora, K.; Rojas-Barrera, I.C.; Andrade, A. *Serratia marcescens* antibiotic resistance mechanisms of an opportunistic pathogen: A literature review. *PeerJ* **2023**, *11*, e14399. [CrossRef]

16. Kim, B.N.; Ryu, J.; Kim, Y.S.; Woo, J.H. Retrospective Analysis of Clinical and Microbiological Aspects of *Klebsiella oxytoca* Bacteremia over a 10-Year Period. *Eur. J. Clin. Microbiol. Infect. Dis.* **2002**, *21*, 419–426.
17. Steverink, D.W.; Soede, N.M.; Bouwman, E.G.; Kemp, B. Semen Backflow after Insemination and Its Effect on Fertilisation Results in Sows. *Anim. Reprod. Sci.* **1998**, *54*, 109–119. [CrossRef]
18. Morrell, J.M.; Wallgren, M. Alternatives to Antibiotics in Semen Extenders: A Review. *Pathogens* **2014**, *3*, 934–946. [CrossRef] [PubMed]
19. Waberski, D.; Luther, A.M.; Grunther, B.; Jakel, H.; Henning, H.; Vogel, C.; Peralta, W.; Weitze, K.F. Sperm Function in Vitro and Fertility after Antibiotic-Free, Hypothermic Storage of Liquid Preserved Boar Semen. *Sci. Rep.* **2019**, *9*, 14748. [CrossRef] [PubMed]
20. Menezes, T.D.A.; Mellagi, A.P.G.; Oliveira, G.D.S.; Bernardi, M.L.; Wentz, I.; Ulguim, R.D.R.; Bortolozzo, F.P. Antibiotic-free extended boar semen preserved under low temperature maintains acceptable in-vitro sperm quality and reduces bacterial load. *Theriogenology* **2020**, *149*, 131–138. [CrossRef] [PubMed]
21. Jakel, H.; Scheinpflug, K.; Muhldorfer, K.; Gianluppi, R.; Lucca, M.S.; Mellagi, A.P.G.; Bortolozzo, F.P.; Waberski, D. In Vitro Performance and in Vivo Fertility of Antibiotic-Free Preserved Boar Semen Stored at 5 Degrees C. *J. Anim. Sci. Biotechnol.* **2021**, *12*, 9. [CrossRef] [PubMed]
22. Mann, T.; Lutwak-Mann, C. Biochemistry of Seminal Plasma and Male Accessory Fluids; Application to Andrological Problems. In *Male Reproductive Function and Semen*; Springer: New York, NY, USA, 1981; pp. 269–336.
23. Prieto-Martinez, N.; Bussalleu, E.; Garcia-Bonavila, E.; Bonet, S.; Yeste, M. Effects of Enterobacter Cloacae on Boar Sperm Quality During Liquid Storage at 17 Degrees C. *Anim. Reprod. Sci.* **2014**, *148*, 72–82. [CrossRef] [PubMed]
24. Wolff, H.; Panhans, A.; Stolz, W.; Meurer, M. Adherence of *Escherichia Coli* to Sperm: A Mannose Mediated Phenomenon Leading to Agglutination of Sperm and *E. coli. Fertil. Steril.* **1993**, *60*, 154–158. [CrossRef]
25. Diemer, T.; Huwe, P.; Michelmann, H.W.; Mayer, F.; Schiefer, H.G.; Weidner, W. *Escherichia Coli*-Induced Alterations of Human Spermatozoa. An Electron Microscopy Analysis. *Int. J. Androl.* **2000**, *23*, 178–186. [CrossRef] [PubMed]
26. Yamamoto, T.; Ariyoshi, A.; Amako, K. Fimbria-Mediated Adherence of *Serratia marcescens* Strain US5 to Human Urinary Bladder Surface. *Microbiol. Immunol.* **1985**, *29*, 677–681. [CrossRef]
27. Hejazi, A.; Falkiner, F.R. *Serratia marcescens. J. Med. Microbiol.* **1997**, *46*, 903–912. [CrossRef]
28. Bonet, S.; Delgado-Bermúdez, A.; Yeste, M.; Pinart, E. Study of boar sperm interaction with Escherichia coli and Clostridium perfringens in refrigerated semen. *Anim. Reprod. Sci.* **2018**, *197*, 134–144. [CrossRef]
29. Paschoal, A.F.L.; Luther, A.M.; Jakel, H.; Scheinpflug, K.; Muhldorfer, K.; Bortolozzo, F.P.; Waberski, D. Determination of a Cooling-Rate Frame for Antibiotic-Free Preservation of Boar Semen at 5 Degrees C. *PLoS ONE* **2020**, *15*, e0234339. [CrossRef]

Article

Phenotypic and Genotypic Characteristics of Antimicrobial Resistance in *Citrobacter freundii* Isolated from Domestic Ducks (*Anas platyrhynchos domesticus*) in Bangladesh

Tarana Ahmed [1,2,*], Md. Saiful Islam [1], Najmul Haider [3,4], Linzy Elton [5], Badrul Hasan [6], Mohammad Nuruzzaman [7], Md. Tanvir Rahman [1], S. M. Lutful Kabir [1] and Md. Shahidur Rahman Khan [1,*]

1   Department of Microbiology and Hygiene, Bangladesh Agricultural University, Mymensingh 2202, Bangladesh; dvm41257@bau.edu.bd (M.S.I.); tanvirahman@bau.edu.bd (M.T.R.); lkabir79@bau.edu.bd (S.M.L.K.)
2   Department of Livestock Services, Ministry of Fisheries & Livestock, Government of the Peoples Republic of Bangladesh, Krishi Khamar Sarak, Farmgate, Dhaka 1215, Bangladesh
3   School of Life Sciences, Keele University, Staffordshire ST5 5BG, UK; n.haider@keele.ac.uk
4   The Royal Veterinary College, University of London, Hertfordshire AL9 7TA, UK
5   Centre for Clinical Microbiology, Department of Infection, Division of Infection and Immunity, Royal Free Campus, University College London, London NW3 2PF, UK; linzy.elton@ucl.ac.uk
6   Department of Jobs, Precincts and Regions, AgriBio, Centre for AgriBioscience, Bundoora, VIC 3083, Australia; badrul.hasan@agriculture.vic.gov.au
7   Ministry of Public Administration, Dhaka 1205, Bangladesh; mnzaman27@gmail.com
*   Correspondence: tarana_ahmed07@yahoo.com (T.A.); msrkhan001@yahoo.com (M.S.R.K.)

Citation: Ahmed, T.; Islam, M.S.; Haider, N.; Elton, L.; Hasan, B.; Nuruzzaman, M.; Rahman, M.T.; Kabir, S.M.L.; Khan, M.S.R. Phenotypic and Genotypic Characteristics of Antimicrobial Resistance in *Citrobacter freundii* Isolated from Domestic Ducks (*Anas platyrhynchos domesticus*) in Bangladesh. *Antibiotics* **2023**, *12*, 769. https://doi.org/10.3390/antibiotics12040769

Academic Editor: Nahla O. Eltai

Received: 16 March 2023
Revised: 13 April 2023
Accepted: 15 April 2023
Published: 17 April 2023

Abstract: Antimicrobial resistance (AMR) in *Citrobacter freundii* poses a serious challenge as this species is one of the sources of nosocomial infection and causes diarrheal infections in humans. Ducks could be the potential source of multidrug-resistant (MDR) *C. freundii*; however, AMR profiles in *C. freundii* from non-human sources in Bangladesh have remained elusive. This study aimed to detect *C. freundii* in domestic ducks (*Anas platyrhynchos domesticus*) in Bangladesh and to determine their phenotypic and genotypic antibiotic susceptibility patterns. A total of 150 cloacal swabs of diseased domestic ducks were screened using culturing, staining, biochemical, polymerase chain reaction (PCR), and matrix-assisted laser desorption/ionization time-of-flight (MALDI-TOF) to detect *C. freundii*. Phenotypic and genotypic antibiotic susceptibility patterns were done by the disk diffusion method and PCR, respectively. In total, 16.67% (25/150) of the samples were positive for *C. freundii*. *C. freundii* isolates showed a range of 20% to 96% resistance to cefotaxime, gentamicin, levofloxacin, ciprofloxacin, cotrimoxazole, tetracycline, ampicillin, and cephalexin. More than 60% of the isolates were phenotypically MDR, and the index of multiple antibiotic resistance ranged from 0.07 to 0.79. Genes encoding resistance to beta-lactams [bla_{TEM-1}-88% (22/25), bla_{CMY-2}-56% (14/25), bla_{CMY-9}-8% (2/25), and $bla_{CTX-M-14}$-20% (5/25)], sulfonamides [$sul1$-52% (13/25), $sul2$-24% (6/25)], tetracyclines [$tetA$-32% (8/25) and $tetB$-4% (1/25)], aminoglycosides [$aacC4$-16% (4/25)], and fluoroquinolones [$qnrA$-4% (1/25), $qnrB$-12% (3/25), and $qnrS$-4% (1/25)] were detected in the isolated *C. freundii*. To the best of our knowledge, this is the first study in Bangladesh to detect MDR *C. freundii* with their associated resistance genes from duck samples. We suggest addressing the burden of diseases in ducks and humans and associated AMR issues using the One Health approach.

Keywords: *C. freundii*; antimicrobial resistance; multidrug resistance; resistance genes; ducks; public health; Bangladesh

1. Introduction

Antimicrobial resistance (AMR) is a major global issue that jeopardizes human, animal, and environmental health [1]. If nothing is done to curb AMR by 2050, it is expected to inflict hundreds of millions of fatalities worldwide, enormous financial consequences, and a significant decline in animal production [2,3]. *Citrobacter freundii* has become more resistant as a direct result of the widespread usage of antibiotics with a broad spectrum of activity [4]. Antimicrobials such as fluoroquinolones, aminoglycosides, nitrofurantoins, carbapenems, and cephalosporins are the typical classes of antibiotics used to treat infections caused by *C. freundii* [5]. However, the concern is rising because *C. freundii* has developed resistance to multiple antibiotics. Moreover, it is possible that low-virulent *Citrobacter* spp., which are able to survive in the host for a long time, could impact the evolution of pathogens by accumulating genes that code for resistance to multiple classes of antimicrobials [6]. The acquisition of resistance genes that confer resistance to multiple antibiotic classes from external sources, such as the environment or other bacteria, can cause multidrug resistance in *Citrobacter* species [7].

Antimicrobial-resistant bacteria may be present in duck droppings, contaminating the environment [8]. Ducks are possible carriers of important antimicrobial-resistant pathogens that might spread to humans because of their interactions with humans [9]. Humans can acquire *C. freundii* from ducks through contact with infected eggs, raw or undercooked meat, and duck carcasses at the slaughterhouse [10]. *C. freundii* infections can be fatal, with death rates ranging from 33 to 48% overall, including 30% mortality in children [11]. The central nervous system of survivor infants may be severely affected, resulting in extreme mental impairment, convulsions, and hemiparesis [12].

The poultry industry in Bangladesh has grown and become a successful agricultural business that makes a significant contribution to the country's overall gross domestic product and provides a valuable source of protein [13,14]. Duck farming is a profitable livestock sector around the world because of the eggs, meat, and feathers it produces [15]. In Bangladesh, duck farming is important in its rural economy, second only to chicken production [16]. Ducks are typically reared in small-scale farming, either indoors or outdoors, or in an integrated farming system in Bangladesh [17], where they come or stay in close contact with humans. However, the greatest barrier to large-scale duck farming in Bangladesh is infectious disease outbreaks, including duck viral enteritis, duck viral hepatitis, avian influenza, botulism, duck cholera, etc. [18,19].

One of the bacterial pathogens found in duck droppings is *Citrobacter* spp. However, very little is known about the role of *Citrobacter* spp. as the source of infections in duck populations. *C. freundii* is the most prevalent among all *Citrobacter* species that causes infections in humans and animals [20]. *C. freundii* was isolated from young ducks having salpingitis [21]. The most common symptoms of *C. freundii* infection in ducks are discharge from nostrils, leg weakness, whitish diarrhea, recumbency, headshaking, and even sudden death [22].

In Bangladesh, the issue is compounded by the fact that poultry farmers come into direct contact with ducks during the rearing process, particularly when raising domestic ducks (*Anas platyrhynchos domesticus*) in their own homes. This direct interaction between humans and ducks increases the risk of transmission of *C. freundii* from domestic ducks to children, creating an even greater cause for concern. Human cases of *C. freundii* have been recorded in Bangladesh, India, and other Asian countries [23–25]; however, *C. freundii* cases in non-humans are not well described in these regions. In fact, this bacterium has not been well characterized in animals from any South Asian countries. The aim of this study was to detect *C. freundii* from cloacal swabs of ducks and determine their phenotypic and genotypic antibiotic resistance patterns to elucidate their potential negative impacts on human health.

2. Results

2.1. Occurrence of C. freundii Isolates

In the polymerase chain reaction (PCR) test, 25 of 150 samples were positive for *Citrobacter* spp. (16.67%, 95% CI: 11.55–23.45%) (Table 1). In MALDI-TOF analysis, all the *Citrobacter* spp. were detected as *C. freundii*. The occurrence of *C. freundii* in cloacal swabs of ducks was higher but not significant in the Kishoreganj district (22%, 95% CI: 12.75–35.24%)

compared to that of Mymensingh (18%, 95% CI: 9.77–30.80%) and Netrokona (10%, 95% CI: 4.35–21.36%) districts (Table 1).

Table 1. Prevalence of *C. freundii* isolated from cloacal swabs of ducks from different districts of Bangladesh.

Locations	No. of Samples Collected	No. of Positive Isolates (%)	95% CI (%)	*p*-Value
Mymensingh	50	9 (18 [a])	9.77–30.80	
Netrokona	50	5 (10 [a])	4.35–21.36	0.261
Kishoreganj	50	11 (22 [a])	12.75–35.24	
Overall	150	25 (16.67)	11.55–23.45	

Values with different superscripts differ significantly ($p < 0.05$) within the variable under assessment, CI = Confidence interval.

2.2. Phenotypic Antibiogram Profiles of Isolated C. freundii

In the antibiotic susceptibility test (AST), *C. freundii* isolates showed the highest resistance to cephalexin (96%, 95% CI: 80.46–99.80%), followed by ampicillin (76%, 95% CI: 56.57–88.50%), azithromycin (56%, 95% CI: 37.07–73.33%), tetracycline (44%, 95% CI: 26.67–62.93%), cotrimoxazole (40%, 95% CI: 23.40–59.26%), ciprofloxacin and levofloxacin (36%, 95% CI: 20.25–55.48%), gentamicin (24%, 95% CI: 11.50–43.43%), cefotaxime (20%, 95% CI: 8.86–39.13%), ceftriaxone and ceftazidime (12%, 95% CI: 4.17–29.96%), and fosfomycin (4%, 95% CI: 0.21–19.54%) (Figure 1). In addition, 100% of the isolates exhibited sensitivity to nitrofurantoin and chloramphenicol (Figure 1).

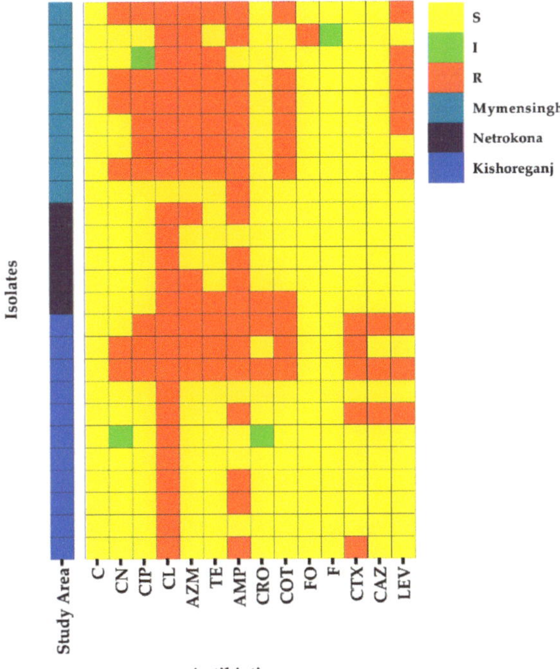

Figure 1. Heatmap showing the distribution of antibiogram profiles in 25 *C. freundii* isolates detected from cloacal swabs of ducks, LEV = levofloxacin; CAZ = ceftazidime; CTX = cefotaxime; F = nitrofurantoin; FO = fosfomycin; COT = cotrimoxazole; CRO = ceftriaxone; AMP = ampicillin; TE = tetracycline; AZM = azithromycin; CL = Cephalexin; CIP = ciprofloxacin; CN = gentamycin; C = chloramphenicol, S = Sensitive, I = Intermediate, R = Resistant.

In bivariate analysis, we observed very high positive significant correlations between resistance patterns against cotrimoxazole and tetracycline ($p < 0.001$); cotrimoxazole and ciprofloxacin ($p < 0.001$); tetracycline and ciprofloxacin ($p < 0.001$); tetracycline and azithromycin ($p < 0.001$); ceftazidime and cefotaxime ($p < 0.001$); cotrimoxazole and azithromycin ($p < 0.001$); cotrimoxazole and gentamycin ($p < 0.001$); levofloxacin and tetracycline ($p < 0.001$); azithromycin and ciprofloxacin ($p < 0.001$); levofloxacin and ciprofloxacin ($p < 0.001$); tetracycline and gentamycin ($p < 0.01$); ceftazidime and ceftriaxone ($p < 0.01$); levofloxacin and cotrimoxazole ($p < 0.01$); and levofloxacin and gentamycin ($p < 0.01$) (Supplementary Table S1). We also found moderate-to-lower significant positive correlations between resistance patterns of *C. freundii* isolates against different antibiotics (Supplementary Table S1).

2.3. MDR and MAR Profiles of C. freundii

The majority of the *C. freundii* isolates (15/25, 60%, 95% CI: 40.74–76.60%) were phenotypically MDR and showed a multiple antibiotic resistance (MAR) index of more than 0.2. Fourteen antibiotic resistance patterns were observed, and eleven of them were MDR. The most common MDR pattern was no. 3 (gentamicin-ciprofloxacin-cephalexin-azithromycin-tetracycline-ampicillin-cotrimoxazole-levofloxacin), which was 26.67% (4/15) of the MDR *C. freundii* isolates. One isolate was resistant to 11 antibiotics (out of 14 tested antibiotics) from seven different classes (out of ten classes) (Table 2). Moreover, the MAR index of *C. freundii* isolates varied from 0.07 to 0.79 (Table 2).

Table 2. Phenotypic multidrug resistance profiles and multiple antibiotic resistance index profiles of *C. freundii* isolated from cloacal swabs of ducks.

Pattern No.	Antibiotic Resistance Patterns	No. of Antibiotics (Classes)	No. of MDR Isolates (%)	MAR Index
1	CN, CIP, CL, AZM, TE, AMP, CRO, COT, CTX, CAZ, LEV	11 (7)	1 (6.67)	0.79
2	CIP, CL, AZM, TE, AMP, CRO, COT, CTX, CAZ, LEV	10 (6)	1 (6.67)	0.71
3	CN, CIP, CL, AZM, TE, AMP, COT, LEV	8 (7)	4 (26.67)	0.57
4	CN, CIP, CL, AZM, TE, AMP, COT, CTX	8 (7)	1 (6.67)	0.57
5	CIP, CL, AZM, TE, AMP, COT, LEV	7 (6)	1 (6.67)	0.50
6	CIP, CL, AZM, TE, AMP, COT	6 (6)	1 (6.67)	0.43
7	CL, AZM, TE, AMP, COT, CRO	6 (5)	1 (6.66)	0.43
8	CL, AMP, LEV, CTX, CAZ	5 (3)	1 (6.67)	0.36
9	CL, AZM, AMP, FO	4 (4)	1 (6.67)	0.29
10	CL, AZM, TE, LEV	4 (4)	1 (6.67)	0.29
11	CL, AZM, AMP	3 (3)	2 (13.33)	0.21
12 *	CL, AMP	2 (2)	4 *	0.14
13 *	CL	1 (1)	5 *	0.07
14 *	AMP	1 (1)	1 *	0.07

MDR = multidrug-resistant, MAR = multiple antibiotic resistance, LEV = Levofloxacin; CAZ = ceftazidime; CTX = cefotaxime; FO = fosfomycin; COT = cotrimoxazole; CRO = ceftriaxone; AMP = ampicillin; TE = tetracycline; AZM = azithromycin; CL = Cephalexin; CIP = ciprofloxacin; CN = gentamycin; * Non-multidrug-resistant.

2.4. Genotypic Resistance Profiles of C. freundii Isolates

Upon PCR analysis, out of 25 *C. freundii* isolates, beta-lactamase genes bla_{TEM-1}, bla_{CMY-2}, bla_{CMY-9}, and $bla_{CTX-M-14}$ were detected in 88% (95% CI: 70.04–95.83%), 56% (95% CI: 37.07–73.33%), 8% (95% CI: 1.42–24.97%), and 20% (95% CI: 8.86–39.13%) of the isolates, respectively (Figure 2). Genes conferring resistance to sulfonamides [*sul1*-52% (13/25), 95% CI: 33.49–69.97%; *sul2*-24% (6/25), 95% CI: 11.49–43.43%], tetracyclines [*tetA*-32% (8/25), 95% CI: 17.21–51.59%; *tetB*-4% (1/25), 95% CI: 0.21–19.54%], fluoroquinolones [*qnrA*-4% (1/25), 95% CI: 0.21–19.54%; *qnrB*-12% (3/25), 95% CI: 4.17–29.96%; *qnrS*-4% (1/25), 95% CI: 0.21–19.54%], and aminoglycosides [*aacC4*-16% (4/25), 95% CI: 6.40–34.65%] were also detected in the isolated *C. freundii*. No isolates harbored bla_{SHV-1}, $bla_{CTX-M-1}$, $bla_{CTX-M-2}$, *tetC*, and *aacC2* genes (Figure 2).

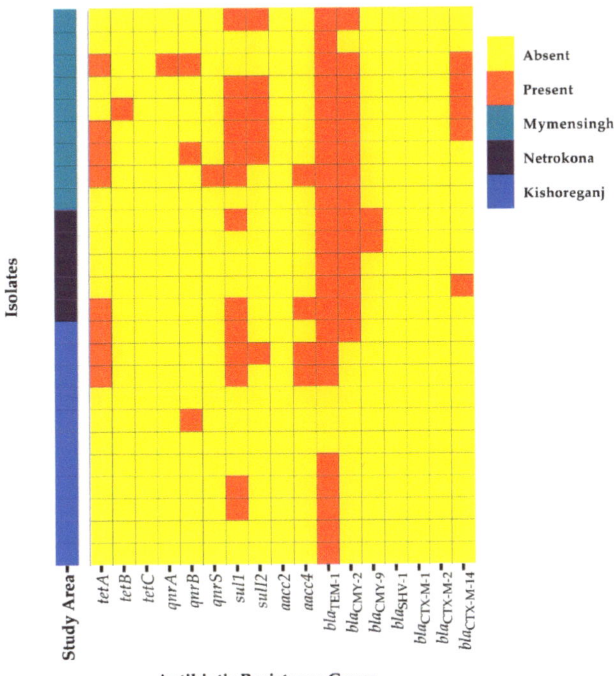

Figure 2. Heatmap showing the distribution of various antibiotic resistance genes of *C. freundii* isolated from cloacal swabs of ducks in Bangladesh.

In the bivariate analysis, a high positive significant correlation was observed between the presence of antibiotic resistance genes *aac4* and *tetA* (Pearson correlation coefficient, $\rho = 0.636$; $p = 0.001$), *qnrA* and *qnrB* ($\rho = 0.553$; $p = 0.004$), and *sul1* and *sul2* ($\rho = 0.540$; $p = 0.005$) (Supplementary Table S2). Moreover, moderate-to-low positive significant correlations were also observed between the presence of resistance genes *tetA* and *sul1* ($\rho = 0.487$; $p = 0.013$), *qnrS* and *aacC4* ($\rho = 0.468$; $p = 0.018$), bla_{CMY-2} and $bla_{CTX-M-14}$ ($\rho = 0.443$; $p = 0.026$), $bla_{CTX-M-14}$ and *sul2* ($\rho = 0.421$; $p = 0.036$), *aac4* and *sul1* ($\rho = 0.419$; $p = 0.037$), bla_{TEM-1} and bla_{CMY-2} ($\rho = 0.417$; $p = 0.038$), *tetB* and $bla_{CTX-M-14}$ ($\rho = 0.408$; $p = 0.043$), and *qnrA* and $bla_{CTX-M-14}$ ($\rho = 0.408$; $p = 0.043$) (Supplementary Table S2).

2.5. Comparison of Phenotypic and Genotypic Resistance Profiles of Isolated C. freundii

In bivariate analysis, a positive significant correlation was observed between phenotypic and genotypic resistance profiles of *C. freundii* isolates against tetracyclines ($\rho = 0.846$; $p < 0.001$), sulfonamides ($\rho = 0.784$; $p < 0.001$), aminoglycosides ($\rho = 0.521$; $p = 0.008$), and fluoroquinolones ($\rho = 0.417$; $p = 0.038$). Given that all the isolates were phenotypically resistant to at least one beta-lactam antibiotic (constant variable), we could not compute the Pearson correlation coefficient to show the correlation between phenotypic and genotypic resistant profiles of *C. freundii* isolates against beta-lactams. However, 88% similarity was exhibited between beta-lactam antibiotics ($n = 25$) and beta-lactamase genes ($n = 22$) based on the phenotypic (disk diffusion) and genotypic (PCR) assays.

3. Discussion

Ducks have the potential to harbor hazardous bacteria that can cause zoonotic diseases in humans, including salmonellosis, *E. coli* infections, cholera, psittacosis, and others [9]. In our study, about 16.67% of the duck samples harbored *C. freundii*. The detection of *C. freundii* in this study suggests that ducks have the potential to transfer this organism to humans, posing an important threat to public health. In Bangladesh, poultry farmers, surrounding people, and the environment are directly exposed to domestic ducks during the process of rearing them. As a result, people in direct contact with ducks, both on farms and in houses, are at an increased risk of being infected with *C. freundii* from ducks. Importantly, *C. freundii* isolates have the ability to cause bacteremia in humans, indicating a high risk to human health [4]. A previous study in South Korea reported that *C. freundii* bacteremia was the major risk factor for a higher mortality rate in hospitalized patients (aged $\geq$ 15 years) [26]. Duck droppings can contaminate the agricultural environment and can be a source of infections for crop farmers and other people exposed to the contaminated environment [27].

Over time, *C. freundii* has the tendency to develop resistance to different classes of broad-spectrum antibiotics, and a major emerging issue is the rapid spread of this antibiotic resistance [26]. The result of the antibiotic susceptibility test showed that *C. freundii* isolates showed resistance to different classes of broad-spectrum antibiotics. For example, resistance was found to multiple classes of antibiotics, including beta-lactams, sulfonamides, fluoroquinolones, tetracyclines, aminoglycosides, macrolides, and glycopeptides, and these antibiotics are used extensively in both human and veterinary medicine. Our study indicates that domestic ducks foraging in different environmental niches could be a potential carrier of antibiotic resistance. Similar to our study, Olaiton et al. [8] also showed that *C. freundii* isolated from duck droppings showed resistance to tetracyclines, aminoglycosides, beta-lactams, and sulfonamides.

Cephalosporins are still reliable antibiotics for the treatment of *C. freundii* infections, and the rapid emergence of cephalosporin resistance in *C. freundii* is considered a global health problem [4]. In this study, *C. freundii* isolates showed phenotypic resistance to ampicillin, cephalexin, cefotaxime, ceftriaxone, and ceftazidime. In addition to that, several genotypes associated with cephalosporin resistance were found. For example, bla_{CMY-2} (56%), bla_{CMY-9} (8%), and $bla_{CTX-M-14}$ (20%) are considered clinically associated biomarkers [28]. The widespread distribution of beta-lactam genotypes in humans, animals, and the environment indicates an immediate need for improvement in the treatment of infectious diseases [29]. Resistance to beta-lactam antibiotics in an organism may also be developed due to the acquisition of other beta-lactamase genes, e.g., bla_{TEM} and bla_{CMY}, in humans and animals [29,30]. The presence of these clinical biomarkers in *C. freundii* isolated from ducks is a clear indication of potential widescale spread from human *C. freundii* isolates or other members of the *Enterobacteriaceae* family.

Sulfonamides are a significant class of synthetic bacteriostatic antibiotics that are still widely used to treat bacterial infections in veterinary medicine [31]. In our study, 40% of the *C. freundii* isolates were resistant to the sulfonamide drug called cotrimoxazole. Resistance to sulfonamides in gram-negative bacteria, including *C. freundii*, typically results from the acquisition of one of two genes, *sul1* or *sul2*, which encode for dihydropteroate synthase forms that are not inhibited by the drug [32]. Sulfonamide resistance genes *sul1* and *sul2* were detected in 52% and 24% of the *C. freundii* isolates, respectively. One possible explanation for the increased occurrence of sulfonamide resistance genes among *C. freundii* isolates collected from the cloaca of ducks is the widespread abuse of this kind of antibiotic in the poultry industry. The presence of sulfonamide-resistant *C. freundii* and their corresponding genes in ducks should be concerning because these resistant bacteria have the potential to be transferred to humans via direct or indirect contact [33].

Resistance to fluoroquinolone antibiotics is an urgent problem in both human and veterinary medicine worldwide. More than 44% of *C. freundii* isolates showed phenotypic resistance to fluoroquinolone antibiotics, either to ciprofloxacin or to levofloxacin, which is an important public health concern as these drugs are widely used for humans and large animals. In general, the fluoroquinolone group of antibiotics is a reliable antimicrobial agent for the treatment of *Citrobacter* infections, especially *C. freundii* bacteremia [4]. In this study, we detected a higher rate of fluoroquinolone resistance gene *qnrB* (12%) than that of other *qnr* (*qnrA* and *qnrS*) genes, which is not unusual. The *qnrB* genes, which encode proteins liable for reduced susceptibility to fluoroquinolones, are by far the most common and diverse subfamily of *qnr* genes [34]. These antibiotic-resistant bacteria and their resistance genes have the potential to spread to humans via the food chain [35]. Given that these resistant isolates can be passed from ducks to humans, the use of fluoroquinolones in poultry needs to be closely regulated in order to avoid any further resistance to fluoroquinolones.

Tetracyclines are commonly used as one of the first-line antibiotics against a wide variety of non-life-threatening infections, including *C. freundii* infections [36]. The development of antibiotic resistance in *C. freundii* limits the treatment options. We report 44% of the *C. freundii* isolates were phenotypically resistant to tetracycline. We also detected tetracycline resistance genes *tetA* (8/25) and *tetB* (1/25) in the *C. freundii* isolates; however, none for the *tetC* gene. The detection of *tetA* and *tetB* genes in tetracycline-resistant *C. freundii* isolates demonstrates that the active efflux system was the initial mechanism of tetracycline resistance in ducks [37,38].

Aminoglycosides are among the most effective antibiotics available for treating serious infections [39]. In the present study, 24% of the *C. freundii* isolates were phenotypically resistant to the aminoglycoside antibiotic named gentamicin. Moreover, we detected the aminoglycoside resistance gene *aacC4* in 16% of the *C. freundii* isolates. The presence of this gene with significance to public health in ducks highlights the necessity of conducting an additional investigation into duck reservoirs for AMR. One of the important causal agents of community-acquired sepsis is *C. freundii* [40]. Therefore, the presence of aminoglycoside resistance in *C. freundii* may limit the treatment options for community-acquired sepsis because combining an aminoglycoside with a beta-lactam antibiotic and metronidazole is a common and effective experimental treatment for community-acquired sepsis [41].

This study found that the phenotypic and genotypic resistance profiles of *C. freundii* isolates against tetracyclines, sulfonamides, aminoglycosides, and fluoroquinolones were significantly correlated ($p < 0.05$). However, some isolates exhibited phenotypic susceptibility in the presence of resistance genes, and some isolates appeared phenotypic resistant but did not harbor resistant genes. Over time, several random mutations can occur in the gene sequence of an antibiotic resistance gene, rendering it inactive and transforming it into a resistance pseudogene that does not show the expected resistance characteristics [42]. Moreover, observed variations in susceptibility may be explained by heteroresistance processes such as random tandem gene amplification, uncommon mutation, and environmental manipulation of resistant genes [43]. The antibiotic may also act as a modulator, causing antibiotic-resistant genes to express poorly in vitro [44]. Therefore, whole-genome sequencing-based analyses have the potential to provide a precise genotype-to-phenotype resistance link.

Infections developed by MDR bacteria (with a high MAR index) have the potential to have severe repercussions for both human and animal health [45]. Humans and animals alike are at risk due to the spread of MDR *Citrobacter* spp. [46]. Infections caused by MDR *C. freundii* have fewer treatment options. In this study, 60% of the isolates were MDR, indicating an alarming issue in ducks and humans. The transfer of resistance genes from one resistant bacterium to another can lead to the development of MDR in bacteria that normally respond to the related classes of antibiotics [7]. Moreover, 60% of the isolates had a MAR index greater than 0.2, suggesting that these bacterial strains originated from a high-risk source of contamination in an area where antibiotics are frequently used [47]. Ducks can spread these pathogens to humans through the food chain or through direct contact, and they can also spread them to the environment through polluted water or feed [48,49].

4. Materials and Methods

4.1. Study Area

From January 2020 to January 2022, the present study was carried out in Mymensingh (24.7539° N, 90.4073° E), Netrokona (24.8103° N, 90.8656° E), and Kishoreganj (24.4260° N, 90.9821° E) districts of Bangladesh (Figure 3). The locations were chosen because of the high density of ducks in these districts, with massive wetland areas.

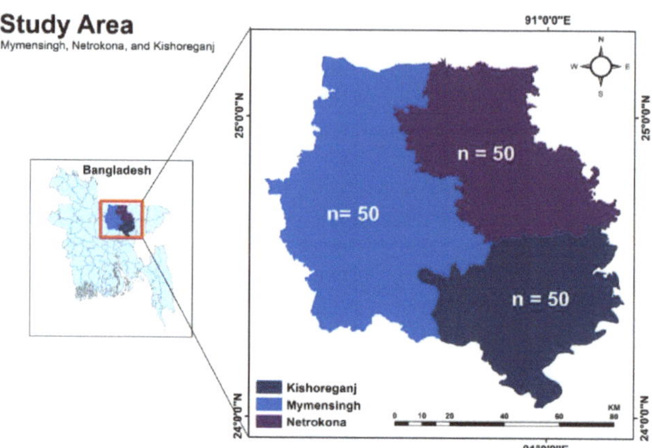

Figure 3. Maps showing study areas in Bangladesh with the number of samples collected from each district.

4.2. Sample Collection and Processing

We collected a total of 150 cloacal swabs of diseased domestic ducks (*Anas platyrhynchos domesticus*) (50 from each district) from different households (10 households from each district, each household reared domestic ducks with a range of 50–100). The diseased ducks had several symptoms, such as whitish or greenish diarrhea, leg weakness, headshaking, and even sudden death. Moreover, the ducks were reared in a scavenging or semi-scavenging system. The cloacal swab was transferred aseptically to a test tube containing Luria–Bertani (LB) broth (HiMedia, Maharashtra, India). The samples were transferred to the lab in a cold chain and enriched the bacteria by incubating the test tube contents of the samples aerobically for 18 to 24 h at 37 °C.

4.3. Isolation and Molecular Detection of Citrobacter spp.

One loopful (1–2 μL) of the overnight growth culture was streaked on a xylose-lysine deoxycholate (XLD) agar (HiMedia, India) plate, and the plate was then incubated aerobically for 24–48 h at 37 °C. For pure colonies, subcultures were done following the same procedures. Colonies showing yellow color with a black center on the XLD agar plate were assumed to be *Citrobacter* spp. isolates. Finally, we screened the single pure colonies for further confirmation by Gram's staining technique and corresponding biochemical tests (such as urease, citrate, catalase, motility, H_2S test, methyl red, Voges–Proskauer, and sugar fermentation tests) [50].

We performed a PCR test for the final confirmation of isolated *Citrobacter* spp. targeting a 16S rRNA gene (F: AGAGTTTGATCMTGGCTCAG, R: TACGGYTACCTTGTTACGACTT) as previously mentioned [51]. For PCR, the DNA from the bacterial genome was extracted by following the boiling and freeze–thawing methods described earlier [52,53]. Each extracted DNA was then amplified using a total of 20 μL of the final volume, including 10 μL of the master mix (2×) (Promega, Madison, WI, USA), 1 μL of each primer (forward and reverse) (100 pmol) (Macrogen, Republic of Korea), 2 μL of nuclease-free water, and 6 μL of genomic DNA. Amplified PCR products were examined on a 1.5% agarose gel (Invitrogen, Waltham, MA, USA) using a gel electrophoresis apparatus (Nippon Genetics, Tokyo, Japan). Amplicon products were observed under an ultraviolet trans-illuminator (Biometra, Göttingen, Germany) after being stained with ethidium bromide. The amplicon size was checked using a 1 kb DNA ladder (Promega, Madison, WI, USA).

4.4. Detection of C. freundii by MALDI-TOF Mass Spectrometry

Citrobacter spp. isolates that were positive in PCR assay were then subjected to matrix-assisted laser desorption/ionization time-of-flight (MALDI-TOF) mass spectrometry to detect *C. freundii* isolates. The MALDI-TOF analysis was done in the QC Laboratory, Dhaka, Bangladesh, and was followed by the procedures previously described by Kolínská et al. [54].

4.5. Antibiotic Susceptibility Test

4.5.1. Phenotypic Analysis

According to the Clinical and Laboratory Standards Institute [55], the AST of isolated *C. freundii* was done by a Kirby–Bauer disk diffusion test [56]. In this study, 14 antibiotics from ten classes were chosen based on their availability in Bangladesh: fluoroquinolones (ciprofloxacin—5 μg, levofloxacin—5 μg), aminoglycosides (gentamicin—10 μg), tetracyclines (tetracycline—30 μg), macrolides (azithromycin—15 μg), cephalosporins (ceftriaxone—30 μg, cephalexin—30 μg, cefotaxime—30 μg, ceftazidime—30 μg), penicillins (ampicillin—10 μg), glycopeptides (chloramphenicol—30 μg), sulfonamides (cotrimoxazole- 25 μg), phosphonic acids (fosfomycin—200 μg), nitrofurantoin (nitrofurantoin—100 μg) (HiMedia, India). A multidrug-resistant (MDR) isolate was characterized as one that is resistant to three or more antibiotic classes [57]. We enumerated the multiple antibiotic resistance (MAR) index by the following formula [47]: MAR = w/v; here, w = number of antibiotics to which an isolate is resistant, v = total number of antibiotics used in this study.

4.5.2. Genotypic Analysis

Genes conferring resistance to beta-lactams (bla_{TEM-1}, bla_{CMY-2}, bla_{CMY-9}, bla_{SHV-1}, $bla_{CTX-M-1}$, $bla_{CTX-M-2}$, and $bla_{CTX-M-14}$), sulfonamides (*sul1* and *sul2*), tetracyclines (*tetA*, *tetB*, and *tetC*), fluoroquinolones (*qnrA*, *qnrB*, and *qnrS*), and aminoglycosides (*aacC2* and *aacC4*) were tested by a simplex PCR assay (Table 3).

Table 3. Primers used in the present study for detecting *Citrobacter* spp. and different antibiotic resistance genes in *C. freundii* isolates from cloacal swabs of ducks.

Factors	Target Genes	Primer Sequences (5′-3′)	Annealing Temp.	Amplicon Size (bp)	References
Beta-lactamase	bla_{TEM-1}	F-CAGCGGTAAGATCCTTGAGA R-ACTCCCCGTCGTGTAGATAA	55	643	[58]
	bla_{CMY-2}	F-TGGCCGTTGCCGTTATCTAC R-CCCGTTTTATGCACCCATGA	55	870	[58]
	bla_{CMY-9}	F-TCAGCGAGCAGACCCTGTTC R-CTGGCCGGGATGGGATAGTT	55	874	[58]
	bla_{SHV-1}	F-GGCCGCGTAGGCATGATAGA R-CCCGGCGATTTGCTGATTTC	55	714	[58]
	bla_{CTXM-2}	F-GGCGTTGCGCTGATTAACAC R-TTGCCCTTAAGCCACGTCAC	55	486	[58]
	$bla_{CTX-M-1}$	F-AACCGTCACGCTGTTGTTAG R-TTGAGGCTGGGTGAAGTAAG	55	766	[58]
	$bla_{CTX-M-14}$	F-GCCTGCCGATCTGGTTAACT R-GCCGGTCGTATTGCCTTTGA	55	358	[58]
Tetracyclines	*tetA*	F-GCGCCTTTCCTTTGGGTTCT R-CCACCCGTTCCACGTTGTTA	55	831	[58]
	tetB	F-CCCAGTGCTGTTGTTGTCAT R-CCACCACCAGCCAATAAAAT	55	723	[58]
	tetC	F-TTGCGGGATATCGTCCATTC R-CATGCCAACCCGTTCCATGT	54	1019	[58]
Fluroquinolones	*qnrA*	F-TCAGCAAGAGGATTTCTCA R-GGCAGCACTATTACTCCCA	55	670	[59]
	qnrB	F-ATGACGCCATTACTGTATAA R-GATCGCAATGTGTGAAGTTT	53	680	[59]
	qnrS	F-ACGACATTCGTCAACTGCAA R-TAAATTGGCACCCTGTAGGC	54	428	[58]
Sulfonamides	*sul1*	F-TCACCGAGGACTCCTTCTTC R-CAGTCCGCCTCAGCAATATC	55	331	[58]
	sul2	F-CCTGTTTCGTCCGACACAGA R-GAAGCGCAGCCGCAATTCAT	55	435	[58]
Aminoglycosides	*aacC2*	F-GGCAATAACGGAGGCAATTCGA R-CTCGATGGCGACCGAGCTTCA	55	450	[58]
	aacC4	F-ACTGAGCATGACCTTGCGATGCTCTA R-TACCTTGCCTCTCAAACCCCGCTT	55	436	[58]

4.6. Statistical Analyses

4.6.1. Descriptive Analysis

We used Microsoft Excel 2013 (Los Angeles, CA, USA) for data entry and the Statistical Package for Social Science (SPSS 25, IBM, Chicago, IL, USA) and GraphPad Prism 8.4.2 (GraphPad Software, Inc., Avenida De La Playa La Jolla, CA, USA) for the data analysis. We performed the chi-square test for relatedness (with a Z-test for proportion) to understand the variations in the prevalence of *C. freundii* among sampling sites. Statistical significance was defined as a *p*-value of less than 0.05. The binomial 95% confidence interval (CI) was calculated following the Wilson and Brown Hybrid method [60]. Using GraphPad Prism, we created the heatmap to show the distribution of phenotypic and genotypic antibiotic resistance profiles of *C. freundii* isolates.

4.6.2. Bivariate Analysis

We performed bivariate analysis in SPSS to determine whether resistance patterns in pairs of antibiotics and antibiotic resistance genes from isolated *C. freundii* were correlated. Moreover, the correlation between phenotypic and genotypic resistance profiles of *C. freundii* isolates against different classes of antibiotics was determined using bivariate analysis. A statistically significant *p*-value was less than 0.05 ($p < 0.05$).

5. Conclusions

Citrobacter spp. was found to be present in more than 16% of the samples collected from ducks in Bangladesh. We reported a range of 20% to 96% resistance in *C. freundii* isolates to different important antibiotics. The detection of genes encoding resistance to various classes of antibiotics in *C. freundii* isolated from ducks indicates a significant risk to human health due to the widespread presence of antibiotic resistance and their associated resistance genes in *C. freundii*. Given the close relationship between ducks, water, and other environmental components, there is a concern about the spreading of antibiotic-resistant *C. freundii* to humans. This may pose a potential human health risk. *C. freundii* should be characterized more elaborately using whole genome sequencing. However, further studies using the One Health approach and developed tools and high technologies are helpful in understanding the potential impact of MDR *C. freundii* in humans and animals and to minimize the risk of the emergence of MDR *C. freundii* in both animals and humans.

Supplementary Materials: The following supporting information can be downloaded at: https://www.mdpi.com/article/10.3390/antibiotics12040769/s1, Table S1: Pearson correlation coefficients assessing correlation between pairs of antibiotics to which *C. freundii* isolates showed resistance; Table S2: Pearson correlation coefficients assessing correlation between pairs of antibiotic resistance genes in *C. freundii* isolates from cloacal swabs of ducks.

Author Contributions: T.A., M.S.R.K. and M.N. contributed to the study conception and design. T.A. investigated the study. T.A. and M.S.I. analyzed the data. T.A. and M.S.I. prepared the first draft manuscript. M.S.I., N.H., L.E., B.H., M.N., M.T.R., S.M.L.K. and M.S.R.K. reviewed the draft manuscripts. All authors have read and agreed to the published version of the manuscript.

Funding: The study was funded by 'Bangladesh Agricultural Research Council' through 'National Agricultural Technology Program Phase II' project for in Country Ph.D. scholarship (P149553).

Institutional Review Board Statement: The institutional ethics committee (Animal Welfare and Experimentation Ethical Committee) at Bangladesh Agricultural University in Mymensingh gave its approval to the procedures and related protocols in this research (AWEEC/BAU/2020[10]).

Informed Consent Statement: Not applicable.

Data Availability Statement: The datasets used and/or analyzed during the current study are available from the corresponding author upon request.

Acknowledgments: We acknowledge the Department of Livestock Services of the Ministry of Fisheries and Livestock of Bangladesh for supporting this research.

References

1. Islam, M.; Nayeem, M.; Hasan, M.; Sobur, M.; Ievy, S.; Rahman, S.; Kafi, M.; Ashour, H.M.; Rahman, M. Virulence determinants and multidrug resistance of *Escherichia coli* isolated from migratory birds. *Antibiotics* **2021**, *10*, 190. [CrossRef] [PubMed]
2. Tawyabur, M.; Islam, M.; Sobur, M.; Hossain, M.; Mahmud, M.; Paul, S.; Hossain, M.T.; Ashour, H.M.; Rahman, M. Isolation and characterization of multidrug-resistant *Escherichia coli* and *Salmonella* spp. from healthy and diseased turkeys. *Antibiotics* **2020**, *9*, 770. [CrossRef]
3. Urmi, M.R.; Ansari, W.K.; Islam, M.S.; Sobur, M.A.; Rahman, M.; Rahman, M.T. Antibiotic resistance patterns of *Staphylococcus* spp. isolated from fast foods sold in different restaurants of Mymensingh, Bangladesh. *J. Adv. Vet. Anim. Res.* **2021**, *8*, 274–281. [CrossRef]
4. Liu, L.H.; Wang, N.Y.; Wu, A.Y.J.; Lin, C.C.; Lee, C.M.; Liu, C.P. *Citrobacter freundii* bacteremia: Risk factors of mortality and prevalence of resistance genes. *J. Microbiol. Immunol. Infect.* **2018**, *51*, 565–572. [CrossRef] [PubMed]
5. Wang, J.T.; Chang, S.C. Citrobacter Species. 2016. Available online: http://www.antimicrobe.org/b93.asp (accessed on 25 February 2023).
6. Pepperell, C.; Kus, J.V.; Gardam, M.A.; Humar, A.; Burrows, L.L. Low-virulence *Citrobacter* species encode resistance to multiple antimicrobials. *Antimicrob. Agents Chemother.* **2002**, *46*, 3555–3560. [CrossRef]
7. Sommer, M.O.; Dantas, G. Antibiotics and the resistant microbiome. *Curr. Opin. Microbiol.* **2011**, *14*, 556–563. [CrossRef]
8. Olaitan, J.O.; Shittu, O.B.; Akinliba, A.A. Antibiotic resistance of enteric bacteria isolated from duck droppings. *J. Appl. Biosci.* **2011**, *45*, 3008–3018.
9. Elmberg, J.; Berg, C.; Lerner, H.; Waldenström, J.; Hessel, R. Potential disease transmission from wild geese and swans to livestock, poultry and humans: A review of the scientific literature from a One Health perspective. *Infect. Ecol. Epidemiol.* **2017**, *7*, 1300450. [CrossRef]
10. Lessenger, J.E. *Diseases from Animals, Poultry, and Fish, in Agricultural Medicine*; Springer: New York, NY, USA, 2006; pp. 367–382.
11. Owoseni, M.; Okoh, A. Assessment of chlorine tolerance profile of *Citrobacter* species recovered from wastewater treatment plants in Eastern Cape, South Africa. *Environ. Monit. Assess.* **2017**, *189*, 1–12. [CrossRef]
12. Doran, T.I. The role of *Citrobacter* in clinical disease of children. *Clin. Infect. Dis.* **1999**, *28*, 384–394. [CrossRef]
13. Islam, M.S.; Sabuj, A.A.M.; Haque, Z.F.; Pondit, A.; Hossain, M.G.; Saha, S. Seroprevalence and risk factors of avian reovirus in backyard chickens in different areas of Mymensingh district in Bangladesh. *J. Adv. Vet. Anim. Res.* **2020**, *7*, 546. [CrossRef]
14. Mili, S.A.; Islam, S.; Al Momen Sabuj, A.; Haque, Z.F.; Pondit, A.; Hossain, G.; Hassan, J.; Saha, S. A Cross-Sectional Seroepidemiological Study on Infectious Bursal Disease in Backyard Chickens in the Mymensingh District of Bangladesh. *Vet. Med. Int.* **2022**, *2022*, 9076755. [CrossRef]
15. Rajput, D.S.; Singh, S.P.; Sudipta, G.; Nema, R.P. Duck farming, fascinating option in India. *J. Vet. Sci. Technol.* **2014**, *5*, 181.
16. Dey, R.K.; Khan, M.S.R.; Nazir, K.H.M.N.H.; Islam, M.A.; Belal, S.M.S.H. Epidemiological investigation on Duck Salmonellosis in some selected areas of Bangladesh. *Bangladesh J. Vet. Med.* **2016**, *14*, 149–160. [CrossRef]
17. Islam, M.A.; Howlider, M.A.R.; Alam, M.A.; Heyamet, M.A.; Debnath, M. Present status, problem and prospect of duck farming in rural areas of Mymensingh district, Bangladesh. *Asian J. Med. Biol. Res.* **2016**, *2*, 202–212. [CrossRef]
18. Hoque, M.A.; Skerratt, L.F.; Rahman, M.A.; Rabiul Alam Beg, A.B.M.; Debnath, N.C. Factors limiting traditional household duck production in Bangladesh. *Trop. Anim. Health Prod.* **2010**, *42*, 1579–1587. [CrossRef]
19. Khatun, A.; Giasuddin, M.; Islam, K.M.; Khanom, S.; Samad, M.A.; Islam, M.R.; Noor, M.; Bhuiyan, J.U.; Kim, W.I.; Eo, S.K.; et al. Surveillance of avian influenza virus type A in semi-scavenging ducks in Bangladesh. *BMC Vet. Res.* **2013**, *9*, 196. [CrossRef]
20. Bai, L.; Xia, S.; Lan, R.; Liu, L.; Ye, C.; Wang, Y.; Jin, D.; Cui, Z.; Jing, H.; Xiong, Y.; et al. Isolation and characterization of cytotoxic, aggregative *Citrobacter freundii*. *PLoS ONE* **2012**, *7*, e33054. [CrossRef]
21. Bisgaard, M. Salpingitis in web-footed birds: Prevalence, aetiology and significance. *Avian Pathol.* **1995**, *24*, 443–452. [CrossRef]
22. Haruna, E.S.; Usman, M.; Ahmed, S.; Shaibu, J.S.; Makinde, A.A.; Lombin, L.H.; Henton, M.M. Isolation of *Citrobacter murliniae* from clinically ill and dead quail, ducks and chickens. *Vet. Rec.* **2004**, *154*, 119. [CrossRef]
23. Rahman, A.; Shamsuzzaman, S.M.; Dola, N.Z. Antimicrobial Susceptibility Pattern and Virulence Genes Detection in *Citrobacter freundii* Isolated from Patients of a Tertiary Care Hospital, Bangladesh: Antimicrobial susceptibility pattern and virulence genes detection in *Citrobacter freundii*. *Int. Arab. J. Antimicrob. Agents* **2022**, *12*, 1. [CrossRef]
24. Mohanty, S.; Singhal, R.; Sood, S.; Dhawan, B.; Kapil, A.; Das, B.K. *Citrobacter* infections in a tertiary care hospital in Northern India. *J. Infect.* **2007**, *54*, 58–64. [CrossRef]
25. Farjana, N.E.; Islam, M.A.; Zerin, T.; Begum, M.A. Bacterial association in urinary tract infection and their drug resistance among patients in Rajshahi, Bangladesh. *Int. J. Community Med. Public Health* **2021**, *8*, 2144. [CrossRef]
26. Kim, B.N.; Woo, J.H.; Ryu, J.; Kim, Y.S. Resistance to extended-spectrum cephalosporins and mortality in patients with *Citrobacter freundii* bacteremia. *Infection* **2003**, *31*, 202–207. [CrossRef]

27. Jones, B.A.; Grace, D.; Kock, R.; Alonso, S.; Rushton, J.; Said, M.Y.; McKeever, D.; Mutua, F.; Young, J.; McDermott, J.; et al. Zoonosis emergence linked to agricultural intensification and environmental change. *Proc. Natl. Acad. Sci. USA* **2013**, *110*, 8399–8404. [CrossRef]

28. Sidjabat, H.E.; Paterson, D.L.; Qureshi, Z.A.; Adams-Haduch, J.M.; O'Keefe, A.; Pascual, A.; Rodrçguez-Bano, J.; Doi, Y. Clinical features and molecular epidemiology of CMY-type β-lactamase–producing *Escherichia coli*. *Clin. Infect. Dis.* **2009**, *48*, 739–744. [CrossRef]

29. Islam, M.; Sobur, M.; Rahman, S.; Ballah, F.M.; Ievy, S.; Siddique, M.P.; Rahman, M.; Kafi, M.; Rahman, M. Detection of blaTEM, blaCTX-M, blaCMY, and blaSHV Genes Among Extended-Spectrum Beta-Lactamase-Producing *Escherichia coli* Isolated from Migratory Birds Travelling to Bangladesh. *Microb. Ecol.* **2022**, *83*, 942–950. [CrossRef]

30. Ngaiganam, E.P.; Pagnier, I.; Chaalal, W.; Leangapichart, T.; Chabou, S.; Rolain, J.M.; Diene, S.M. Investigation of urban birds as source of β-lactamase-producing Gram-negative bacteria in Marseille city, France. *Acta Vet. Scand.* **2019**, *61*, 1–7. [CrossRef]

31. Daeseleire, E.; Van Pamel, E.; Van Poucke, C.; Croubels, S. Veterinary Drug Residues in Foods. In *Chemical Contaminants and Residues in Food*, 2nd ed.; Woodhead Publishing: Cambridge, UK, 2017; pp. 117–153.

32. Botts, R.T.; Apffel, B.A.; Walters, C.J.; Davidson, K.E.; Echols, R.S.; Geiger, M.R.; Guzman, V.L.; Haase, V.S.; Montana, M.A.; La Chat, C.A.; et al. Characterization of four multidrug resistance plasmids captured from the sediments of an urban coastal wetland. *Front. Microbiol.* **2017**, *8*, 1922. [CrossRef]

33. Hasan, M.S.; Sultana, M.; Hossain, M.A. Complete genome arrangement revealed the emergence of a poultry origin superbug *Citrobacter portucalensis* strain NR-12. *J. Glob. Antimicrob. Resist.* **2019**, *18*, 126–129. [CrossRef]

34. Ribeiro, T.G.; Novais, Â.; Branquinho, R.; Machado, E.; Peixe, L. Phylogeny and comparative genomics unveil independent diversification trajectories of qnrB and genetic platforms within particular *Citrobacter* species. *Antimicrob. Agents Chemother.* **2015**, *59*, 5951–5958. [CrossRef]

35. Samtiya, M.; Matthews, K.R.; Dhewa, T.; Puniya, A.K. Antimicrobial Resistance in the Food Chain: Trends, Mechanisms, Pathways, and Possible Regulation Strategies. *Foods* **2022**, *11*, 2966. [CrossRef]

36. Truong, R.; Tang, V.; Grennan, T.; Tan, D.H.S. A systematic review of the impacts of oral tetracycline class antibiotics on antimicrobial resistance in normal human flora. *JAC Antimicrob. Resist.* **2022**, *4*, dlac009. [CrossRef]

37. Roberts, M.C. Tetracycline resistance determinants: Mechanisms of action, regulation of expression, genetic mobility, and distribution. *FEMS Microbiol. Rev.* **1996**, *19*, 1–24. [CrossRef]

38. Roy, K.; Islam, M.S.; Paul, A.; Ievy, S.; Talukder, M.; Sobur, M.A.; Ballah, F.M.; Khan, M.S.R.; Rahman, M.T. Molecular detection and antibiotyping of multi-drug resistant *Enterococcus faecium* from healthy broiler chickens in Bangladesh. *Vet. Med. Sci.* **2022**, *8*, 200–210. [CrossRef]

39. Krause, K.M.; Serio, A.W.; Kane, T.R.; Connolly, L.E. Aminoglycosides: An Overview. *Cold Spring Harb. Perspect. Med.* **2016**, *6*, a027029. [CrossRef]

40. Ferranti, M.; Cicogna, G.T.; Sattin, A.; Alaibac, M. *Citrobacter freundii* sepsis in an immunosuppressed patient with pemphigus vulgaris. *BMJ Case Rep. CP* **2018**, *11*, e227091. [CrossRef]

41. Norskov-Lauritsen, N.; Sandvang, D.; Hedegaard, J.; Fussing, V.; Mortensen, K.K.; Sperling-Petersen, H.U.; Schonheyder, H.C. Clonal origin of aminoglycoside-resistant *Citrobacter freundii* isolates in a Danish county. *J. Med. Microbiol.* **2001**, *50*, 636–641. [CrossRef]

42. Davis, M.A.; Besser, T.E.; Orfe, L.H.; Baker, K.N.; Lanier, A.S.; Broschat, S.L.; New, D.; Call, D.R. Genotypic-phenotypic discrepancies between antibiotic resistance characteristics of *Escherichia coli* isolates from calves in management settings with high and low antibiotic use. *Appl. Environ. Microbiol.* **2011**, *77*, 3293–3299. [CrossRef]

43. Urmi, U.L.; Nahar, S.; Rana, M.; Sultana, F.; Jahan, N.; Hossain, B.; Alam, M.S.; Mosaddek, A.S.; McKimm, J.; Rahman, N.A.; et al. Genotypic to Phenotypic Resistance Discrepancies Identified Involving β-Lactamase Genes, blaKPC, blaIMP, blaNDM-1, and blaVIM in Uropathogenic *Klebsiella pneumoniae*. *Infect. Drug Resist.* **2020**, *13*, 2863–2875. [CrossRef]

44. Depardieu, F.; Podglajen, I.; Leclercq, R.; Collatz, E.; Courvalin, P. Modes and modulations of antibiotic resistance gene expression. *Clin. Microbiol. Rev.* **2007**, *20*, 79–114. [CrossRef]

45. Talukder, M.; Islam, M.S.; Ievy, S.; Sobur, M.A.; Ballah, F.M.; Najibullah, M.; Rahman, M.B.; Rahman, M.T.; Khan, M.F.R. Detection of multidrug-resistant *Salmonella* spp. from healthy and diseased broilers having potential public health significance. *J. Adv. Biotechnol. Exp.* **2021**, *4*, 248–255. [CrossRef]

46. Liu, L.; Qin, L.; Hao, S.; Lan, R.; Xu, B.; Guo, Y.; Jiang, R.; Sun, H.; Chen, X.; LV, X.; et al. Lineage, Antimicrobial Resistance and Virulence of *Citrobacter* spp. *Pathogens* **2020**, *9*, 195. [CrossRef]

47. Krumperman, P.H. Multiple antibiotic resistance indexing of *Escherichia coli* to identify high-risk sources of fecal contamination of foods. *Appl. Environ. Microbiol.* **1983**, *46*, 165–170. [CrossRef]

48. Kang, X.; Wang, M.; Meng, C.; Li, A.; Jiao, X.; Pan, Z. Prevalence and whole–genome sequencing analysis of *Salmonella* reveal its spread along the duck production chain. *Poult. Sci.* **2022**, *101*, 101993. [CrossRef]

49. Fong, I.W. Animals and Mechanisms of Disease Transmission. In *Emerging Zoonoses: A Worldwide Perspective*; Fong, I.W., Ed.; Springer: Toronto, ON, Canada, 2017; pp. 15–38.

50. Brenner, D.J.; O'Hara, C.M.; Grimont, P.A.; Janda, J.M.; Falsen, E.; Aldova, E.; Ageron, E.; Schindler, J.; Abbott, S.L.; Steigerwalt, A.G. Biochemical identification of *Citrobacter* species defined by DNA hybridization and description of *Citrobacter gillenii* sp. nov.(formerly *Citrobacter* genomospecies 10) and *Citrobacter murliniae* sp. nov.(formerly *Citrobacter* genomospecies 11). *J. Clin. Microbiol.* **1999**, *37*, 2619–2624. [CrossRef]
51. Hashim, M.H.; AlKhafaji, M.H. Isolation and identification of *Citrobacter freundii* from chicken meat samples using cultural and molecular techniques. *Iraqi J. Sci.* **2018**, *59*, 1216–1224.
52. Ievy, S.; Islam, M.; Sobur, M.; Talukder, M.; Rahman, M.; Khan, M.F.R.; Rahman, M. Molecular detection of avian pathogenic *Escherichia coli* (APEC) for the first time in layer farms in Bangladesh and their antibiotic resistance patterns. *Microorganisms* **2020**, *8*, 1021. [CrossRef]
53. Islam, M.S.; Paul, A.; Talukder, M.; Roy, K.; Sobur, M.A.; Ievy, S.; Nayeem, M.M.H.; Rahman, S.; Nazir, K.N.H.; Hossain, M.T.; et al. Migratory birds travelling to Bangladesh are potential carriers of multi-drug resistant *Enterococcus* spp., *Salmonella* spp., and *Vibrio* spp. *Saudi J. Biol. Sci.* **2021**, *28*, 5963–5970. [CrossRef]
54. Kolínská, R.; Španělová, P.; Dřevínek, M.; Hrabák, J.; Žemličková, H. Species identification of strains belonging to genus *Citrobacter* using the biochemical method and MALDI-TOF mass spectrometry. *Folia Microbiol.* **2015**, *60*, 53–59. [CrossRef]
55. Clinical and Laboratory Standards Institute (CLSI). *Performance Standards for Antimicrobial Susceptibility Testing, M100-S30*; Clinical and Laboratory Standards Institute: Wayne, PA, USA, 2020.
56. Bauer, A.T.; Kirby, W.M.M.; Sherris, J.C.; Turck, M. Antibiotic susceptibility testing by a standardized single disc method. *Am. J. Clin. Pathol.* **1966**, *45*, 149–158. [CrossRef]
57. Sweeney, M.T.; Lubbers, B.V.; Schwarz, S.; Watts, J.L. Applying definitions for multidrug resistance, extensive drug resistance and pandrug resistance to clinically significant livestock and companion animal bacterial pathogens. *J. Antimicrob. Chemother.* **2018**, *73*, 1460–1463. [CrossRef]
58. Chen, S.; Zhao, S.; White, D.G.; Schroeder, C.M.; Lu, R.; Yang, H.; McDermott, P.F.; Ayers, S.; Meng, J. Characterization of multiple-antimicrobial-resistant *Salmonella* serovars isolated from retail meats. *Appl. Environ. Microbiol.* **2004**, *70*, 1–7. [CrossRef]
59. Azeez, D.A.; Findik, D.; Hatice, T.Ü.R.K.; Arslan, U. Plasmid-mediated fluoroquinolone resistance in clinical isolates of *Escherichia coli* in Konya, Turkey. *Cukurova Med. J.* **2018**, *43*, 295–300. [CrossRef]
60. Brown, L.D.; Cai, T.T.; DasGupta, A. Interval estimation for a binomial proportion. *Stat. Sci.* **2001**, *16*, 101–133. [CrossRef]

www.ingramcontent.com/pod-product-compliance
Lightning Source LLC
La Vergne TN
LVHW071938160726
843515LV00011B/2644

MDPI AG
Grosspeteranlage 5
4052 Basel
Switzerland
Tel.: +41 61 683 77 34

Antibiotics Editorial Office
E-mail: antibiotics@mdpi.com
www.mdpi.com/journal/antibiotics